Bionics by Examples

Werner Nachtigall · Alfred Wisser

Bionics by Examples

250 Scenarios from Classical to Modern Times

Springer

Werner Nachtigall
Scheidt
Germany

Alfred Wisser
Natural Sciences and Technology III
Department 8.3
Saarland University
Saarbrücken
Germany

This book was originally published in German (2013) under the title "Bionik in Beispielen"

ISBN 978-3-319-37514-4 ISBN 978-3-319-05858-0 (eBook)
DOI 10.1007/978-3-319-05858-0

Springer Cham Heidelberg New York Dordrecht London

Printed on acid-free paper

Springer is part of Springer Science+Business Media (www.springer.com)

To Martha and Christa.

„The manner in which BIONICS will mark its greatest contribution is through the revolutionary impact of a whole new set of concepts …"

J.E. Steele (1960)

Preliminary remarks

Of all good matters are three.

In the second edition of the book: "*BIONIK - Grundlagen und Beispiele für Ingenieure und Naturwissenschaftler*" written by W. Nachtigall that appeared in 2002, the field of bionics was introduced for the first time in a wide summary, delimited and classified. In the meantime, the bionics has developed in the way of its proceeding as well as according to the large number of its attempts.

On the one hand , it was therefore necessary to summarise the scientific proceeding – from understanding over abstraction towards technical realisation – and to support it. This happened in 2009 with the second book of the bionics trilogy: "*Bionik als Wissenschaft*", also written by W. Nachtigall.

On the other side the danger exists that bionics is reduced in public, but also in the bio- and technical sciences, on always the same, though highly significant, but also amply wide-stepped examples, like the lotus effect or the shark scale effect. However, bionic research and technical application run at many places, often not so spectacularly, but just already on a very wide base. Therefore, it seemed necessary to present an example collection arranged after bionic sub-areas. It should reflect the huge number of the approaches, already presented, work out the wide base of bionics, and thus sensitise the public opinion to the importance of this discipline. The third book in this trilogy should take over these duties.

Thus, the appealed trilogy presents itself as follows:

1.) *BIONIK - Grundlagen und Beispiele für Ingenieure und Naturwissenschaftler (1998)*
2.) *BIONIK als Wissenschaft - Erkennen - Abstrahieren – Umsetzen (2010)*
3.) *BIONIK in Beispielen - 250 illustrierte Ansätze (2013)*

Since W.N. has dealt many years of his life with technical biology and bionics (and now, as well-meaning people say, has a lot of time for the study of literature as an emeritus), the herewith-completed trilogy should also show a final summary from his view. It should help - as many of his preceding publications - to fix the "right learning from nature to technology"farther in science and society. We always understand bionic work in the strict scientific sense. The representation closes also bionics as a science from dubious methodical attempts and esoteric fare dodgers.

Our book should also strengthen the term "Bionics". We feel the trend to replace the well-introduced term "Bionics" in the German language area with the Anglo-Saxon term "Biomimetics". This happens without need, so to speak, in advanced obedience what concerns a blurred aimed „international harmonisation", because the term of "bionics" would be taken also negatively in English speaking countries. However, "Mimetic" (μίμησις) means "imitation", and we do not want to copy the nature, but just work out her principles and integrate this “lege artis” into the engineer's sciences. There is no shorter and clearer (and in addition very well introduced) term, which expresses this way of proceeding, as the term "bionics".

Thus manifold correspondences, reprints, magazine articles and book representations as well as notes, recordings and report volumes have been sift through for this final volume by conferences and congresses and own works and for the newer aspects of course in vast manner the Internet. W.N. has carried out many bionics meetings as founder and long-standing leader of the Society for Technical Biology and Bionics (GTBB) and published the Saarbrücken BIONA-reports first as editor, then as a co-editor with A.W. as editor. Meanwhile the number of the bionics conferences has increased. The literature is already extensive, so that we had to select typical examples. This happened according to the following criteria:

1.) Attempts, which are already transformed in products capable of market or available or, nevertheless, patents.
2.) Attempts, which are not transformed yet, however, open promising new territories.
3.) Attempts of rather technical-biological kind, which show, however, a certain power for later conversion.

First, it was our intention to consider only newer works, selected according to the subtitle “Newer examples”. Then, however, many attempts, which one expects in such an example collection, would necessarily be neglected. The exceptionally known lotus effect, for example, would have been 10 years ago absolutely a candidate for this subtitle. Today it belongs already to the classical period, just as the up to now most important development in this field, the evolution strategy. Therefore, can we leave out both bionic examples? Finally, from these and similar reasons we have forced ourselves, then to take up already the most important "pre-classical" and "classical" examples with. Thus, the collection has become also a sort of guide by the history of the development of the bionics. However, at least in the segment "modern age", new examples predominate, partly also those which have been known only shortly before the publication of this book or of which we were told by the authors.

For the arrangement of the examples, we have taken over the proven and extensively accepted gradation from the first mentioned book. Nevertheless, we have introduced a further aspect on the border to the biotechnology in addition, namely the technical use of organic materials including energy plants.

The number of the present bionic attempts is worldwide around some thousands. 250 examples given here explain indeed less than 10 % of all attempts. Anyway, it was tried to indicate attempts from the whole area of bionics. However, personal predilections and focuses cannot be completely dammed. Moreover, there is a line of branches, which have "moulted" long ago to independent disciplines. They are represented here only with few typical examples (cf. the table of contents).

For the representation, we have used a uniform box pattern, for every example a single side. This pattern encloses generally: Heading, two pictures (biological and technical), the principle, biology, abstraction, application, and literature. Because only maximum seven lines are available for the bigger boxes, the distillation of a longer work on the most essential connections was often an osseous work, particularly as these seven lines should nicely fill the completely small box in the flush setting also. However, this has the advantage for the reader that he finds the real essentials of a bionic attempt in brief. The literature given can serve for the more detailed checking up. They come, where possibly, from simple accessible magazines. Pictures and occasional quotations come, as a rule, from the first-cited work.

We have split the work on the book in the following way: W.N. has met the first choice, has written the texts, and has selected the pictures; A.W. has taken all duties, which deal with the layout of the pictures and pages, the Internet search for the newest examples up to the project management. Our last corporate work was „Biologisches Design – Systematischer Katalog für bionisches Gestalten", which appeared in the Springer publishing company.

In the initial-phase, Dr. habil. Claus Ascheron of the Springer publishing company has accompanied the project and in the final phase, it has been in charge of Dr. Dipl.-Phys. Vera Spillner of the Springer/Spektrum publishing company. The authors thank the mentioned for the pleasant cooperation.

Saarbrucken, in autumn, 2013

(Prof. Dr. Werner Nachtigall) (Dr. Dipl.-Biol. Alfred Wisser)

CONTENTS

About the Author

Among other things, Prof. emeritus Dr. rer. nat. Werner Nachtigall studied biology and technical physics. It was always an important request for him to bring together bio-sciences and technical disciplines. He created the two terms "Technical Biology and Bionics" (TBB) and worked with his work-groups in this area for several decades, he also established the "Society for Technical Biology and Bionics", whose long-time chairman he was. He is author of approximately 300 original-publications and approximately four dozens of books. Also, he has received several prices and he is an elected member of several academies and societies. Internationally, he is regarded as one of the most important promoters of the bionics because of his long-time and successful activity to fix the TBB in science and society world-wide.

Dr. rer. nat. Alfred Wisser, former co-worker in a work-group of Prof. Nachtigall, is zoologist with the main emphasis on "functional morphology of insects". For example, basic findings of the morphology of the wing-joint of dipterans are decreased of his studies. As editor of the BIONA-reports for Technical Biology and Bionics, specialist for scanning-electron-microscopy and data processing as well as project-manager he already has, together with Prof. Nachtigall, realized several reports and book-projects, like the book *"Biological Design"* (Springer publishing house). Among other things, he has also taken on the picture-layout and the project-management for the present book.

About the Authors

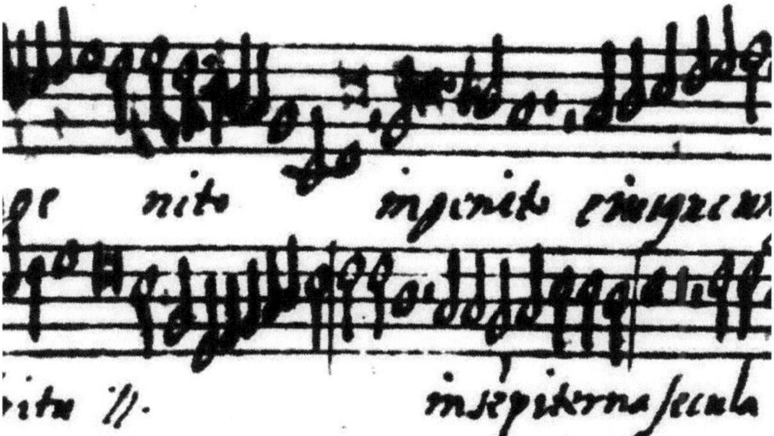

"PREHISTORY"

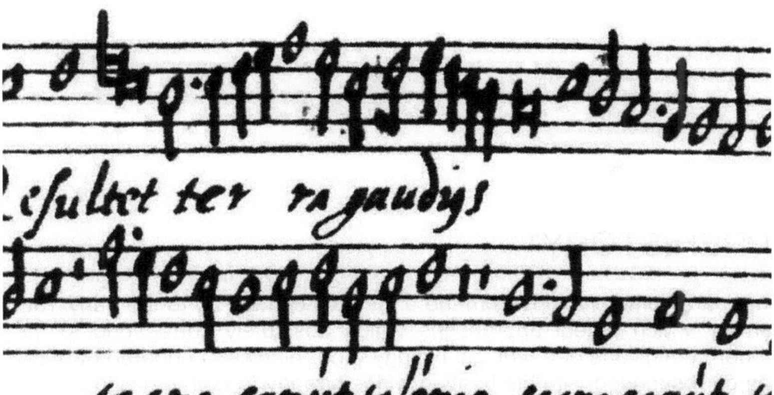

W. Nachtigall and A. Wisser, *Bionics by Examples*,
DOI 10.1007/978-3-319-05858-0_1

"PREHISTORY"

There are some citations from the early 16th century up to the early 19th century which reveal approaches which absolutely correspond to the modern definitions of „technical biology" and "bionics". Leonardo da Vinci has corresponded with his beating wings already to both definitions. The spectrum ranges from Borelli's first model experiment in science history to Cayley's "aeronautical" experiments in the twenties of the 19th century and beyond it.

Different downstroke and upstroke in beating wings

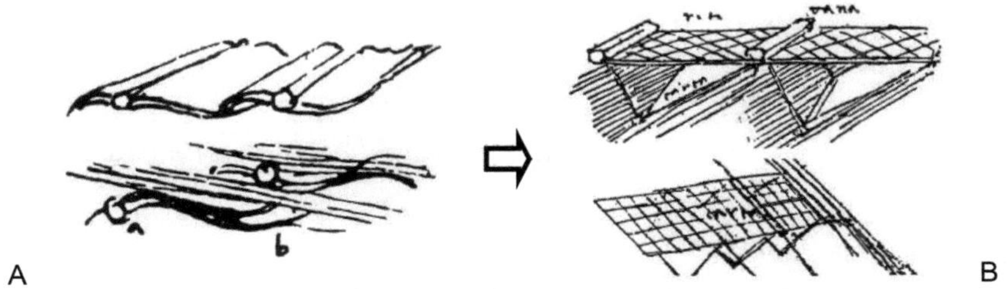

The morphological and functional heterogeneity of the bird's wing is transformed in an analogous flap system.

BIOLOGY:
According to the Leonardo da Vinci's imagination, the widely overlapping feathers of a bird's wing close the gaps during downstroke (A, top), such generating a cushion of compressed air from which the downwards beating wing pushes off. With the impact against it, the feather cascade opens and leaves gap-shaped slits between the single feathers (A, bottom). Thus, a flow arises without great drag production.

PRINCIPLE:
The principal morphological function of the beating wing behaves according to the opinion of Leonardo in such a way that it undergoes changes in their form during down- and upstroke. This arises from the fact that, due to the unsymmetrical overlapping of feathers, a pressure from below leads to a closing of the slits, so that the feathers have an effect as a compact surface. At pressure from upside (upstroke) the same overlapping geometry leads to an opening of the slits.

TECHNOLOGY:
As a technical realization, Leonardo has provided a system of flaps, analogously to the automatically changing wing-cascade. These are arranged in such a way, that they are closed during downstroke until they hit one another (B, top), such connecting to a uniform surface; during upstroke they open in contrast (B, bottom). They are made of oil-drenched linen-pieces, braced between osier stakes.

LITERATUR: Leonardo da Vinci (1505): Sul volo degli uccelli. Firenze. – Gibbs-Smith, C. H. (1967): Leonardo da Vincis aeronautics. Science Museum, London.

The underwater hulk of Baker's galleon

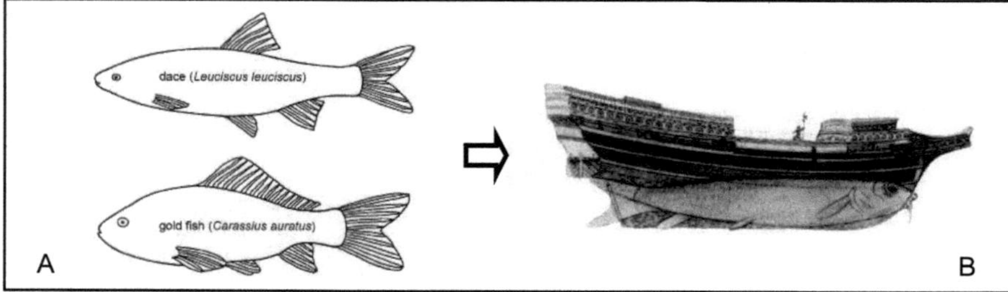

A very early example shows that natural abstraction, not copy of nature, leads to a technically usable solution.

BIOLOGY:
The bodies especially of fast-swimming fishes altogether appear flow-optimized, however, one doesn't know exactly until today, which body-form for specific conditions of swimming and circulation is the best possible. Therefore it would make little sense to shape, for example, the underwater-hulk of a ship after the form-model of a particular fish. Admittedly, "the fish-form" at itself can give stimulation.

HISTORY:
After wishes and planning of John Hawkins, the ship construction master Matthew Baker drafted a new galleon-type from 1576 that distinguishes after all by the reduction of the water-drag and, together with other innovations, should achieve higher speed as well as better mobility and course-stability. For the optimization of the underwater-hulks, he studied the flow-optimized form of fishes (A).

RESULT:
The picture B shows the drawing of one of his representations, a galleon with a plotted fish „with codfish head and mackerel tail“, that is a generally abstracted shape of a fast-swimming, big sea fish. „For the first time M. Baker has integrated in the construction of ship hulks also practical nature studies“ (Soyener). Among other things, England became a dominating maritime authority by these changeovers, too.

LITERATUR: Baker, M. (ca. 1590): Fragments of ancient Shipwrightry. Zit. nach Soyener (1993). – Soyener, J. K. (1993): Die Elisabethanische Galeone. Der Einfluss von Schiffsbau und Bewaffnung auf die Seegefechtstaktik in der zweiten Hälfte des 16. Jh. Das Logbuch 29(4), 160–165.

Study of nature and the first model experiment

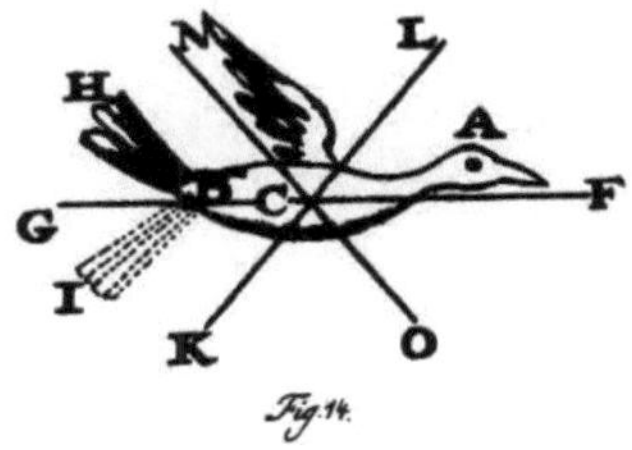

A

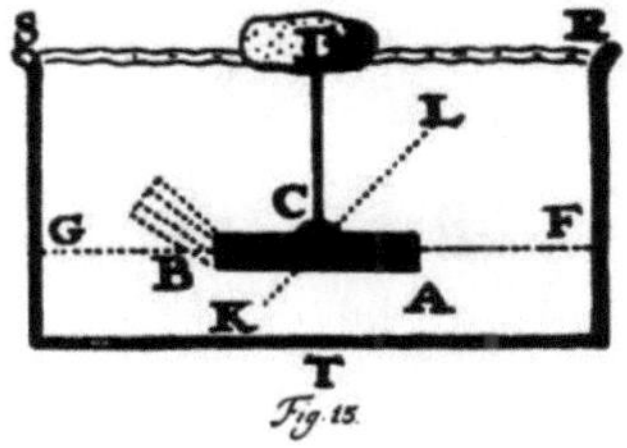

B

Bionic procedure means exploration → abstraction → transformation. Borelli's model was the very earliest abstraction.

BIOLOGY:
Birds steer in the horizontal as in the vertical plane by deflecting their tail. If, for example, they want to climb from the horizontal-flight, they "use the elevator": They tilt the tail upward (A). Thus, a positive tilting moment is created around the lateral axis that presses the body-section behind the centre of mass downwards and therefore raising the head. In consequence, the trajectory points diagonally upward; the bird ascends.

EXPERIMENT, FIRST:
From J.A. Borelli, 1608 -1679, teacher of mathematics in Milan, in 1685 the book "De motu animalium" appeared, in which the locomotion of animals is explained. In order to understand the above described effect of the bird-tail, he made an experiment. In a water-dish, a stone, at which a tail-like, diagonally upward directed paper-sheet was glued (B), was lowered from a cork. When moving in direction F, the stone with the front-edge was tilted upwards.

ABSTRACTION ACCORDING TO MODEL:
This experiment is known as the first bionic model experiment. In a model made analogously to a natural form one can work out the principle after which a natural process - which could not be examined itself so easily - expires. So abstraction of the indispensable inter-step for bionic action is carried out. In the present example: "Generation of a head-raising pitching moment in nature as well as in the model-experiment."

LITERATUR: Borelli, J. M. (1685): De motu animalium (Über die Fortbewegung von Tieren; *Opus postumum*). Angeli Barnabi. 2. Aufl. Ludg. Batav. Neudruck Leipzig 1927.

Cayley's meadow-buck-beard-parachute

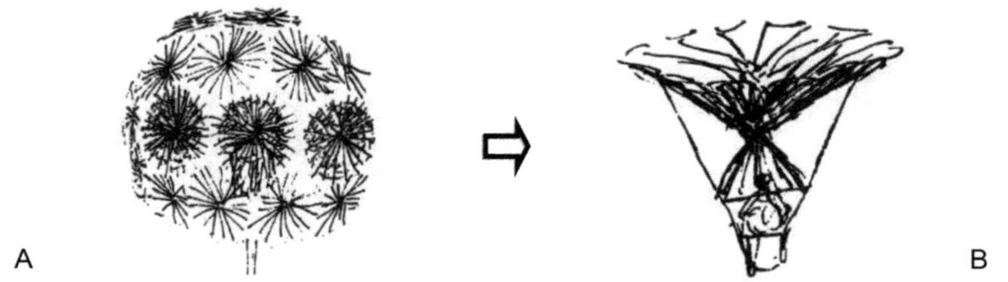

Not blueprints but principles mean pursuing bionics. Cayley's parachute was an original-bionic invention.

BIOLOGY:
The meadow-buck-beard's diaspors, *Tragopogon oriental* (A), belong to the "parachute gliders", as well as the corresponding spread facilities of the dandelion, *Taraxacum officinale*, or, with connected pappus-processes, of the valerian, *Valeriana officinalis*. Two typical, functional peculiarities provide stability: the deep-set mass centre and the easily upward bent, "dihedral" form of the pappus-parachute.

PRINCIPLE:
The little nut lies, as a result of the long-stretching pappus-stem, far from the pappus shield, and thus the mass centre is positioned downwards. Thereby the restoring moment is enlarged, too. When the diaspore is tilted by a gust of wind from its stable situation, the sloping form of the shield gives bigger drag on the downwards tilted side and therefore an additional restoring moment. Thus, the two principles add up.

TECHNOLOGY:
Sir George Cayley, 1773 -1857, founder of the science of "aeronautics", (flight-physics) studied natural forms many times to initiate technologies, so for example bodies of dolphins and also woodpeckers as models for steerable balloons. After the two meadow-buck-beard-principles, in 1829, he drew (B) an auto-stable parachute-form that could not turnover any more when sinking and so should increase the security at a jump.

LITERATUR: Cayley, G. (ab 1809): On aerial navigation, parts I, II & III. Nicholsons Archiv 24, 25 & 26. – Gibbs-Smith, C. H. (1962): Sir George Cayleys aeronautics. Science Museum, London.

It was hard to let loose the „form model of nature“

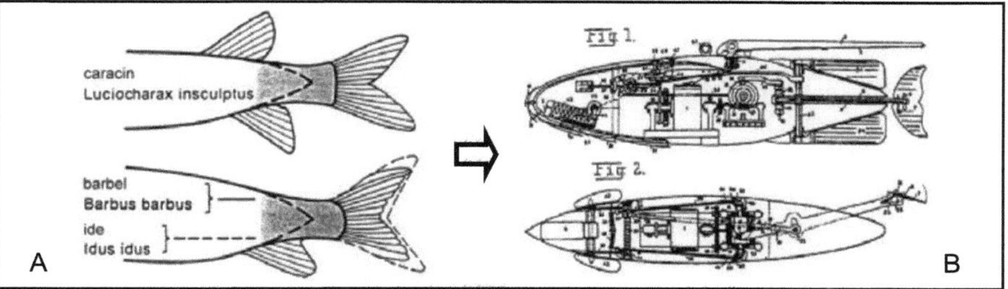

At the beginning, nature-form (through neglecting of the function-concept) became inadequately important.

BIOLOGY AS UNREFLECTED MODEL:
Fast-swimming bigger fishes have characteristic body form, for instance mackerels, salmon, tunas, sharks or others fast swimmers. Also, the tail flukes with their half-moon shapes are quite similar (A). The impressive swimming-performances of these fish-forms were known. Therefore, there was a tendency to give early swimming-robots (B) a meticulous fish-form and fin-design.

MORPHOLOGICAL, NOT FUNCTIONAL SIMILARITY:
In the year 1905, C. Lie has given a device, which should pull a rope under water, the shape of a fish and called it, moderately logical, a "pilot-fish". The propulsion-generating tail fin had an outline, as one knows approximately from the codfish. At the same location as a fish, the device had back, ventral and pectoral fins, which were used for lateral- and depth-control.

ADEQUACY:
The wire dragging device doesn't need to have the form of a fish; according to its Reynolds-numbers, other forms should be more appropriate. It doesn't have to possess a tail fin-drive either. A drive would be more manoeuvrable over a pivot-propeller. Only the position of the lateral- and depth-control behind, respectively in front of the mass centre should be chosen in such a meaningful manner, as realized by a fish.

LITERATUR: Lie, C. (1905): Vorrichtung zum Lotsen von Schiffen. Kaiserliches Patentamt Nr. 21315.

W. Nachtigall and A. Wisser, *Bionics by Examples*,
DOI 10.1007/978-3-319-05858-0_2

„EARLY HISTORY“

From the middle of this 19th century, there are increasingly more tendencies for a „learning from nature“, that lead into technical realizations or at least stimulate these. It is little known that barbed wire and ferro-concrete are bionic inventions. The inventor of the ferro-concrete, J. Monier, lives in the term „Moniereisen“ of the mason-guild. Besides naively appearing attempts of realization still in the 20th century, there are also more and more attempts with a firm physical basis, for example the flipper propulsions. The series reaches as far as to the zeppelins of the twenties and the wartime.

Barbed wire is a bionic invention

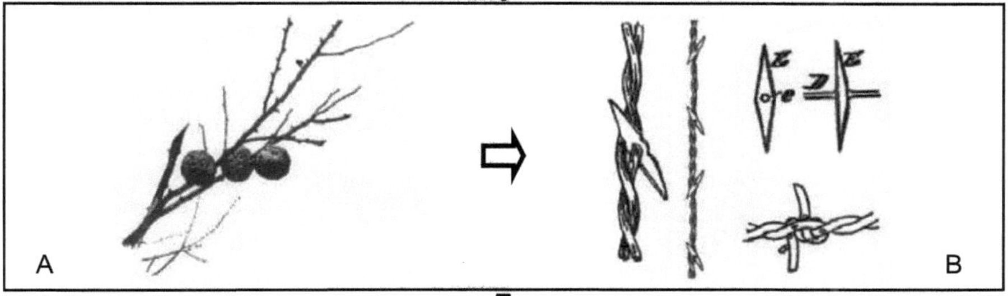

Sometimes, a bionic background is in an invention, that nobody has remembered any more: Barbed wire.

BIOLOGY:
Since the begin of the stock farming, the North America's farmers has known, that animals are not to be persuaded by anything to penetrate through the prickly hedges of the Osage-orange, *Maclura pomifera*. Therefore, such hedges (A) were planted as separations of the grassland and pen-fence. However, the increasing of the borders of the meadow-areas requested for engineering solutions especially since the hedges of *Maclura* only grow very slowly.

PRINCIPLE:
With the first patent of a „thorny wire" (Mike Kelly, 1868), the sharp thorns close to the basis of an Osage-branch were imitated. In fact, they were built of duplicate-sharpened, thorn-like metal-platelets between two wire-strings (B). The nature model is documented in the letter of patent: *„My invention relates to imparting to fences of wires a character approximating that of the thorn hedge."* But the wire was still quite expensive.

DEVELOPMENT:
Since the Kelly-wire was quite extensive because of the complicated handicraft, it was replaced by automatically manufactured products. Today, it is only used in the military area where money doesn't play any role. The Glidden-wire, that was announced to the patent in 1874, was more practicable. This sharp metal-platelets, tied-in by hand, were replaced (B) by mechanically plaited double-thorns. Today, there are up to 200 barbed wire-patents.

LITERATUR: Kelly, M. (1868): Improvement in fences. US Patent No. 74379. – Glidden, J. F. (1874): Improvement in wire fences. US Patent No. 157124.

Ferro-concrete is a bionic invention

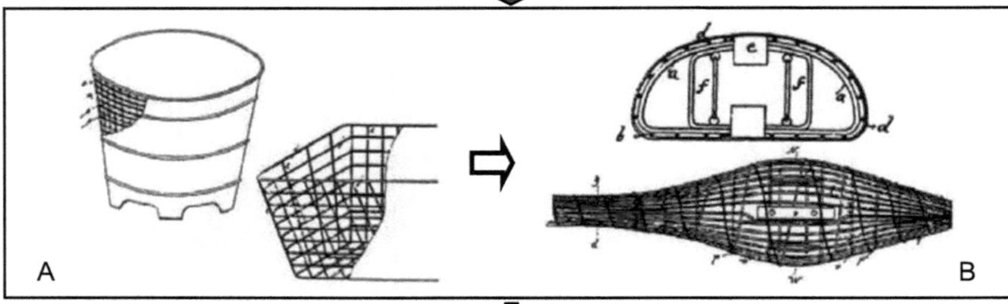

One can always speak of a bionic invention when natural models have given the essential inspiration.

PROBLEM AND BIOLOGICAL MODEL:
The Parisian „horticulteur et paysachiste" Joseph Monier had been angry about the high costs and the risks of fracture of the ceramic or stony plant pots. On the other hand, he had observed that the linked sclerenchyma structure of a decomposed opuntia leaf gives solidity to the leaf mass. From it the ideas originated, patented in 1867, to produce flower pots in an inexpensive manner in form of multi-component constructions.

PRINCIPLE:
In the plant world the parenchyma mass consist of cells standing narrowly side by side, which are stabilized by the turgor pressure, but still sensitive to traction. On the other hand, the connected structure of the sclerenchyma, forming multiple lignified, long-stretched bundles of cells, is not very stabile against lateral pressure; however, it is insensitive against fraction forces. The combination of the two materials combines in an ideal way compression- with tensile-strength.

TECHNOLOGY:
The pressure-proof parenchyma mass keeps the tensile-proof network of sclerenchyma at distance. For the last one, a wire-skeleton is analogue in the technical range, for a flower pot a wire-basket (A), for a sleeper an accordingly formed winding of wire (B), for the first one the matrix of the poured, solidifying cement. Therefore, a sleeper or a bucket of cement with wire-matrix-filler combines compression with tensile strength: Ferro-concrete.

LITERATUR: Monier, J. (1867): Nouveau système de caisses et bassins mobiles et portatifs au fer et ciment applicable à l horticulture. Prevet Français Nr. 77165. – Monier, J. (1880): Verbindung von Metallgerippen mit Cement. Kaiserliches Patentamt Nr. 14673, Kl. 80.

Importance of analogy considerations: ferro-concrete

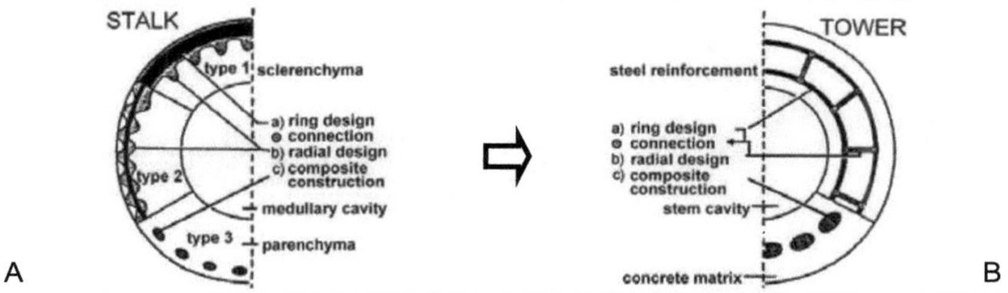

The importance of an analogy-research was highlighted the first time by W. Rasdorsky at the example of the ferro-concrete.

TECHNICAL DESCRIPTION AND BIOLOGICAL UNDERSTANDING:
W. Rasdorsky, who critically regarded S. Schwendeners vision of double-T-beams, got the idea „from lectures, heard in the years from 1906 to 1907, about the iron concrete structure" that the plant is to be understood as a composite construction, „in which the rows of sclerenchyma correspond to the iron armouring, the cells of parenchyma to the concrete matrix" (probably it was meant: the cement matrix): the right way to a functional understanding.

PRINCIPLE OF ANALOGY-RESEARCH:
With the above-mentioned quotation from the year 1911, the heuristic role of the analogy-research, so important today, already had been shown very early: „Therefore, between the technical composite constructions and the plant-organs, there is an extensive analogy" of the whole construction-principle. Giesenhagen notes in 1912 that leaves „with their strengthening tissues form a grid like the iron insert in a ferro-concrete mass (A, B)".

IMPORTANCE OF ANALOGY-RESEARCH:
The early analogy-research led not only to a correct understanding of the morphological structure, but also influenced the manner of view of the following generation of researchers. Bachman compared in 1922 the configuration of the tensile fibres within the bamboo with „a reinforcement of the mainly stressed outer layer during flexion (like the ferro-concrete). Bower referred in 1923: „Ordinary herbaceous plants are constructed on the same principle."

LITERATUR: Rasdorsky, W. (1911): Bull. de la Société des Naturalistes de Moscou, Sect. Biol. 4, 351–405. – Rasdorsky, W. (1928): Ber. d. Dtsch. Bot. Ges. 46, 48–104. – Giesenhagen, K. (1912): Handwörterbuch d. Naturwiss. 2, 1–35. – Bachmann, F. (1922): Jb. Wiss. Bot. 61, 372.

Naive suggestions for realizations lead into a void

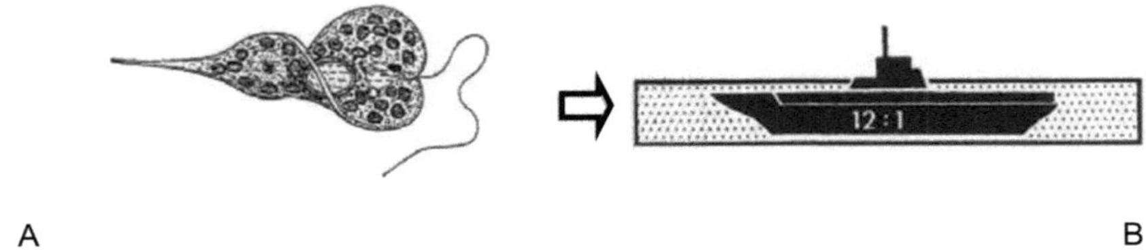

One must give up suggestions for realizations, which don't include the physical-technical bases.

BIOTECHNICAL COMPARISON CAN LEAD TO IDEAS, BUT ...
R.H. Francé, one of the early proponents of a „learning from nature for technology", for example, already tried in his 1919-work „Die technischen Leistungen der Pflanzen", to overcome borders of disciplines. So, he proposed to build a „rotating submarine" after the model of the „water-screw-form" of a unicellular green-alga. Representatives of the type *Phacus* and others actually move helically.

... BIOLOGY AND TECHNOLOGY ARE NOT ALWAYS COMPARABLE:
A comparison is sense-empty in this case since the physical prerequisites for the movement of a 1/10 mm long green-alga (A) and a 30 m long submarine (B) are completely different. Not only the lengths have differed about the factor $3 \cdot 10^5$ also the speeds differ in similar manner. So the Reynolds-numbers, which mark the characteristics of the water flow, differ approximately in the relation of $1{:}10^6$.

TECHNOLOGY:
A submarine cannot rotate screw-like together with its team through the water. One cannot transfer the mechanism of movement of the small green-alga to a 30 m long submarine either. So, the proportions of pressure - and friction drag are example-wise completely different. A submarine could not swim in such a way. The same is valid vice-versa: Small algae with submarine-form and -drive would not make progress in the water.

LITERATUR: Francé, R. H. (1919): Die technischen Leistungen der Pflanzen. Veit & Co, Leipzig.

Technical and physical principles as basis

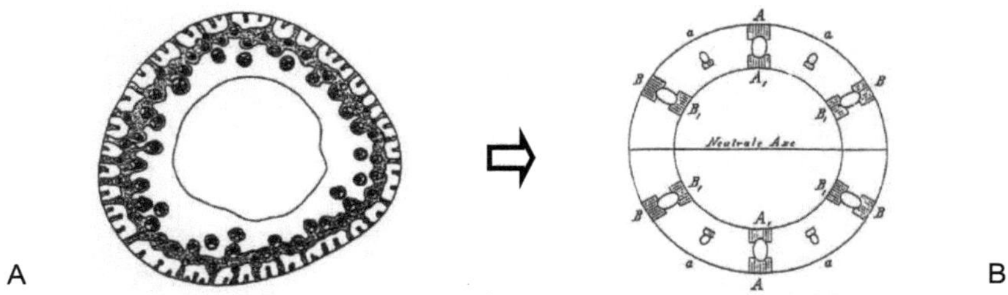

New or rediscovered technical or physical knowledge was already used for early principal explanations.

BIOLOGY:
Cross-sections through plant-stems or -stalks, for example of the guts-reed, *Cladium mariscus* (A), normally are marked by annular load-bearing structures of merged sclerenchyma. These often form centrifugal and centripedal bulgings. Such structures show common enough cross-section-forms that resemble technical double-T-steel-profiles. Accordingly, then they were also described as „biological double-T-profiles."

PRINCIPLE:
One knows about technical double-T-beams since their introduction into the concrete- and railroad-constructions of this 19th century, that due to their typical cross-sections they show an especially high moment of inertia and are relatively resistant to bending and torsion (B). This is especially true if they are combined to radial complexes: So technical biology can explain the so-being of biological structures by analogue conditions.

EXPLANATION BY AN OTHER FIELD:
S. Schwendener, botanist and biomechanics in the late 19th century, was stimulated „by the viewing iron bridges and railway stations with their numerous double-T-profiles" to understand rigid plant-stalks as systems of such profiles. In the year 1888, he wrote in a book: „The plant undoubtedly constructs after the same rules like the engineers only that their technology is a very much finer and more perfect."

LITERATUR: Schwendener, S. (1874): Das mechanische Prinzip im anatomischen Bau der Monocotylen ... Engelmann, Leipzig. – Schwendener, S. (1888): Über Richtungen und Ziele der mikroskopisch botanischen Forschung. Naturwiss. Wochenschrift, Berlin.

„Fish-propellers" of the tail fin-type

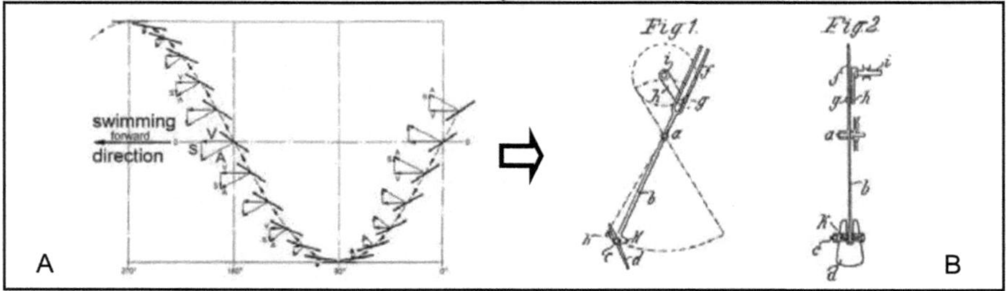

The reference „prototype nature“ is often given in older patent-writings, but not in newer once.

BIOLOGY:
The efficiency of the tail fin-drive of fishes has been regarded as outstanding since ages ago. In the patent-writing of Zdenko Knight of Limbeck is written: „Object of the present invention is a fin-propeller for ships, which is mounted instead of a screw-propeller on the rear-ship and includes an onward driving movement corresponding to the tail fin of fish, through deflections to starboard and port side.“

PRINCIPLE:
The thrust-production in a to and fro swinging fin (A) happens in the way that it is put at a positive angle of attack relative to its path through the water. So a lift component A is created at a right angle to the direction of the beat in addition to a drag component in the beat-direction. This one can be separated into a thrust-component V in swimming direction and a lateral force-component S vertically to it.

TECHNOLOGY:
The drive consists of an eccentric-fork-system (B), that lets a driving-rod b swing to and fro. At the end of the rod, a fin is fixed, that is eccentrically built, so that its surface is always orientated against the water pressure d. It beats against the block h and the opposite block and regulates itself in such a way, that it always generates thrust, at the reversion-points the side of attack changes.

LITERATUR: Ritter von Limbeck, Z. (1903): Fischpropeller für Schiffe. Kaiserliches Patentamt Nr. 153810, Kl. 65 f.

„Wave-propeller" with an elastic fin

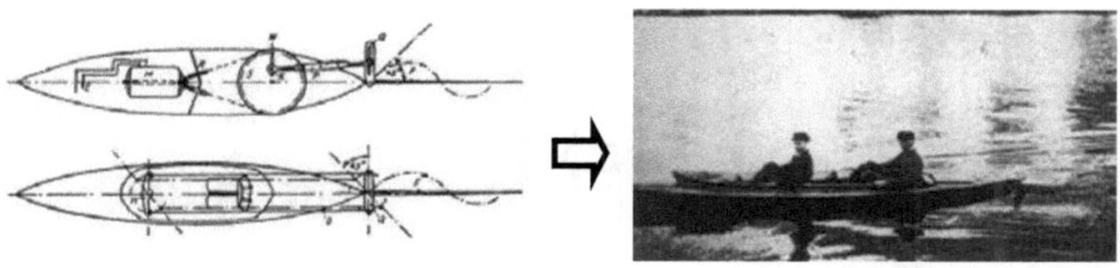

A B

There are few examples in the area of early bionic realization from the step of idea / model to full-scale-implementation.

BIOLOGY AND TECHNOLOGY:
After the study of the propulsion-mechanisms of fishes, the engineer H. Schramm, who had given his book „The undulation as propulsion-factor in nature and technology“ the apodictic subtitle „Thoughts of an engineer about the problem swinging propulsion in technology and biology“ took a chance for realization of models into full-scale-implementations. Men carrying boats with undulating stern-fins were built and tested.

PRINCIPLE:
Long stern-fins (A) were put into horizontal-swinging. It was recognized on that occasion, „... that the elasticity of the fin should not be continuously the same but that, at the point of attack of the fin-pressure-force, the front part of the fin, flexible connected to the boat-stern, should be strong and of little elasticity and to its rear end it should be continuously soft and flexible.“ - An analogy to the fish.

TECHNOLOGY:
Five such fins were tested. The best one showed efficiencies that were higher than those of ship-propellers. Within certain boundary conditions, a screw-propeller attains efficiency of 0.52 and contrary to this, the undulating propeller 0.78. The basic approach was enlarged to a two-men-canoe (B). With the best fin, coming like a fish-fin to a hair-sharp edge at the end, especially high static thrust has been achieved.

LITERATUR: Schramm, H. (1927): Die Schwingung als Vortriebsfaktor in Natur und Technik. Gedanken eines Ingenieurs über das Problem der schwingenden Propulsion in Technik und Biologie. De Gruyter, Leipzig.

Test for the patent-office: Francé's salt shaker

Will the patent-office consider an „invention, which the nature has already made," patentable? The test was positive.

PROBLEM:
Raoul H. Francé, the discoverer of the microscopic life in the soil („Edaphon"), tried to multiply microorganisms from an even inoculation of a soil area. So he looked for a method for even inoculation, „strewing quite evenly, each square millimetre with one dozen of the smallest life-germs." However, all the spreading devices usual in the 20-er years like conventional salt shakers (A), powder-shakers for small children, atomizers or strainers, scattered quite unevenly.

PRINCIPLE:
„I found the solution of the problem in the capsule of the poppy. Everyone knows them, everyone knows, that the holes, grouped in circles below their covers, are used for scattering the small poppy-grains, but never somebody has thought that here an invention of the plant is given, which surpasses the our ones. A poppy-capsule, filled with the grains of the earth, scattered very much more evenly as I had succeeded with other devices until then."

PATENT:
„With a bold decision, I wanted to have certainty. I drew a shaker for salt, for powders or other medical purposes after the model of the poppy-capsules (B) and declared this as invention to the pattern-protection. One didn't deny me the protection; an invention was done. After short, I got confirmed that by the patent-office..." Francé held himself only, sympathetic modest, for a „miserable copier of the nature."

LITERATUR: Francé, R. H. (1919): Die technischen Leistungen der Pflanzen. Veit & Co., Leipzig. – Francé, R. H. (ca. 1929): Deutsches Patentamt Nr. 723730.

Airship-constructions of the 20er years

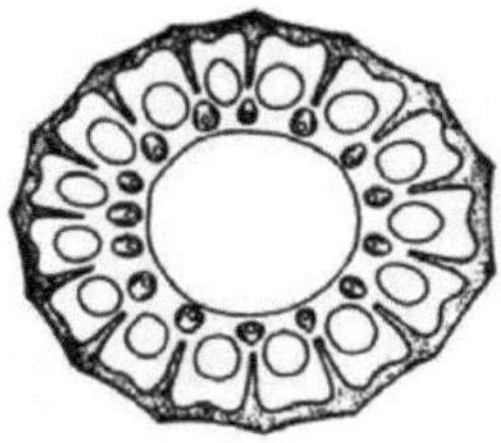

A

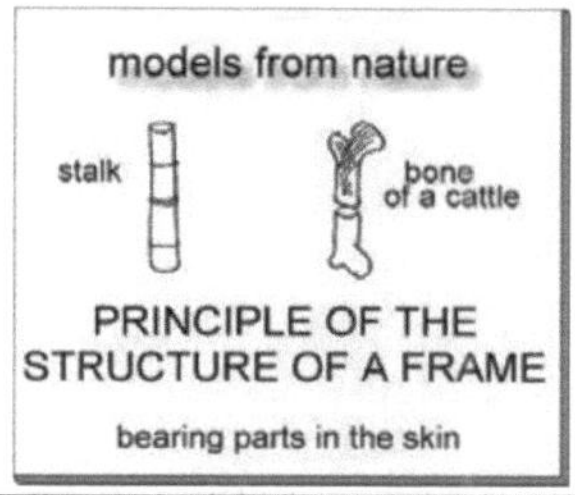

B

The light construction of biological stalks with far outside lying supporting frame supplied stimulation for airship frames.

BIOLOGY:
Grass-stalks and horsetails (A) carry their sclerenchyma stiffening-system far peripheral in form of interconnected ring-structures. Support elements especially surround the vascular bundles, so that they become stiff tubes. These are interconnected by sclerenchyma with their neighbouring tubes. So a type of a ring form results. From time to time effective, radial geared double-T-beams occur, too.

PRINCIPLE:
Through the described arrangement, a big second moment of inertia of these girder structures occurs, an essential reason for their considerable bending and distortion-stiffness. There are numerous types, from the horsetails up to the bloom-plants, which work all after the same principle. Through the outside displacement the supporting system doesn't become so very much died out, so that it would become sensitive against local buckling.

TECHNOLOGY:
For the conception of the aluminium made beam-frames of airships, one studied both stalks and bones „after the model of the nature (B)." Concerning the „principle of the carrier-body-construction" one was interested to localize „bearing parts in the outside-skin." A direct realization of these studies is not verifiable. The qualitative proof however, that already Graf Zeppelin or his followers carried on bionics, is secured.

LITERATUR: Zeichnung aus einer technischen Publikation über Messungen und Rechnungen an Zeppelinmodellen (Bauweisen, Strömungsverhalten, Antriebe etc.) in einem Museum. Nicht näher eruierbar. Skizze B ist eine direkte Nachzeichnung des Ausstellungsblatts.

Bionics in totalitarian systems

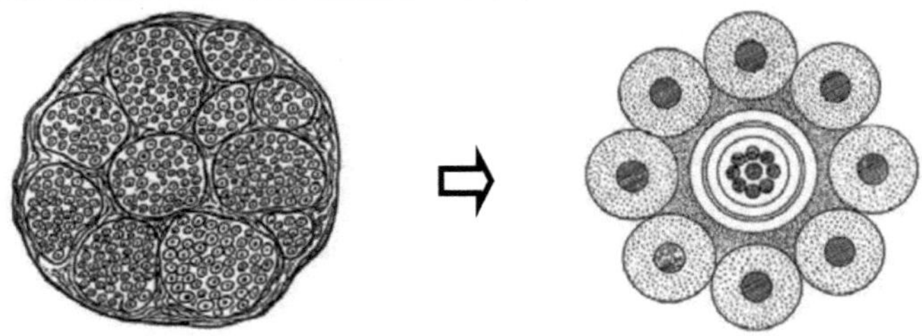

A B

Even dictatorial systems as the national socialism and communism had accepted the „prototype nature."

NATIONAL SOCIALISM:
Into the blood-and-soil-ideology of the third empire, the natural was embedded, though in emotionally exaggerated revaluation. 1939, Alf Gießler, a then "Gauleiter", set natural and technical realities opposite to each other in his book „Biotechnology", for example nerves and cable-cross-sections (A, B): „The high technical proficiency of the northern human being ... has its unequalled model in the processes of the nature itself."

COMMUNISM:
Marx, Engels as well as Lenin referred many times and desiderated on principles of the biosphere: „... that there is our entire reign over the nature ... to recognize her laws and, to be able to apply them right." „Morphological and physiological appearances, form and function, stipulate each other reciprocally." „Darwin has drawn attention to the natural technology ...." „From the living view to the abstract thinking ...."

CRITICISM:
In both cases, applicable statements were turned into the goal of the fulfilment of political demands. This admittedly doesn't devaluate the comparisons, but it relatives them. For example A. Gießler tried to find groups of prototypes in the nature (A) for each technical line from the construction-statics up to the electrotechnology (B). He therefore compared already whole areas to find similarities between nature and technology: early analogy research.

LITERATUR: Gießler, A. (1939): Biotechnik. Quelle & Meyer, Leipzig. – Marx, K.; Engels, F. (Gesammelte Werke, ed. 1962): Dialektik der Natur. Bd. 20, 496 und 611–620. – Lenin, W. I. (Gesammelte Werke, ed. 1964): Bd. 29, 229; Bd. 38, 160. Dietz, Berlin.

Transition to the functional aspects

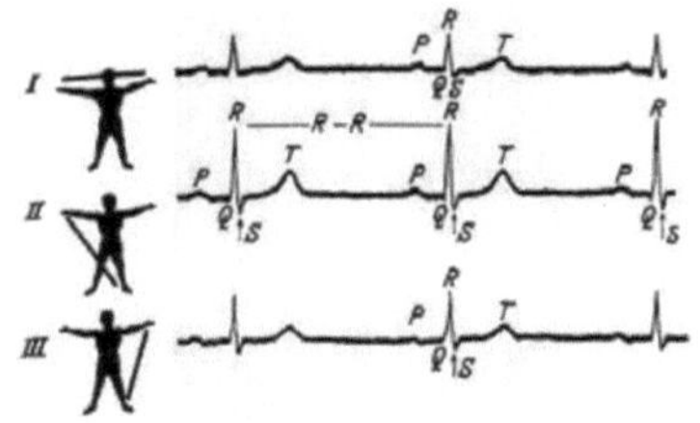

A

B

The consideration of function in biology challenges the technically-physical disciplines as far as at their borders.

BIOLOGY RANGES AMONG THE MEASURING DISCIPLINES:
Nature-enthusiasm or politically motivated super elevation of nature, to early and uncritical transferring and partly naive view-manners have contributed, approximately as far as to the middle of this 20th century, only here and there something to the deepening of the cross-relationships. This changed as soon, as the measuring biological disciplines were able to offer questions and results that were equal to those of the technical sciences.

BIOLOGY BEGINS TO ASK BORDERLINE QUESTIONS:
If biology hardly was capable previously to fit in suitably suggestions from technology, the medal now turned back quickly. The questions, which the biologists and human-physiologists asked the technicians and physicists, became gradually so difficult, that the physical-technical disciplines had to resign at first. Their interest awakened this on the other side, they became interested and this led to continuing developments.

EXAMPLE:
As in the late 19th century, human-physiologists began to examine the electrocardiogram of the human heart (A) and to appraise it already clinically, the sensitivities and time-functions of the used measuring instruments (e.g. mercury-drop-galvanometer) were inappropriate. The demand for fast indicators without intolerable phase-shift led to the construction of sensitive rotating-mirror-galvanometers (B).

LITERATUR: Rein, H.; Schneider, M. (1971): Einführung in die Physiologie des Menschen. 16. Aufl. Springer, Berlin. – Lueger, O. (Hrsg.) (1904): Lexikon der gesamten Technik und ihrer Hilfswissenschaften. 2. Aufl. Deutsche Verlags-Anstalt, Stuttgart/Leipzig.

In the architecture especially the function counts

A   B

Architecture-bionics is at risk to be lost in form-similarities („biomorphic constructions"). However, the function counts.

FORM-SIMILARITIES IN BIOLOGY AND TECHNOLOGY:
That pure form-similarity doesn't have anything to do with bionics can be seen in the Russian architecture of the post-war era, that sometimes exhausts itself in nature imitation and so has little to do, with the new Russian architecture of the twenties. 1985, J.S. Lebedev writes in his book: „In the last two to three decades, we became eyewitnesses of a new creative impulse in the architecture." But which one?

FORM-PROTOTYPYES IN NATURE:
Continuing Lebedev: „Forms of the nature, like ocean mussels, petals, turtle-tanks, bended and pleated plant-leaves became converted into *forms* (highlighting by the authors) of the built environment." If the hollyhock-fruit (A) was the model for a „hollyhock-high-rise"(B) on the basis of a pure form-similarity, this is pure architecture-*design*. If however, one compares the possibilities of the arrangement of the part-fruits, function comes into play.

CRITICISM OF THE PURE FORM-MODEL:
Nature-forms are always functional, then however only with reference to their respective tasks. The intestine has a series of functions, from the transportation up to the re-absorption. If one shapes a „biomorphic" building in intestine-form (F. Kiesler, an American architect of the 60th years, s. p. 91), such functions don't play any role, and new technical functions are not deducible from the form. The building therefore is not bionic.

LITERATUR: Lebedev, J. S. (1983): Architektur und Bionik. Verlag für Bauwesen, Berlin. – Nachtigall, W. (2003): Bau-Bionik. Springer, Berlin.

Trabeculae of bone-spongiosa and isostatic ribs

A

B

The formation of stress-trajectories in small spongiosic trabeculae leads to analogue frame structures in buildings.

BIOLOGY:
The spongy structure of trabeculae for example in neck and head of a femur (Femur) is not oriented randomly. The single trabeculae orient themselves in relation to the trajectories of stress that is in direction of the main pressure- and tension-stresses. The lines of the same stress are at right angles to each other. In space the lines form areas (A) of the same stress, which, in each case, also are vertically to each other (comp. p. 119).

PRINCIPLE:
Frame elements which run in line with the trajectories of stress in the plain or in space, so line-shaped or shell-like, relieve themselves mutually of bending stresses which may become highly dangerous. Biological or technical materials endure pressure-forces to a certain extent, but also traction-forces to a high extent. Therefore, it makes sense to accept higher pressure- or also tension-stresses for the benefit of reduction of bending stress.

TECHNOLOGY:
One can dissolve a bearing area, for example a floor-blanket, into girders, that follow the directions of trajectories of pressure- and of traction-stresses in each case. Altogether, the frame system (B) then can become lighter than a normal-blanket. The architect P. Nervi comments in his patent-writing about concrete-elements with isostatic ribs: „One frequently finds examples in the nature, and that of the bone trabeculae ... is classic."

LITERATUR: Kummer, B. (1962): Funktioneller Bau und funktionelle Anpassung des Knochens. Anat. Anzeiger 110, 261–293. – Nervi, P. et al. (1950): Perfezionamento nella costruzione di solai, volte ... Italienisches Patent Nr. 455678.

"Frühgeschichte"

W. Nachtigall and A. Wisser, *Bionics by Examples*,
DOI 10.1007/978-3-319-05858-0_3

"CLASSICAL PERIOD"

Shortly before the 2nd world war, modern bionics begins to develop. A centre of these approaches - from the present-day point of view already to be named "classical" - develops at the Technical University of Berlin, where the botanist G. Helmcke and the aerodynamicist H. Hertel work together in questions of structure and locomotion of animals, and where later on I. Rechenberg and his companions develop the evolution-strategy. In Munich and then in Saarbrücken, W. Nachtigall brought in approaches of the technical biology, that lead into bionic developments. With the discovery of the well known Lotus-Effect® at the begin of the 90ies the circle is closing.

From where does the term "Bionik" come?

"... it is not through the solution of specific problems or the design of particular devices."

⇨ "Herein lies part of the motivation behind the generation of bionics."

Usually, the word BIONICS is composed from "BIOlogy" and "techNICS." This doesn't correspond to the history.

ORIGIN:
From the 13th until 15th 9. 1960, a congress, sponsored by the Wright Air Development Division, took place in Dayton, Ohio, named: "Bionics symposium. Living prototypes - the key to new technology". Here in the first session, J.E. Keto talked about "Bionics - new frontiers of technology through fusion of the bio- and physio-disciplines". Here, the word "bionics" appeared for the first time. In the end-session, Major J. E. Steele talked about: "How we get there".

DEFINED BY J. E. STEELE?
One reads occasionally that the named author defined the word "bionics". However, this is not to be clearly taken out of the 499-page report. Obviously, the name originated in the preparatory discussion as gripping catchword. The key-sentence in Steele's lecture is: "We have given the name "BIONICS" to the recognition and practice of these methods". It should give a view beyond the fence of the specialization.

PROPHETIC VIEW:
Steele quite clearly saw that a gate had been pushed open with it: *"The manner in which BIONICS will mark its greatest contribution is ... through the revolutionary impact of a whole new set of concepts, a fresh point of view".* W.N. then had put the *TECHNICAL BIOLOGY* that first explores the nature with help of technical-physical procedure aside the *BIONICS*. "Recognizing must go ahead applying" (Max Planck).

LITERATUR: Anonymus (1961): Bionics symposium. Living prototypes – the key to new technology. Wadd Technical Report 60–600, 5. Mar. 1961, 23–899. United States Airforce, Wright-Paterson Airforce Base, Ohio.

"TUB-TUB" in Berlin: Solid physics at the basis!

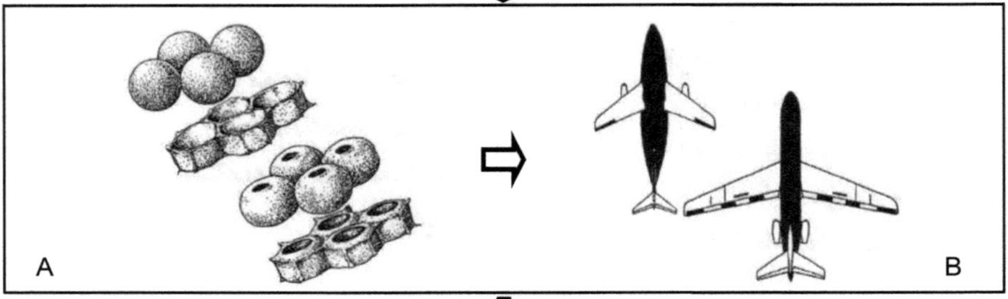

"TUB-TUB" - "Technology and Biology" at the Technical University of Berlin: This was the TU-battle cry of the sixties.

COOPERATION OF BIOLOGY AND TECHNOLOGY:
At the TU Berlin, the biologist G. Helmcke and the aerodynamic H. Hertel in the early 60s met in their effort to overcome the then borders of the disciplines. The story goes that they welcomed themselves with this battle cry on the walks of the TU. Helmcke dealt with the fine-construction of diatoms, Hertel with the flow-mechanics of airplanes and bodies of fishes and of other biological and technical aspects, always on the basis of solid physics.

PRINCIPLE:
To bring together the two disciplines got along in each case under the patronage of an expelled specialist. Neither on the one nor the other side anyone behaved amateurish. Of course, Helmcke, with his proposals for implementation, had to serve himself the loyal help of construction-engineers, Hertel, on the other hand, was dependent on the solid information from fish scientists: Inter-disciplinary cooperation of particular expertises.

EXAMPLES:
Helmcke analyzed the chamber-construction of diatoms (A) and was partner of Otto and Mahnleitner who designed analogous frame constructions of giant-cinema-canvases or of belfries. From Hertel, for example, the design of "thick fuselages for airliners" originated (B), that in comparison with the conventional cigar-like-fuselages can transport more passengers at less fuel-consumption.

LITERATUR: Helmcke, G. (1959): Form und Funktion der Diatomeenschalen. Gesetzmäßigkeiten im Kleinsten. Beitr. Naturkunde Niedersachsens 12, 110–114. – Hertel, H. (1964): Biologie und Technik. Struktur, Form, Bewegung. Krausskopf, Mainz.

Rechenberg's evolution-strategy: Classic bionics

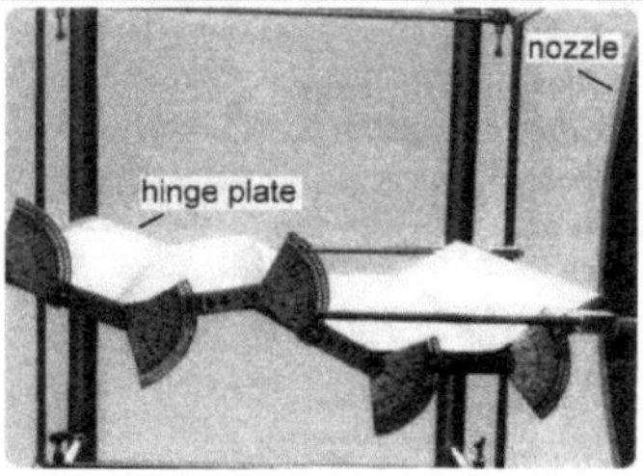

A

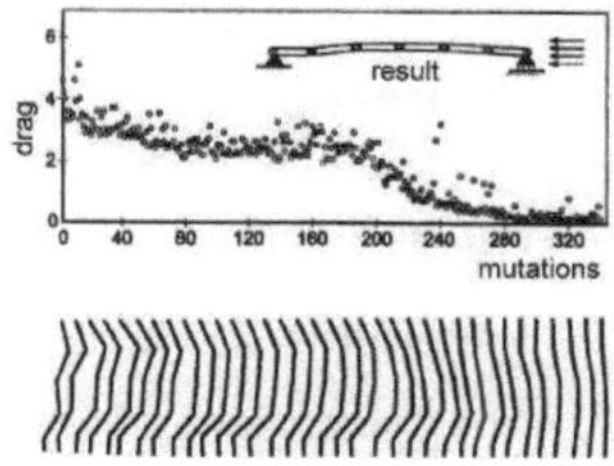

B

The evolution-strategy of Rechenberg, Schwefel and Bienert leads to the theoretical expected optimum.

BIOLOGY:
Mutation, *recombination*, *isolation* and *selection* are parameters of natural evolution. Small accidental alterations of the genetic make-up (mutations) spread over during accidental formation of chromosomes at the division of germ-cells (recombination) to the descendants. They must prove themselves in the evaluation field of the environmental conditions and come differently to reproduction (selection), particularly at barriers (isolation).

PRINCIPLE:
Evolution-strategy works with the 4 named parameters of the natural evolution. If an introduced mutation leads to better reproductive-successes, the respective characteristic-bearer is taken as parent for a new generation, alternatively one goes back to the prior one. Such an evolution-strategic action especially proves itself in particular where there are (still) theories or calculation-procedures available.

EXAMPLE:
In a classic experiment, where one already knows, what will be the result, the evolution-strategy was tested. A hinge plate (A), exposed to a flow, has lowest drag if all parts of this plate are lying in a plane parallel to the flow. Going out from an incidentally folded plate, all angles of attack of the parts of this plate were accidentally changed. After approximately 340 mutation-selection-steps, the expected configuration (B) resulted.

LITERATUR: Rechenberg, I. (1973): Evolutionsstrategie. Optimierung technischer Systeme nach Prinzipien der biologischen Evolution. Frommann-Holzboog, Stuttgart. – Neubearbeitung (1994): Evolutionsstrategie 94. Frommann-Holzboog, Stuttgart.

Optimization of a hinge plate at oblique flow

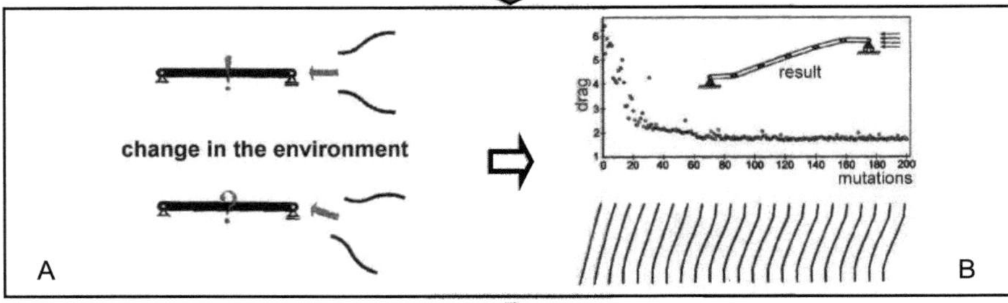

Evolution-strategy leads also to a theoretically expected optimum if the "environment-conditions" change.

BIOLOGY:
The "Experimentum crucis" described at the previous page should confirm, that evolution-strategic optimization actually leads to a theoretically predictable optimum. One could now object, that here the technical boundary conditions were constant. This corresponds to a constant environment in the biology. However, the mutative range now offers survival-chances, too, when the environment-conditions change.

MODIFICATION:
In order to take these objections into account, in the altered experiment described here "the environment was changed." The flat plate was not attacked by parallel flow but under an angle of attack unequally zero (flow inclination; A). A drag-optimum, that is a certain configuration of least-possible drag, is also to be expected here. However, this can be no more the flat plate, because it must offer the flow a favourable phased-in part and phased-out part.

RESULT:
After further 200 mutations, the hinge plate had configured to an approximated sigmoid form. In this form, the phased-in part and phased-out part are now lined up approximately flow-parallel. This is the forthcoming form of lowest drag for the now given boundary-condition (B) (tilt-position to the current). With this variation, the efficiency of the evolution-strategic procedure was shown another time.

LITERATUR: Rechenberg, I. (1973): Evolutionsstrategie. Optimierung technischer Systeme nach Prinzipien der biologischen Evolution. Frommann-Holzboog, Stuttgart. – Neubearbeitung (1994): Evolutionsstrategie 94. Frommann-Holzboog, Stuttgart.

Optimization of spreading a wing (winglets)

A

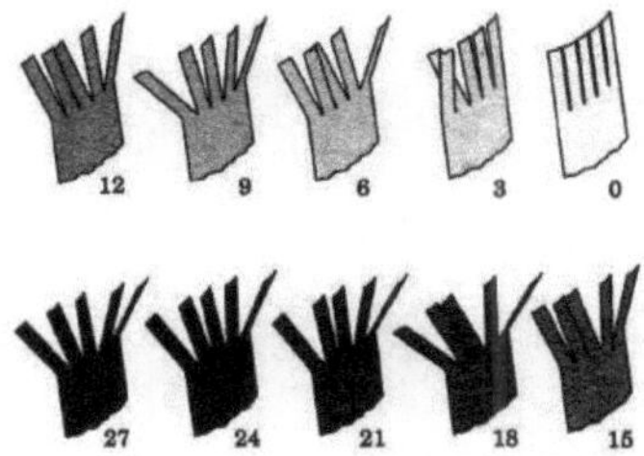

B

The evolution-strategy leads to the same morphological optimal configuration as the biological evolution.

BIOLOGY:
Big, glide- and sail-capable land-birds, like eagles, vultures, storks as well as pelicans (A), possess primary fan out feathers. In optimal-attitude (angle of twist, angle of positioning, angle of attack of the single feathers), these can influence the compensation-flow between lower and upper side, around the wing tip, positively. This attitude takes place passively and automatically because of the type of the elastic fastening of the individual primary feathers.

PRINCIPLE:
If the drag-reducing effect of the cascade of the hand feathers becomes noticeable only with a certain optimal-configuration of their setting parameters, exactly this configuration should come out by using evolution strategy, because it emerged over evolutive processes. Since several parameters are to be taken into account, a multiplicity of combination-possibilities emerges, that cannot be tested individually.

TECHNOLOGY:
In a wind-tunnel, a planar, technical wing (B, 0) similar to the wing of a stork, with 5 adjustable winglets, whose above mentioned adjustment possibilities could be changed in an accidental manner, was examined by evolution strategy. Goal was the evolution of an as small (good) as possible glide ratio $\varepsilon = c_W/c_A$. A configuration very similar the primary feathers (B, 27) emerged, which improved the lift-to-drag ratio of B,0 around 11 % (comp. p. 144).

LITERATUR: Rechenberg, I. (1973): Evolutionsstrategie. Optimierung technischer Systeme nach Prinzipien der biologischen Evolution. Frommann-Holzboog, Stuttgart. – Neubearbeitung (1994): Evolutionsstrategie 94. Frommann-Holzboog, Stuttgart.

Optimization of a quadrant pipe elbow

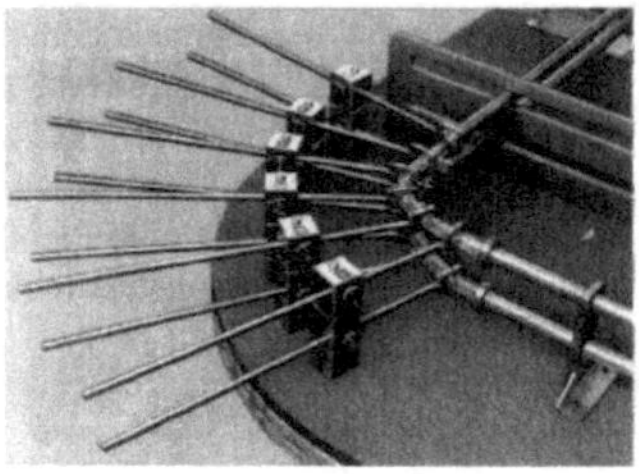

A

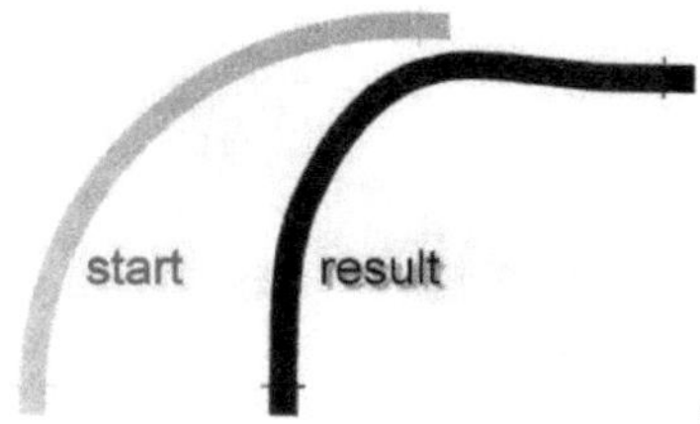

B

Pressure and deflection losses from usual pipe elbows could be decreased markedly by application of evolution-strategy.

PROBLEM:
Quadrant pipe elbows, which turn a flow by 90°, are used very frequently by installers, for example, if tubes are led around wall-corners. Since every pipe elbow has an amount of pressure and deflection loss, the serial arrangement of many pipe elbows can increase drastically the necessary pump-power in a tube-system. From even this reason also a small loss-decrease has a very positive effect.

PRINCIPLE:
For an evolution-strategic action a quadrant pipe elbow, made of an elastic material, which could run from a compressed air-reservoir, was changed mutative in its local curvatures (A) by 6 hand-moved rods. Then the setup was estimated better when the exit-speed due to the more crooked tube was higher. Another experiment was carried out automatically by 10 robot-moved Bowden-cables.

RESULT:
Both experiments led to the same result. The evolutionary reached final form presented a steadier turning to the quadrant in comparison with the original form and on the end a small "swerve" (B). This optimized pipe elbow displayed (with a longitudinal-diameter-relationship from 31:1) a pressure-loss-reduction by 2 and a total deflection loss by 10 percent. This makes it more suitable for example for a vehicle intake manifold (comp. p. 226).

LITERATUR: Rechenberg, I. (1973): Evolutionsstrategie. Optimierung technischer Systeme nach Prinzipien der biologischen Evolution. Frommann-Holzboog, Stuttgart. – Neubearbeitung (1994): Evolutionsstrategie 94. Frommann-Holzboog, Stuttgart.

Efficiency-improvement in a hot water steam-nozzle

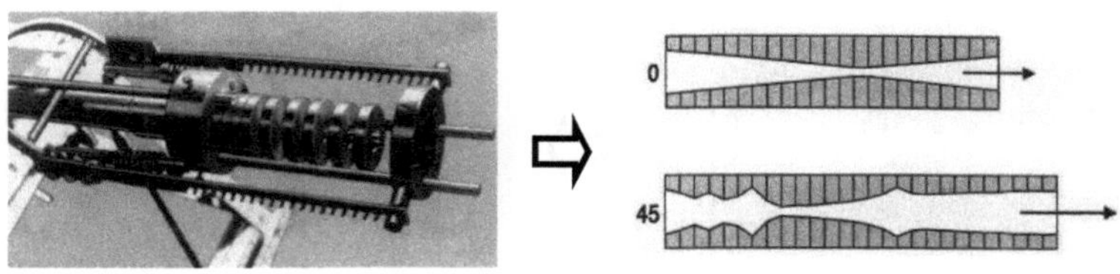

A B

The evolution-strategy leads to an optimal configuration of the nozzle even if there is (still) not any usable theory for it.

PROBLEM:
In a hot water steam-nozzle, temporarily, it comes to steam formation in the closest jet-cross-section, which pushes with the expanding steam the plug of liquid forward. Complex two-phase-currents could not be calculated at the moment of the test, and therefore no usable theory existed for the selection of the optimal form for the nozzle. Therefore, the problem was subjected to an evolution-strategic examination.

PROCEDURE:
The nozzle-form was composed stepless from drilled open segments (A). For this, 330 segments with altogether 10^{60} possible combinations were provided. The procedure was started with a conventional form of a Laval-nozzle (B, 0), with an efficiency of η = 0,55. Then the segments were composed in an accidental sequence to the complete nozzle, and the new form was analyzed again.

SOLUTION:
If the efficiency of the new nozzle-form was better, this was further changed accidentally, if not, one returned back to the prior form, and this was changed further on. By means of the rules of a (1+1)-evolutionary-strategy, over 44 successful inter-forms (not shown), the end form (B, 45) worked at η = 0.80 (improvement by 46 percent). This showed, with its zigzag formed cross-section, a completely unexpected geometry.

LITERATUR: Schwefel, H.-P. (1968): Experimentelle Optimierung einer Zweiphasendüse. Bericht 35 des AEG-Forschungsinstituts Berlin zum Projekt MHD-Staustrahlrohr. – Rechenberg, I. (1994): Evolutionsstrategie 94. Frommann-Holzboog, Stuttgart.

Optimal focussing of an eye lens

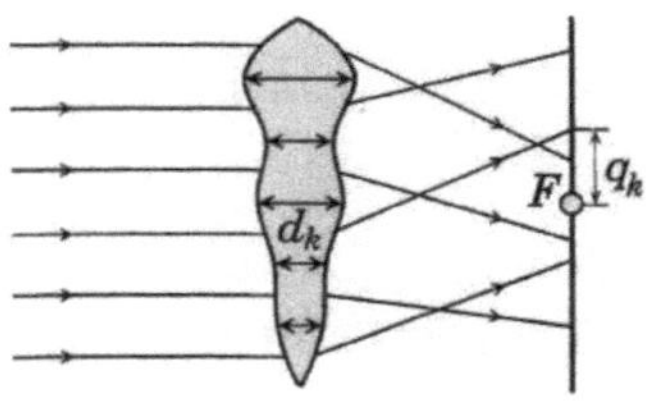

A

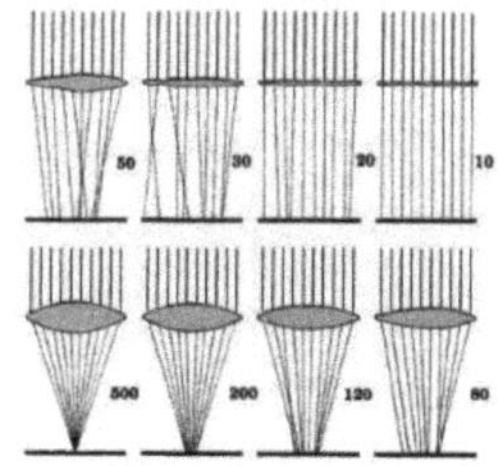

B

Even a complex, highly functional system can develop itself evolution-strategically from a simple, dysfunctional one.

BIOLOGY:
In the course of the evolution, image producing eyes developed by epidermal inversion, completion of the emerging cave with a refractive body and optimization of this body to a high-imaging lens. The question is, whether the conversion of such a body which is dysfunctional with reference to image sharpness can change into a high-imaging lens "by try and error", that means by evolution strategy.

PROCEDURE:
An arbitrarily shaped, deformable and refractive body may have the local thickness d_k that deflects an incident ray with the deviation q_k from the ideal focus (A). The optimization approach then means that the sum of all aberration-squares should tend towards a minimum: $\Sigma\ q_k^2 \rightarrow$ min. (Squaring to compensate negative signs). As soon as this is the case, the refractive body takes the form of an ideal converging lens (B).

RESULT:
The named goal was reached in 500 mutations-selections-steps by using a (1,10)-evolution-strategy, starting from a not all focusing plane-parallel plate (B). After it, the image was and remained ideally sharp. This relativists the many-quoted statement, that one could not imagine the evolution of such a complex organ like the eye after the Darwin evolution-formalism, because it would not exist a "five-percent-eye".

LITERATUR: Rechenberg, I. (1973): Evolutionsstrategie. Optimierung technischer Systeme nach Prinzipien der biologischen Evolution. Frommann-Holzboog, Stuttgart. – Neubearbeitung (1994): Evolutionsstrategie 94. Frommann-Holzboog, Stuttgart.

Remodelling the colour-alteration of peppered moths

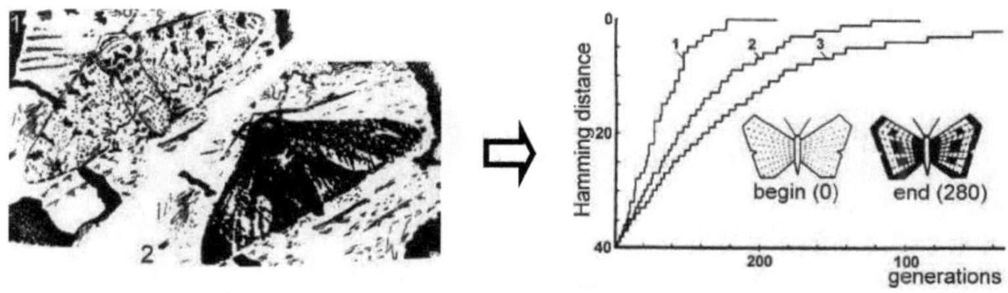

A B

With different evolution-strategies, results of the evolution, for example colour patterns, can be remodelled quite correctly.

BIOLOGY:
In the course of England's industrialization, the originally rather lightened peppered moth of the species *Biston betularia* (A, 1) became more dark coloured (A, 2): "Industry-melanism", proved for 70 out of 780 English butterflies and moths. Darker mutants probably contrast so well from the soot-blackened surroundings and, therefore, were not discovered by birds and less picked of. This created reproductive advantages to the latter.

QUESTION:
If this was an evolutiv process, it should be remodelled by means of the evolution-strategy: Test on their effectiveness. To this, accidental alterations of size and distribution of the darker wing-stains (→ mutation) were generated, tested in their recognisability and taken, if this was reproductive-strategically more favourable, as parents for a new generation (→ selection). Three types of evolution-strategy were applied.

RESULT:
These three types of strategies correspond to the stairway-curves B, 1 (10 individuals, mutation, recombination and selection), B, 2 (10 individuals, mutation and selection) and B, 3 (1 individual, mutation and selection). The basic patterns of the dark form could be shown after 70, 170 and 290 mutation-steps, which show exactly enough the effectiveness of this procedure. As expected, the first-named strategy was the fastest.

LITERATUR: Rechenberg, I. (1973): Evolutionsstrategie. Optimierung technischer Systeme nach Prinzipien der biologischen Evolution. Frommann-Holzboog, Stuttgart. – Neubearbeitung (1994): Evolutionsstrategie 94. Frommann-Holzboog, Stuttgart.

Earthworm and peristaltic working crawling pneu

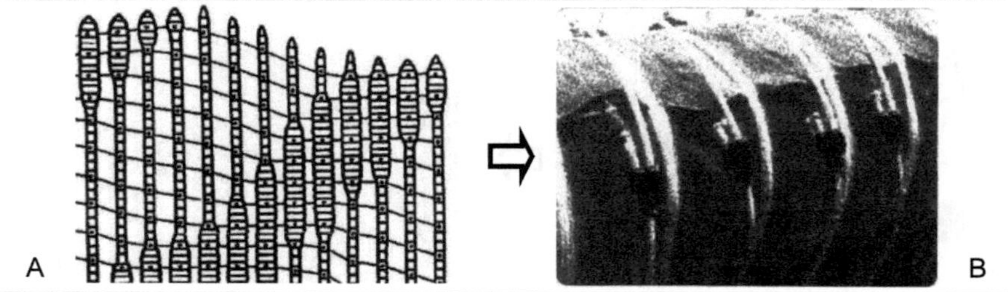

The peristaltic crawl of the earthworm was converted into the movement of phased coupled segments of a crawling hose.

BIOLOGY:
Earthworms contract a body-section, get stuck or become wedged with the ground and then push the section lying before it long-stretched to the front, where again it get stuck with the ground. Then, they contract this section while the up to now contracted section "lets go" and is pulled forward (A). Consequently, contraction-waves seem to proceed over the body. This "peristalsis" is suitable for crawling in narrow tubes.

PRINCIPLE:
Each segment works in the same rhythm of "contracting" and "extending", where in the first case the segmental ring-muscles are active, and in the latter case the segmental longitudinal-muscles. The segments are actuated by the central-nervous system in the correct rhythm. One can therefore imitate such a system, if one equips each single segment movement-technically autonomously, however stimulates it "from outside" in the right phase.

TECHNOLOGY:
By the order of the company FESTO, the Swiss future-workshop PROSPECTIVE CONCEPTS took the first steps to such a pipe-crawler (B). The plastic segments are actuated individually. A certain mean internal pressure (that corresponds to the hydraulics of the segments) provides the necessary stiffness in cooperation with pressure alterations (that correspond functionally to the contractions of the two muscle-groups).

LITERATUR: Festo AG & Co. KG (Hrsg.): Reinhard, A.; Zeller, E.: Prospective Concepts. „L*ai*men" – Einblicke in eine Zukunftswerkstatt. Regenwurm. S. 142–147.

Snake-scales and cross-country ski-covering

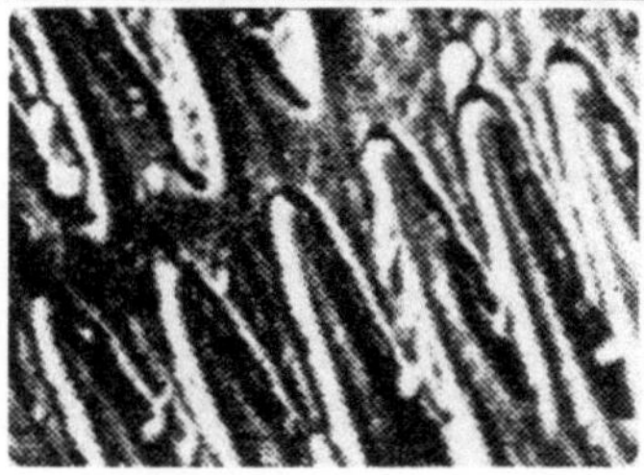

A

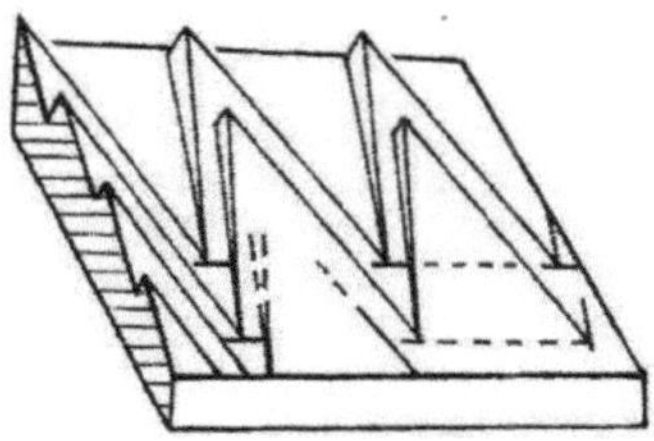

B

The principle of directional-dependent generation of friction leads to a not re-sliding cross-country ski-covering.

BIOLOGY:
Snakes of the *Leimadophys*-type that live on slippery grounds of tropical rainforests have specifically shaped ventral scales. These carry nearly parabolic stampings (A). The snakes move onward after the principle of „sole-crawl". On that occasion, the spinal column and the body-cover are put forward reciprocally against each other. The ventral geometry of the scales makes backward slipping difficult.

PRINCIPLE:
Since the particular form of the ventral scales on the one hand admittedly impedes the irritating backward slipping, on the other hand, however, hardly influences the desired forward gliding, one can derive a physical principle, namely the principle of the "directional-dependent friction-generation." The scale-geometry takes care that the friction between the scale- and the ground-surface depends in a functional manner on the movement-direction.

TECHNOLOGY:
The *Leimadophys*-principle of the "directional-dependent friction-generation" was transferred on a cross-country ski-covering. This carries analogous but in detail differently formed (snow-adjusted) triangular stampings (B). These possess no kernings, so that the foil can jump out of the form without problems. The stampings don't hinder the forward gliding, but clearly decrease the irritating backward slipping at the hillside-ascent.

LITERATUR: Coineau, Y.; Kresling, B. (1987): Les inventions de la nature et la bionique. Hachette, Paris. – Patent Brevet français No. 8301241 (1983). INPI, Paris.

Fin-pump according to the trout-tail fluke

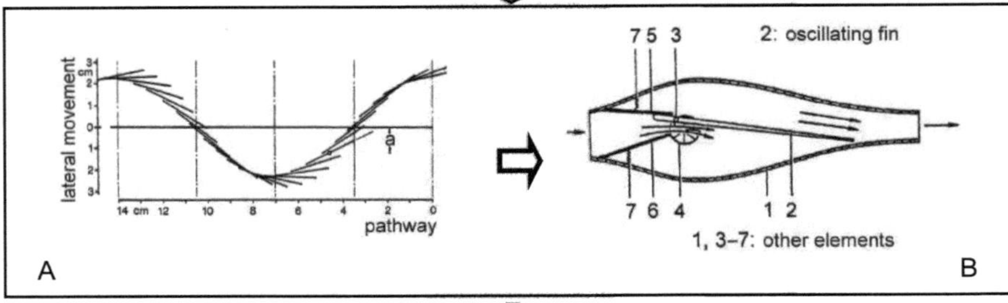

Through the flap of the fins, a fish generates thrust. Reverse consideration: Held tight, it generates a strong water jet.

BIOLOGY:
Most fish and also the whales generate propulsion (thrust) with their flapping tail fluke. They either move their fin to and fro (fish) or up and down (whales). Hereby, due to their intrinsic functional elasticity, they always set with small positive angles of attack against their tracks of movement (A). As a consequence, a thrust-component is generated in every flapping-position. The lateral force-components cancel each other.

PRINCIPLE:
The reversion of the fish-fin-drive-principle leads to the generation of a water jet. If one holds on the fish, now his undulating tail fin transfers power on the surrounding water in the same manner and stirs its motion to the back. So the propulsion-apparatus turns into a pump-apparatus. Many smaller water-animals, for example insect-larvae (ephemera, plecoptera), generate a breath-water-stream with help of undulating little plates.

TECHNOLOGY:
A rectangular plate, that becomes thinner to the back like a fish-fin, locked up into pump-casing swings in low distance to the wall of the casing up and down. It is driven by periodic tilting movements of the front-edge; the rear-edge then swings with the biggest amplitude. Particles, which are going-on, are side-slipped. Such a fin-pump (B) doesn't become blocked and therefore is suitable for strongly polluted waste water.

LITERATUR: Hertel, H. (1964): Biologie und Technik. Struktur, Form, Bewegung. Krausskopf, Mainz. – Affeld, K.; Hertel, H. (1973): Pumpe zum Fördern von Flüssigkeiten mittels schwingender Flächen. DP H 58-654-Ic/59e; Offenlegungsschr. 1703294.

Dolphin-skin-coating for drag-reduction

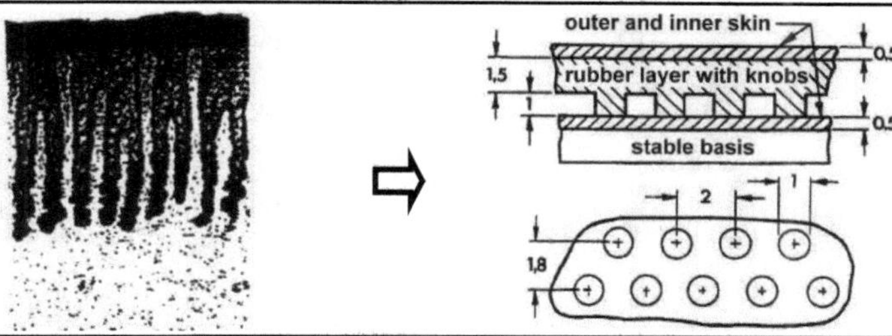

A B

An elastic system with lacunas filled with liquor damps down emerging turbulences and maintains the boundary-layer laminar.

BIOLOGY:
The sub skin (corium) of dolphins is 2-3 mm thick and composed of a loosely meshed tissue that contains fat-cells as well as connective tissue and muscular structures. The outer-skin (epidermis) is about 1.5 mm thick and composed of spongy tissue with a high intra- and above all extracellular, easily movable water-portion of 80 percent. Numerous cones push themselves from the corium through the epidermis (A).

PRINCIPLE:
Locally appearing pressure-dents, as they appear with disturbances of the boundary-layer can lead to turbulent turnovers of this. This leads clearly to higher friction-drag. These dents displace the extracellular water that streams back after termination of the disturbance. Here we have a not-linear, viscous-elastic behaviour with vibration-damping characteristic. At favourable choice of the parameters, disturbances can't build themselves up.

TECHNOLOGY:
Travelling by ship, O. Kramer observed how dolphins playfully pass his ship. He assumed a reduction of drag due to laminar posture of the boundary-layer. According to the knowledge of the construction of the dolphin-skin, he developed a membrane containing rubber knobs with an absorption-liquid included in the lacunae (B, dimensions: mm). Torpedoes and submarines, enveloped by such a membrane, could swim faster with a given drive-power.

LITERATUR: Kramer, O. (1960): The dolphins secret. New Scientist 7, 1118–1120. – Kramer, O. (1960): Boundary layer stabilization by distributed damping. ASNE-Journal 2, 25–33.

Technical Biology is the mother of bionics

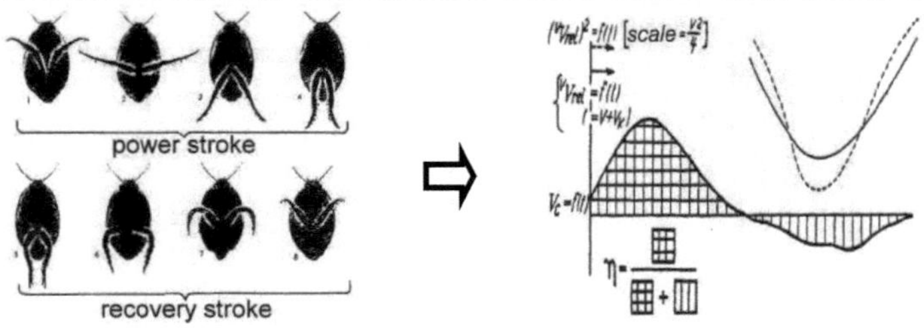

A B

"Recognizing must go ahead applying" (Quotation: Max Planck). Technical Biology creates the necessary knowledge.

PROCEDURE:
At first, as the quotation says, a nature-construction or a method of the nature must be studied in such a way, that they can be recognized and that the results can be formulated. Since technical realization is aimed at (→ Bionik), the appropriate mode of considering is many times characterised by the methods and results of technical physics. The discipline, that explores such bio-systems, is called: "Technical Biology".

EXAMPLE:
Until in the 60er years, this procedure was customary only at certain points. As example for its consistent introduction, W. Nachtigall cites his thesis (1959), in which he had examined swimming of water-bugs (A) with respect to morphological, biomechanical and technic-biological aspects. Key-concepts to the labelling of the swimming-qualities were e.g. "drag coefficient c_W" and "efficiency η" (B), concepts borrowed from technology.

PRESENT-DAY MEANING:
At that time, technically-biological procedure has led to the establishing of the field "physiology of movement“ that was, until then, not even represented in the standards of classification of referencing organs (e.g. “Biological reports"). As a basis for bionic realization, it became the usual mode today. On the one hand their results quite stand for themselves. Technical Biology, on the other hand, acts towards bionics like a picture to its mirror.

LITERATUR: Nachtigall, W. (1960): Über Kinematik, Dynamik und Energetik des Schwimmens einheimischer Dytisciden. Z. Vergl. Physiol. 43, 48–118. – Nachtigall, W. (Hrsg.) (1990): Rundschreiben der Gesellschaft für Technische Biologie und Bionik 1 (Juli).

Books can anchor aspects of view

A

B

The anchoring of a discipline in a broad public is almost just as important as in the scientific community.

MEANING:
A discipline, for example just bionics, then is well positioned, if its principles, procedures and results are recognized in the corresponding scientific community. However, the money to carry it on comes from the taxpayer. It therefore is appropriated to keep a popularized society well acquainted with this discipline. This can happen through lectures and programs; an important role still plays the book, however.

THE BOOK AS A MEDIATOR FOR SCIENCE:
Specialised books are read by experts. One of the most important specialised books in bionics is Rechenberg's "Evolution-strategy". The summarizing book "Bionik" with the subtitle "... for engineers and natural scientists" of W.N. is addressed to the technically trained reader. Public science, earlier, one said "Popularly-science", on the other hand, demands another writing-style, more narrative, however in an objectively perfect type.

BIONICS AS "PUBLIC SCIENCE" IN BOOKS:
Blooming of bionics in the last decades was not possible without lengthy preparation, a sensitization of the society, together with her organizing and financial institutions. Under this point of view, the "public science" brought more than specialised journalism. It is said, that in 1973 W.N's classic book "Phantasie der Schöpfung" (A) and later the specialised work "Das Große Buch der Bionik" (B) have been helpful for this.

LITERATUR: Rechenberg, I. (1973): Evolutionsstrategie. Frommann-Holzboog, Stuttgart. – Nachtigall, W. (1973): Phantasie der Schöpfung. Hoffmann & Campe, Hamburg. – Nachtigall, W.(1998): Bionik. Springer, Berlin. – Nachtigall, W.; Blüchel, K. (2000): Das Große Buch der Bionik. DVA, München.

Automatic form-optimization of flapping wings

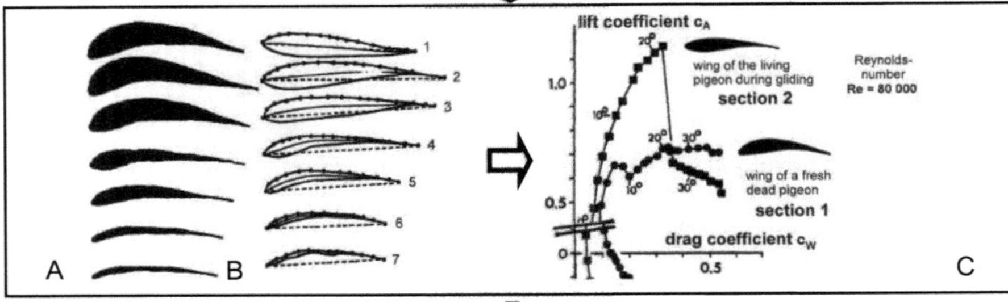

Due to the interaction of elasticity and fluidic pressure wing profiles and angles of attack are optimized automatically.

BIOLOGY:
The flapping wings of birds are cambered and profiled. If one measures the profiles of fresh-dead doves (A) and tests geometrical similar models in the wind-tunnel unfavourable polars arise from these (C). Testing the actual camber of wings of living animals during gliding, one gets different profiles "with S-shape" (B) and favourable polar arise from these (C). The wing-profiles become optimal only in the interaction with the momentary aerodynamic forces.

PRINCIPLE:
A quickly flapping wing of a bird, approximately this of a house-sparrow, alters not only its angle of attack against the airflow but also its geometrical profile parameters like camber, point of maximal thickness etc. from millisecond to millisecond in a manner, that the airflow always generates optimal aerodynamic force components. By neural regulation, the optimal-parameters could not be adjusted so quickly. They arise fully automatically instead.

TECHNOLOGY:
An elastic wing, that optimizes itself intermittently during wing beating in the mentioned manner, are not known in technology until now. Nevertheless, in mini-airplanes ("Micro Air Vehicles, MAVs"), which are driven by flapping wings, the membrane-like wings alter their configuration in their flapping rhythm with respect to the type of the locusts rear-wings. Alone to fin-drives of small-boats as well as with flippers, this principle already is applied.

LITERATUR: Nachtigall, W. (1975): Vogelflügel und Gleitflug. J. Ornithol. 116(1), 1–38. – Bilo, D. (1971): Flugbiophysik von Kleinvögeln. I. Z. Vergl. Physiol. 71, 382.

Wing beat kinematics of flies as basis for MAVs

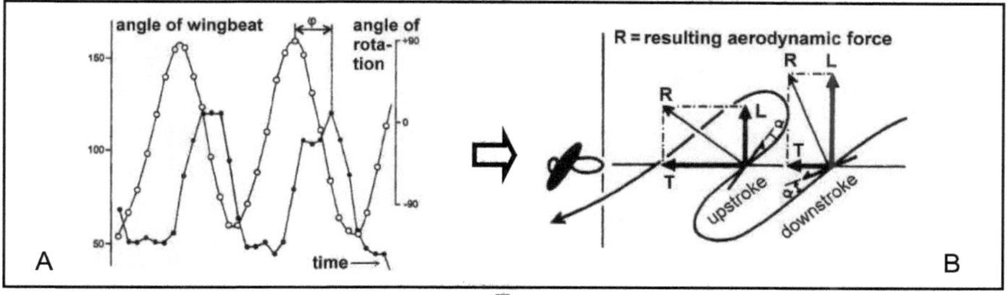

Flapping-rotational-oscillations coupled with an ideal phase-angle generate lift and thrust during down- an upstroke.

BIOLOGY:
Bluebottles as well as hoverflies let the wings swing fore- and backward (flapping oscillation) with approximately 200 s^{-1} from top-behind to front-below and turn them simultaneously approximately around the longitudinal-axis (rotational-oscillation) (A). Near the upper and lower turnover-points the angular velocity of the flapping oscillation is minimal, that of the rotational oscillation is at its maximum. The phase-angle φ between the two oscillations is variable.

PRINCIPLE:
The phasic coupled flapping-rotational-oscillation takes care that the wing during down stroke is blown against the underside with favourable angles of attack, so that lift L as well as thrust T (B) are generated. On the other hand, during upstroke, it is blown against the upper side also under favourable angles of attack and generates lift and thrust, too. Unfavourable force components (negative lift and return drive) are avoided by it in all flapping phases.

TECHNOLOGY:
Today "Micro Air Vehicles" (MAVs) are developed at several places of the world, especially in the USA. They, possibly in whole swarms, especially should take over military spy-tasks. The principles of the drive are slow turning propellers or flapping wings. The last one - if they should develop useful aerodynamic forces and good efficiencies - should be moved analogously to flies kinematics (comp. p. 150, 151).

LITERATUR: Nachtigall, W. (1966): Die Kinematik der Schlagflügelbewegungen von Dipteren. Methodische und analytische Grundlagen zur Biophysik des Insektenflugs. Z. Vergl. Physiol. 52, 155-211. Nachtigall, W. (2003): Insektenflug. Springer.

Bionics and the handling of complex systems

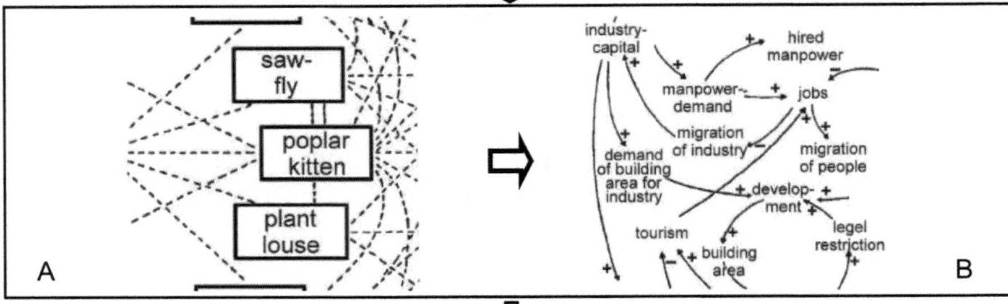

Only relatively late, one discovered nature strategies as models for the management of complex systems in economy.

BIOLOGY:
Biological systems are almost always extraordinarily complex whether one looks at processes in a single cell, the interaction of organs and their subsystems in an organism or ecological systems, like for example an edge of the forest (A). They usually are characterised by a great number of stand-alone elements that are connected by more positive or more negative information-transfer. However, the system is to a certain temporal constancy.

PRINCIPLE:
The complex cross-relationships of the single elements and subsystems, the functioning of which is hold approximately steadily, are not connected in series in form of feedback-free steering-chains. They rather form meshes with diverse information-transfers (A). Everything depends so to speak on everything, but through it well regulated mesh-works are relatively stable against disturbances.

ECONOMY:
The stability-maintaining principle of the negative feedback ("feedback-loop") became only since the thirties-years of this 20. Century applied on living systems: "Where the first feedback was, was the first life" (S. Wagner, physiologist). In going on, one recognized, that a nature-analogous system of regulators and sub-regulators with feedback-loops can protect also economy and administration (B) from disturbing break-ins.

LITERATUR: Dylla, K.; Krätzner, G. (1977): Das biologische Gleichgewicht. Quelle & Meyer, Heidelberg. – Vester, F. (1999): Die Kunst, vernetzt zu denken. Ideen und Werkzeuge für einen neuen Umgang mit der Komplexität. DVA, München.

The military promotes bionic realization

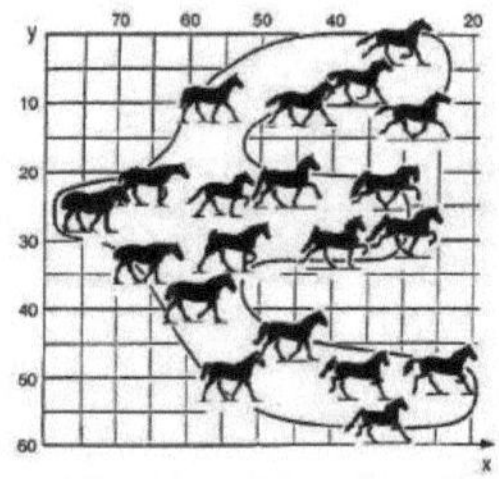

A

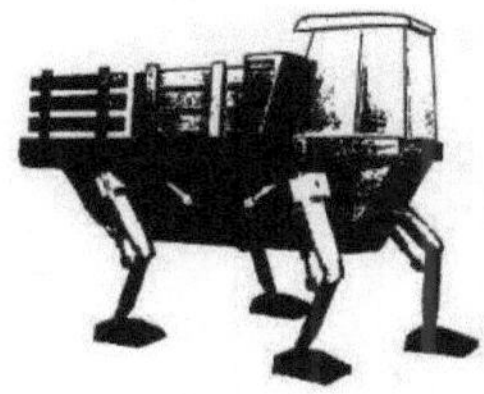

B

Classic example: the development of the Pedipulators according to the principles of the horse-walk. Today, the flies are in.

BIOLOGY:
Horses have a number of gait patterns like trot, gallop, and so on at their disposal. Two easily calculable biomechanical parameters x and y (not defined here) were plotted (A) in a Cartesian diagram. The pattern of a three-pronged fork emerged. Running patterns within this fork were stable; thought-possible running patterns out of the fork-surface unstable. Only with this knowledge, the realization succeeded.

PRINCIPLE:
The biologist M. Hildebrand had worked out 1965 the pattern of the stable running coordination as a "gait-diagram" (A). Based on this, R.A. Liston of the Warren Locomotive Centre in Michigan, 1965, was able to build a stable, four-legged running machine, named Pedipulator (B). It could move better than a machine with wheel-drive in rough terrain. At that time, the Indochina-war had advanced the development strongly.

FURTHER EXAMPLES:
How military demand and support can push bionic developments, this is also shown by Kramer's "dolphin-skin-coating" for submarines and torpedoes. Polaris-submarines partially were equipped with such rubber-coatings that worked saving fuel at given swimming-speed. Recently, micro-air-vehicles for espionage-purposes are conceived by use of older and newer researches about flies (comp. p. 39 151).

LITERATUR: Hildebrand, M. (1965): Symmetrical gaits in horses. Science 150, 701–708. – Liston, R. A. (Hrsg.) (1965): Development of an ambulating quadruped transporter. Dept. of Defence, Electronics Division, General Electrics, Pittsfield, Mass., USA.

The hook and loop fastener "Velcro" - a world-success

A 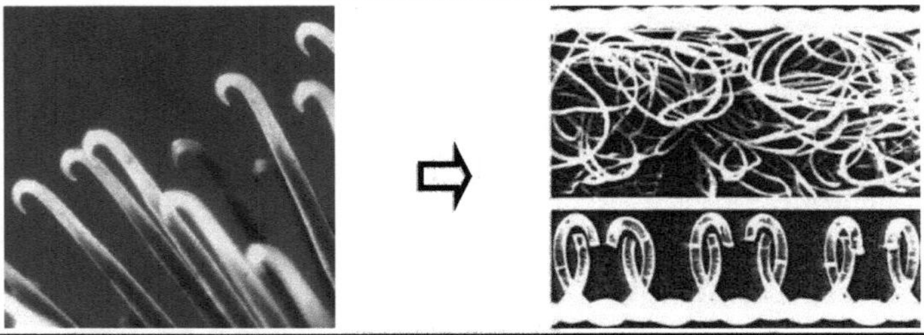 B

Accidental observations can be basis for a commercial world-success. Burrs in the dog-fur were such a case.

BIOLOGY:
Hooked fruits, for example the big burr seeds of the thistle *Arctium lappa*, spread by hooking with the bent ends of their bracts (A) in the fur of wild animals. So they are dragged along for a while. Beside this much known example, there are numerous further examples for epichoric spreading through Velcro backing. The agrimony, *Agrimonia eupatoria*, is spread by barbed bristles of the little nut.

HISTORY:
The inventor of the adhesive fastener, the Swiss engineer George de Mestral, gave in the year 1980 an interview in the newspaper "The Boston Globe" to the journalist Diana Dumanowsky in which, among other things, is reported of de Mestral's forest-walk and his dog, whose fur was full of burrs, about: "Intrigued de Mestral inspected one of the burrs ... The magnification revealed hundreds of tiny hooks ... So the seed of the idea was planted."

NAME AND TECHNOLOGY:
The name "Velcro" comes from the French: "velour" - fluffy material and "crochet" hook. How simply the principle: The development of a working hook and wool-band (B) lasted more than 10 years. The problem of the hook-formation was solved by a looped nylon-ribbon, whose bows were cut open to 2 hooks each. But then, the company "Velcro" should have a start profit of over $ 30 million per year.

LITERATUR: Dumanowsky, D. (1980): The big ZZZRRRR/PPP/velcro: An idea born of a burr stuck on wool. The Boston Globe (June 1).

The gecko-principle; ± analogous realizations

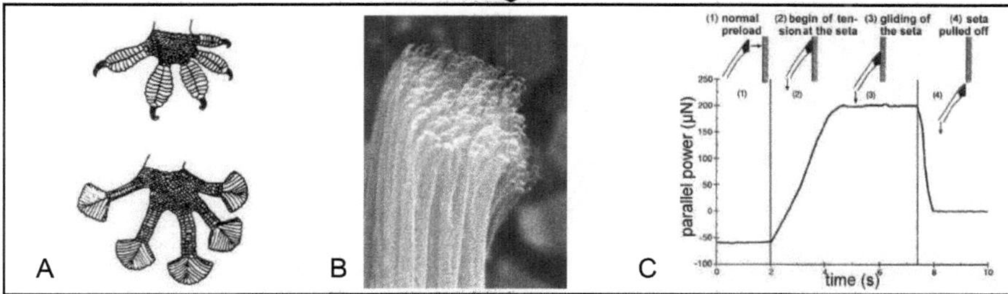

The adhesion of gecko-toes is essentially based on van-der-Waals-forces. Technically, it also goes differently, however.

BIOLOGY:
The toes of the geckos are widened (A) and carry ordered on lamellae dense bundles of fine processes (setae), that dissolve into very fine brush-similar ends (spatulae, B.) The "spatulated" ends of the spatulae are only approximately 0.2 µm thick. With a contact surface split in such a way it can contact closely itself on micro-rough surfaces. One single seta can develop (C) an adhesion force up to 150 µN (comp. p. 192).

PRINCIPLE:
The adhesion is, according to present-day opinion, essentially based on the very close approximation (approximately 10 nm) of the spatulae to the contact surface so that intermolecular forces (van-der-Waal-forces) become efficient. Since finest water coatings always are present, a certain adhesion-effect can also be assumed. Additionally there is a “pulling in” micro-roughness. Suction effects probably don't play any role.

TECHNOLOGY:
A series of teams tried for a long time to use the gecko-effect technically. Meanwhile, one can pull out from plastic-films closely-set long tops that show a certain, but by far not yet practicable adhesion-effect. G. Winkler therefore has followed just another way and used with his climbing system "Gekkomat" suction-disks which allow windows cleaners to climb up at vertical glass-surfaces.

LITERATUR: Autumn, K. et al. (2000): Adhesive force of a single gecko foot-hair. Nature 405, 681–685. – Gorb, S.; Scherge, M. (2001): Biological micro- and nanotribology. Nature's solutions. Springer, Berlin. – Winkler, G. (2000): Gekkomat. www.gekkomat.de.

Mirror-optics of the crayfish-eye and x-ray-telescope

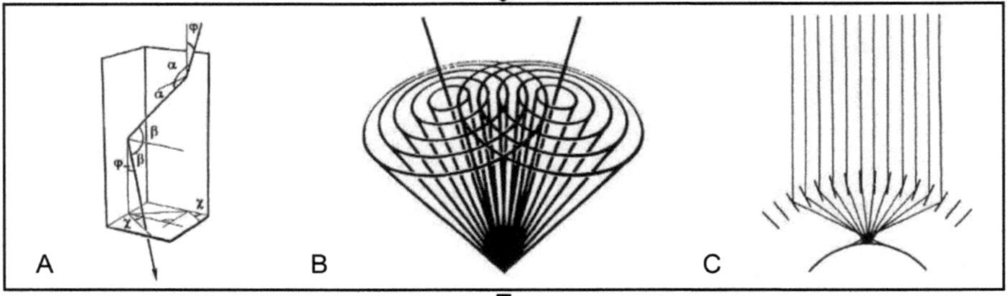

The compound eye of the crayfish works with a kind of mirror-optics, with which one can also focus x-rays.

BIOLOGY:
The single eyes (ommatidia) in the compound eye of the crayfish, *Astacus leptodactylus,* don't have the 6-angular cornea as it is typical for insects but a quadratic one. Therefore in top view, they appear quadratic gridded, as well as cuts through the crystal-cones do. Within the insects these cones bundle the light because of a concentric density stratification. Within the crayfishes a reflective cover of crystals undertakes this function.

PRINCIPLE:
With their "partly coated" inner-surfaces of the crystal cones, the ommatidia reach a doubles-reflection of incident rays of light (A), that leads to a picture-generation according to the type of an "omnidirectional mirror-system using virtual mirror-surfaces". One can imagine (B) the conical bags plugged one into the other as a system in which strongly enough hinder themselves paradoxically. This results in a sharp and light-intense image at the same time.

TECHNOLOGY:
X-rays let themselves reflect at high-polished little steel plates at very small angles of incidence (C). An x-ray-telescope (admittedly very "faint") could be build with it. A new "crayfish-eye-optics" of hemispherical ordered, x-ray-reflective small tubes eliminates this disadvantage and combines extreme wide-angle-qualities with high definition and high "light" intensity. Also x-ray-collimators to manufacture chips are conceivable.

LITERATUR: Vogt, K. (1975): Zur Optik des Flusskrebsauges. Z. Naturforsch. 30c, 691. – Vogt, K. (1980): Die Spiegeloptik des Flusskrebsauges. J. Comp. Physiol. A 135, 1–19. – Chow, M. (1996): The eyes have it. X-ray lens brings finer chips into focus. New Sci. 6, 3 u.18.

Neural nets in biology and computer science

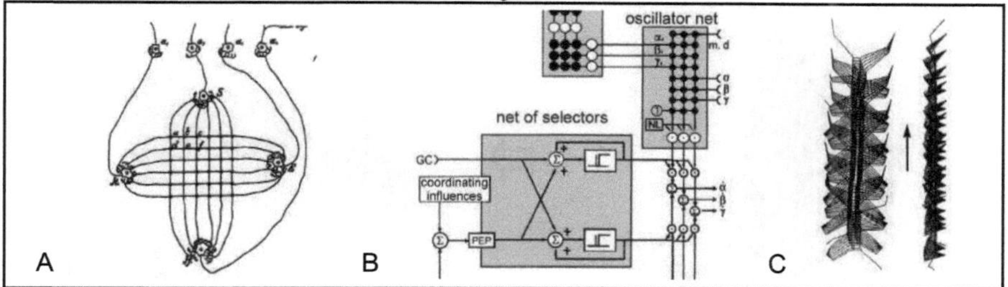

Neural net-structures were suspected already very early. Their principles are largely usual in the computer science today.

BIOLOGY:
Already in the year 1894, the physiologist S. Exner suspected a neuronal network (A) for the optical perception of the human being, at whose "interfaces" information should be stored. In the marine snail *Aplysia*, one knows exactly the interconnection of the neurons in smaller ganglia. From the brain of the housefly, *Musca domestica*, Franceschini abstracted circuits for the area-perception of moving robots.

PRINCIPLE:
Neurons can conduct impulses (in the active state) or not (in the passive state). A neuron becomes conductive if the sum of all inputs exceeds a threshold. Then it falls back into the non-conductive state. In the abstracted neuron network, all elements are uniformly built and ordered in sheets. From the input activities of each of these elements an output activity is determined by an appropriate function.

TECHNOLOGY:
H. Cruse mimicked the movement-process of the rod-locust, *Carausius morosus*, with help of artificial neural networks (B). For running-robots he could generate a "natural" walking-rhythm (that is also insensitive to disturbances) due to the Carausius-principle (C). Neural nets are capable for learning. With their help, one can optimize software for language-recognition, that then adapts itself more and more to the manner of speaking of the secretary.

LITERATUR: Exner, S. (1894): Entwurf zu einer physiologischen Erklärung ... Deuticke, Leipzig. – Franceschini, N. et al. (1996): Biona-report 10, 47–60 – Cruse, H. et al. (1997): Simulation komplexer Bewegungen mit Hilfe künstlicher neuronaler Netze. Neuroforum 2, 9–15.

Termite hills and acclimatisation of buildings

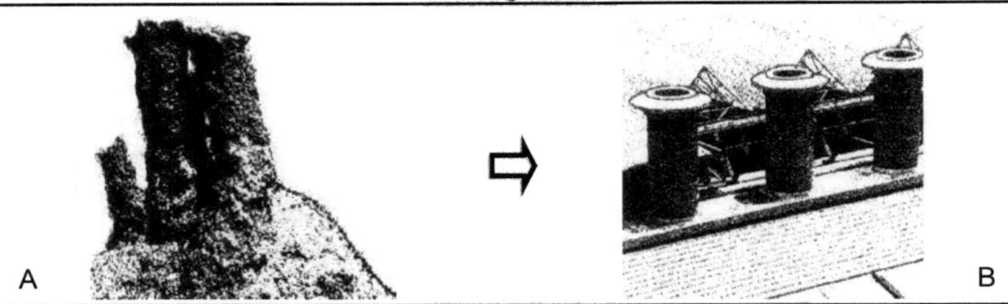

Some termite hills attain solar acclimatisation: Model for temperature and humidity regulation of buildings.

BIOLOGY:
Termite hills are built as closed constructions. They are made of a adamant, hard cured material that nevertheless is gas-permeable. By solar radiation input or by the metabolism-heat, the implied air circulates. Certain types of the genus *Macrotermes* on the other hand work with an open system. They build termites' hills that were completed by chimney-like top parts (A). The latter nature-model was used technically.

PRINCIPLE:
Within closed systems, the implied air that circulates through the cool and moist „cellars" can exchange O_2 and CO_2. Within open systems, the air warms up in the "chimneys" through solar radiation, rises and moves cool and moist air from the "cellar" (this has connection to the groundwater by long corridors) upward. The air flow through the fungus-gardens and the queen-chamber and air-conditions the construction consequently.

TECHNOLOGY:
A big office building in Eastgate, Harare / Zimbabwe was designed by the architect M. Pearce and the engineer-office O. Arup according to the principle of the open termite constructions. It carries numerous round chimney-tops (B). Air ducts form a continuous system, in which hollow ceilings and walls are included. Through the chimney-effect, cool and moist air, which formed a cold-air-lake in the yard at night, is high-sucked.

LITERATUR: Lüscher, M. (1955): Der Sauerstoffverbrauch bei Termiten und die Ventilation des Nestes bei *Macrotermes nataliensis* (Haviland). Acta Tropica 12, 289–307. – Smith, F. (1997): Harare, Zimbabwe. In: Brown, D. J. (Hrsg.): The Arup Journal 1, 2–8.

The polar bear-fur and the translucide insulation

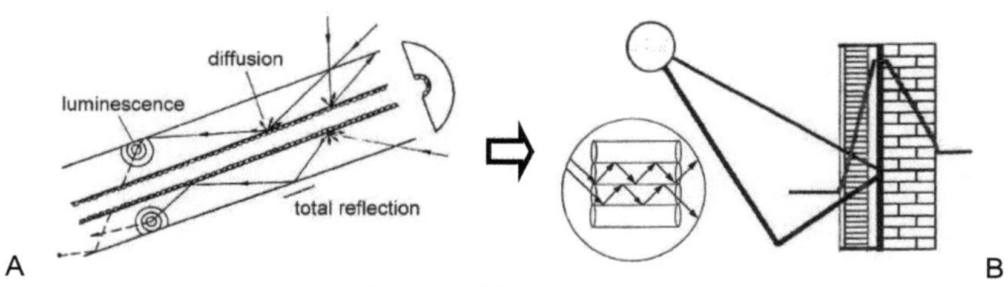

In the hair due to the fibre-optics-principle, also infrared-radiation is led to the black skin that heats it up.

BIOLOGY:
The white hairs of the polar bear possess, in the contrast for example to those of white horses, a reflective central-cylinder. The fur is not uncommon long (in contrast to the also perennial musk-ox), but very densely instead and includes many tiny heat insulating air-bags. The white fur was probably created as adaptation to the white surroundings. However, it also can fill up with solar heat and that even on cloudy days.

PRINCIPLE:
Incident light- and heat-radiations are held between inside of the outside-cover and outside of the central-cylinder through total-reflection (A) and finally reach the skin-surface. (This mechanism is discussed in different manners.) The skin is black and consequently absorbs the radiation. This way, the skin-surface heats up. The heat-loss is held small by the insulating fur that includes a lot of tiny air-bags.

TECHNOLOGY:
The polar bear-mechanism was transformed many times technically. Even if parts of it may appear uncertain: it gave stimulation for the conception of facade-elements according to the principle of transparent insulation. A layer of little glass tubes covered by a glass panel outside leads the radiation over an air space onto a black painted wall. This heats up; the heat diffuses inside (B) (comp. p. 128, 129).

LITERATUR: Tributsch, H. et al. (1990): Light collection and solar sensing through the polar bear pelt. Solarenergy Materials 21, 219–236. – Koon, D. W. (1998): Is polar bear hair fibre optic? Applied Optics 37(5), 3198–3200. – Prospekt Okkalux Kapillarglas GmbH, o. J.

6-legged, insect-analogous running-machine

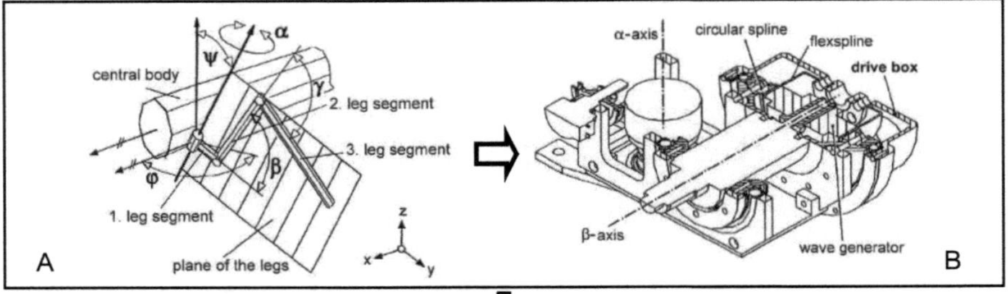

This classic run-machine therefore was so successful because mechanics and control were imitated in detail.

BIOLOGY:
Before the first transfer, the model, the rod-locust *Carausius morosus*, was exactly examined with reference to both the mechanical construction of the running legs and the complex neuro-motor activity of the leg-control very much. As examples, the trochanter-femur-joint (A) have to be mentioned on the one hand and on the other hand circuits of singles-leg-regulators that play together. All these have their technical correspondence.

PRINCIPLES:
The main-limbs, coxa, femur and tibia of the *Carausius*-legs, lie in a plane that can be turned about its α-axis. Another possibility of turning around the δ-axis helps with the adaptation to unevenness of the underground. - The neural leg regulation consists of three stages. A swing-loop lets swing the leg for- and backwards. This is part of a single leg regulator. These on the other hand are parts of the entire control-system of the leg coordination.

TECHNOLOGY:
The trochanter femur joint stimulated the construction of a new β-joint (B) that can transfer big forces with low net weight. For this, a harmonic-drive principle was used, whose external flexspline transmits the flux of force to the internal flange. - The technical regulation was operating quite analogously the three biological stages. Also the tasks of the three hierarchical levels correspond to each other.

LITERATUR: Cruse, H. (1990): What mechanisms coordinate leg movement in walking arthropods? Trends Neuroscience 13, 15–21. – Pfeiffer, F.; Cruse, H. (1994): Bionik des Laufens. Technische Umsetzung biologischen Wissens. Konstruktion 46, 261–266.

The multi-functional honeycomb-principle

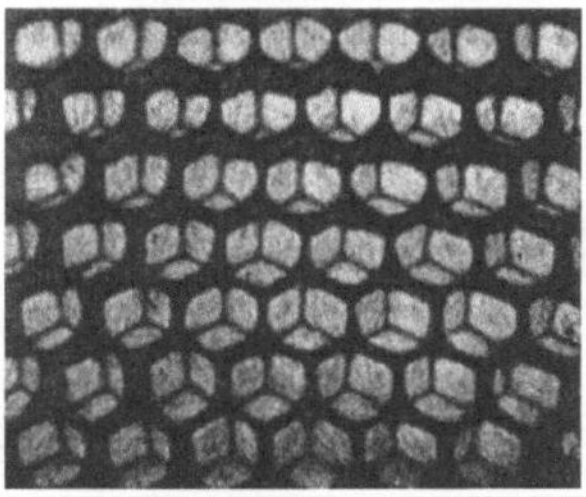
A

B

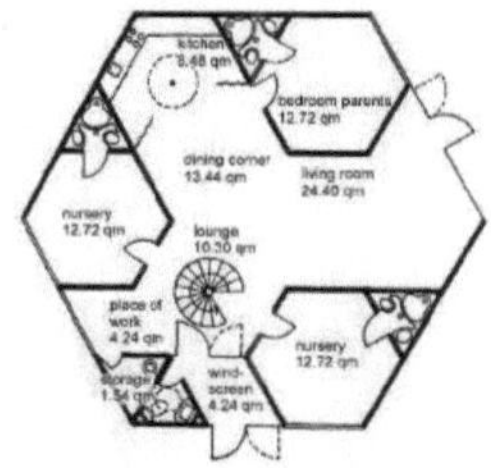

C

The hexagonal honeycombs influenced not only technical procedures of weight-saving and area-utilization.

BIOLOGY:
The wax cells of the honeycombs (A) of the honey-bee, *Apis mellifica*, are with neglectable deviation ideal-forms, that achieve a given room package under given boundary conditions with minimum cost of materials. The cells of opposite sides of a honeycomb are folded in the form of interlaced rhombedodecahedrons. Besides big stability, this also provides for the biggest possible volume of the single-cell (comp. p. 239).

PRINCIPLE AND TECHNICAL REALIZATION:
The construction-principle of the honeycombs, stable space filling with minimum cost of materials, for example, allows flat lightweight-structures in the technical realization. For stabilization, a technical flat honeycomb-system must be stuck with a thin membrane at both sides (B). Space saving hexagonal buildings (C), honeycomb bricks and pressing tools for tires, quite resistant against distortion, also use this principle.

EXAMPLES:
A flat lightweight-structure is the "honeycomb expanded metal" for airplane-wings or by a configuration of pressed material for light door panels (B). Another example is the calotte-configuration, the company Cabasse manufactures highly stiff membranes for large loudspeakers. The company Freiburg & Lausanne AG produces light bricks (Thermo Cellit) with hexagonal honeycomb-structure. Hexagonal "Trelement"-houses and tire structures are further examples.

LITERATUR: Meretz, W. (1963): Die Wabenzelle der Honigbiene. Bull. Math. Biophysics 25, 95–110. – Kat. Fa. Neckermann (ca. 1960): Das Trelement-Haus. – Prosp. „Thermo-Cellit-Backstein", Fa. Freiburg & Lausanne AG. – Mundl, R. (1999): Reifen mit Hexagonallamellen. Pers. Mittg.

Notches without notch stresses

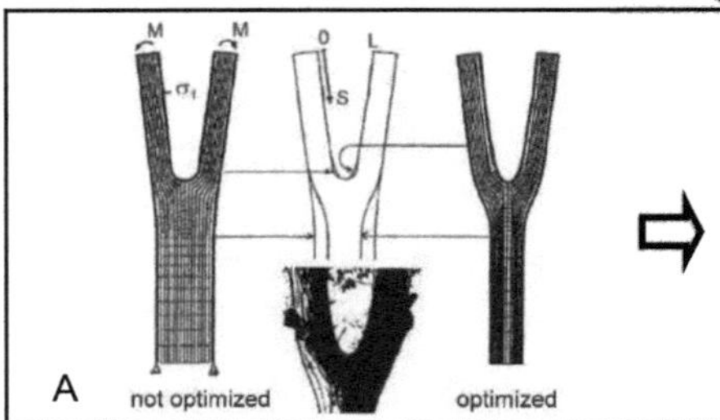

A

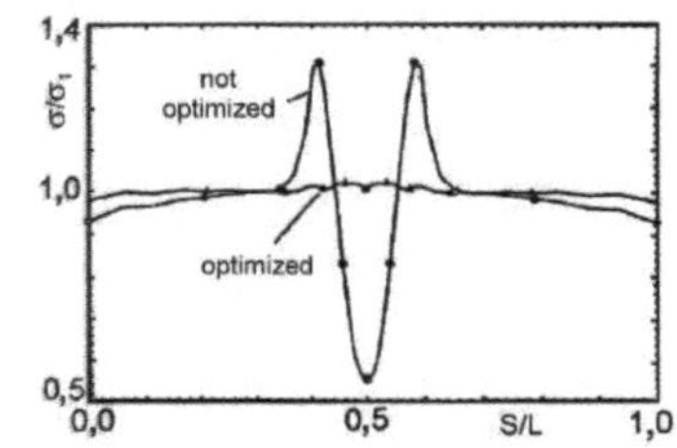

B

The form of tree-forks reduces notch stresses. Many technical realizations, for example beam shoulders, use the principle.

BIOLOGY:
If a forked branch of a tree (A) is shaped, the bottom wedge doesn't have a semicircle like form but a parabolic one. Also the angles of branches that stand vertically to the trunk or analogous changeover of the root bases of trees never are half or quadrant shaped but resemble the named moulding. This can reduce bending stresses that are created in the overlapping region (notch stresses).

PRINCIPLE:
Numerical modelling shows, how the notch stress (Von Mises stress relating to the local tension) plugged over its inner-contour is reduced nearly against zero (B), if one changes (A) a semicircular notch, at first not optimal, analogously to the tree notch by change in the material deposition. The differences don't appear very big, but: "This nature-phenomenon is ... a master-design - a notch without notch stress" (C. Mattheck).

TECHNOLOGY:
Several dozen patented uses have yielded from this nature based form of notches (C. Mattheck), for example: pedicle screws, that (by normal load) no more break, forms of draw shackles, angle brackets, beam shoulders and shoulders of hollow-shafts, that tear no more at the curvature, easier and at the same time more stable crane-claws, motor mountings for vehicles, rims for car-wheels and all many more.

LITERATUR: Mattheck, C. (1992): Design in der Natur. Der Baum als Lehrmeister. Rombach, Freiburg. – Mattheck, C. (2010): Denkwerkzeuge nach der Natur. KIT-Verlag. – Harzheim, L. et al. (1999): Topologieoptimierung ... ATZ 101, 530–539.

Dye-sensitized solar-cells

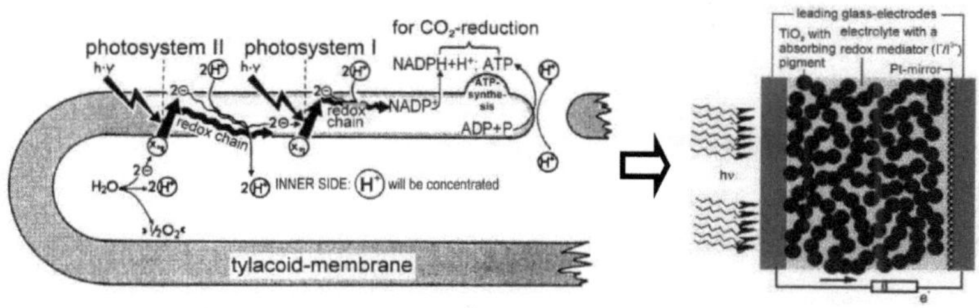

A B

One tries to use this nature-technology, based on chlorophyll-dye, technically for solar utilization.

BIOLOGY:
The green plant produces sugar-substance from CO_2 and H_2O, whereby O_2 is released, with the help of sunlight as an energy donor. On that occasion, the green dye chlorophyll functions as central absorbent. Internally (A), the green plant transports hydrogen (as H^+) in the primarily-reaction and, with it connected, electrons. Energy for synthesis is stored in ATP simultaneously. This is used by sugar-substances in the secondary reaction (CO_2-reduction).

PRINCIPAL TRANSFER-POSSIBILITIES:
The hydrogen is not transported as gas (H_2) but in form of protons (H^+) internally. A solar hydrogen-production over technical dye cells therefore would have to comprise an additional technological step. The electron-transportation over redox cells would be another possibility: solar photovoltaic with help of dye-cells. Both ways have been explored intensively for 2-3 decades, but there are not yet marketable products (comp. p.126, 127).

TECHNOLOGY:
As example for an already longer examined dye-sensitized solar-cell the Grätzel cell (Swiss institute for technology, Lausanne) (B) is mentioned. While in silicon based installations semiconductors simultaneously absorb light and provide for the separation of electric charge, in a dye based cell a mono-molecular dye-layer takes over the light-absorption; after it a semiconductor layer executes the separation of charge.

LITERATUR: Kalyanasundaram, K.; Grätzel, M. (1999): Sensitised solar cells (DYSC), based on nanocrystalline oxide semiconductor films. http://www.epfl.ch/lcp-2/solarcellE.html. (Angabe von Review-Artikeln). – Nachtigall, W. (1997): Vorbild Natur. Springer.

Owl-wing structures make airplanes quieter

A

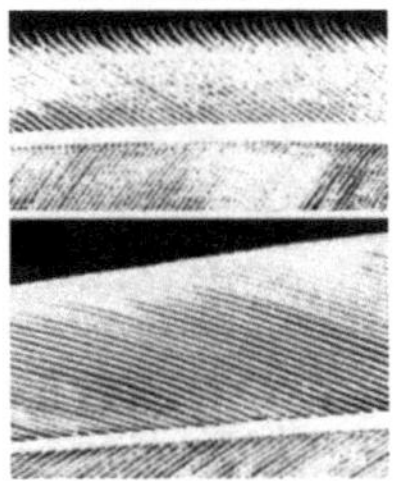

B

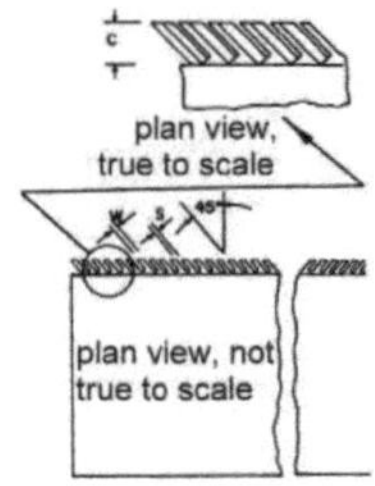

C

Combs at the leading edge of owl-wings help to a "silent" flight. At airplanes, they also work.

BIOLOGY:
The barn owl *Tyto alba* has such a silent flight at her disposal, that approximately 50 percent of its nocturnal approach to food seeking mice is successful. Beside other structures, the owl carries a (A) sound-absorbing appliance at the wing leading edge in shape of a fine feather comb with semicircle shaped upwards bent "teeth". The corresponding, sharp edged feather of the mute swan *Cygnus olor* (B) generates a loud flight sound instead.

PRINCIPLE:
The feather comb hacks emerging coarse turbulences which are connected with loud noise and such generates fine vorticities. This fades away through fuzz and fringes further. Consequently, also the sound production reduces drastically. Moreover the comb of the leading edge makes the emerging boundary layer turbulent and stabilizes it through it, so that the flow doesn't separate even at high angles of attack during the prey-flight.

TECHNOLOGY:
The stabilization of the boundary layer by tripwires or stumbling ribs at Reynolds-numbers of the bird-flight is a well known appliance. Sound diminishing at the owl-wing was published by Kroeger et al. at 1972. However, turbulence combs for noise reduction according to the type of the owl-wings were used successfully already 1971 by Hersh and Hayden at propeller blades and airfoils (C). With very loud ducted fans, the effect is especially conspicuous.

LITERATUR: Kroeger, R. A. et al. (1972): Low speed aerodynamics for ultra-quiet flight. AFFDL-TR-71-75, Dayton, Ohio. – Hersh, A. S.; Hayden, R. E. (1971): Aerodynamic sound radiation from lifting surfaces with and without leading-edge serrations. NACA CR-114370.

Biological and technical micro-mechanics

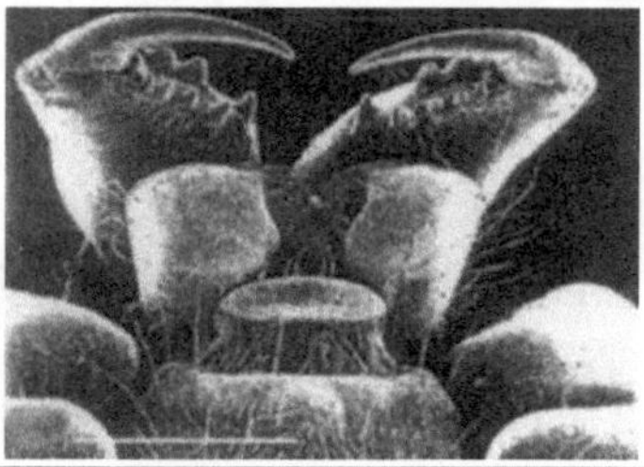

A

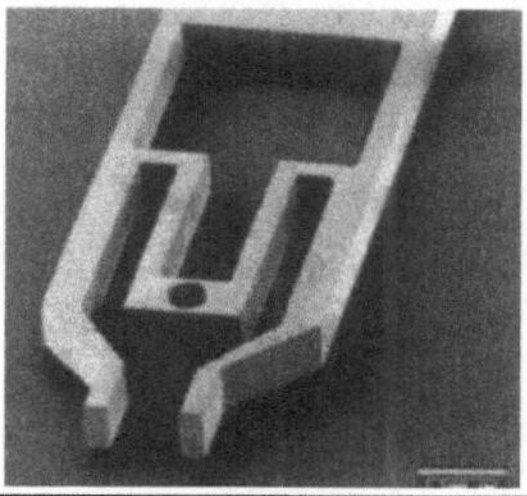

B

Biological micro structures deliver stimulation to the increasingly important questions as mechanical miniaturization.

BIOLOGY:
There is an immeasurable number of working mechanical structures especially within the range of arthropods (insects, spiders, centipedes, shellfishes), which are established in the macroscopic as well as microscopic area. Because of their smallness, they frequently distinguish themselves through specific peculiarities that are due to the lack of space. These amongst others include clamps, holding tools (A) and miniature pumps (comp. p. 197).

PRINCIPLE:
They integrate diverse amendments into their basic material of chitin, so that they, despite of the uniform material basis, are provided with different local elasticity, hardness and other physical-technical parameters. They are moved through the pull of finest muscle ropes or through hydraulic pressure. High mechanical transmission ratio or replacement of an active mechanism by a passive one is usual.

TECHNOLOGY:
Since the 90ies, the technical miniaturization could not use the material-technical peculiarities of biological models so very much, but the function-morphological principles instead. This means, that in the micro range volume-proportional effects of mass and gravity decrease and the influence of lengths- and surface-proportional adhesion- and friction-effects increase. This is reflected for example in the construction of a miniature gripper (B).

LITERATUR: Schilling, C. et al. (1995): Klein, aber komplex: Der Beitrag bionischer Forschung für die Mikrosystemtechnik. In: Nachtigall, W. (Hrsg.): BIONA-report 9, 51–65. – Nachtigall, W. (1993): Biomech. Mikrosysteme. VDI Technol. Analyse Bionik, 93–98.

Spider-threads and "artificial spider-silk"

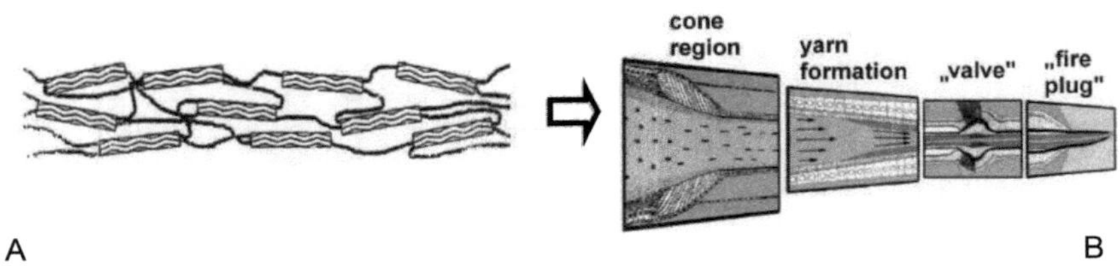

Spider-threads surpass steel-strings in some mechanical qualities. Furthermore they are extremely light and thin.

BIOLOGY:
In the comparison to hairs (diameters ≤100 μm), single spider-threads are extremely thin (0.5–5 μm). Their breaking length amounts to 80 km. They are three times as elastic as nylon with 31 % of elongation without strain residue. Their tensile strength achieve 10^9 N m^{-2} contrary to 10^6 N m^{-2} for rubber, their specific rapture energy is 10^5 J kg^{-1} contrary to ≤ 10^4 J kg^{-1}. In the spinning nozzle, the threads leaving the spinning gland are formed and stretched (A).

PRINCIPLE:
The bio-polymer "spider-silk" contains, if it is formed in the spinning nozzle, amorphous amino acid-chains with crystallized proteins. In the native (stretched) thread, the crystal volume share amounts about 25 percent. The molecular axis of the polymer chains is (at the central threads of the spider *Eriophora fuliginea*) parallel oriented to the macroscopic fibre-axis (A). Such molecular qualities provide for the striking mechanical parameters.

TECHNOLOGY:
For "artificial spider-silk", a high-tech-product of the future, there is a big demand world-wide. It could show its mechanical qualities in ropes for giant parachutes or bullet proof vests, for example. For its industrial production it is important to imitate the principles of both the typical molecular construction and the specific formation-process in the spinning nozzle (B). As yet, this has only partially succeeded (comp. p. 74).

LITERATUR: Gosline, J. M. et al. (1999): The mechanical design of spider-silks: From fibroin sequence to mechanical function. J. Exp. 202, 3295–3303. – Vollrath, F.; Knight, D. P. (2001): Liquid crystalline spinning of spider-silk. Nature 410, 541–548.

Composite-material of favourable fracture toughness

A

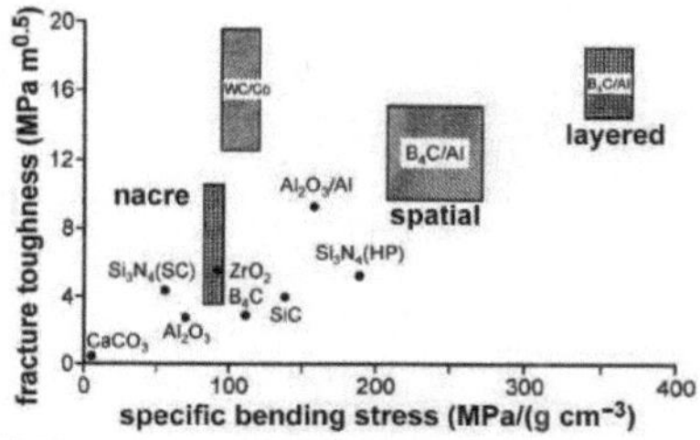

B

The nacre of the mollusc shells combines strength, hardness and a favourable fracture behaviour using crack stoppers.

BIOLOGY:
Mother-of-pearl consists of aragonite - "bricks" with layer-thicknesses of 150-500 nm, separately by thinner layers of an organic polymer material of 20-250 nm thicknesses (A). The "bricks" represent disk like single crystals that are oriented uniformly in the total layer. The organic matrix of amino polysaccharides, also chitin, is enveloped by a protein that improves the adhesion to the crystals of aragonite (comp. p. 182).

PRINCIPLE:
The crystal-structure results in hardness, the connecting matrix tenacity, the combination of both breaking strength and self repairing. Micro breaking cracks end at neighbouring aragonite-"bricks"; the matrix slowly fills the cracks again so that a secondary hardening occurs afresh. The breaking stress of mother-of-pearl amounts to approximately 190 MPa. The fracture toughness amounts to approximately 8 MPa $m^{-0.5}$, the specific bending stress 90 MPa/(g cm^{-3}).

TECHNOLOGY:
The two last-named parameters are plotted in the diagram (B). As is recognizable, this combination cannot be reached by the combination of technical composite materials. The same is valid for the fracture energies. This of the single layer amounts to 1 kJ m^{-2} and that between neighbouring layers to 0.1 kJ m^{-2}. This combination, together with the break-stop-behaviour and the ability for self-repairing of micro-breaks, cannot be imitated technically yet.

LITERATUR: Jackson, A. P. et al. (1988): The mechanical design of nacre. Proc. Roy. Soc. London B 234 (1277), 415–440. – Kamat, S. et al. (2000): Structural basis for the fracture toughness of the shell of the conch *Strombus gigas*. Nature 405(6790), 1036–1040.

Shark scales and riblet foils

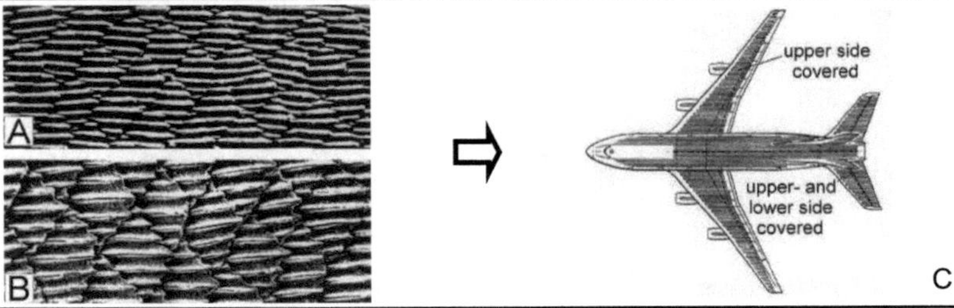

Sharks decrease their friction drag with the help of grooved scales. In airplanes riblet foils cause the same effect.

BIOLOGY:
Fast swimming high sea sharks, as for example the scalloped hammerhead *Sphyrna lewini* (A) or the velvet belly *Etmopterus spinax* (B), carry coatings of densely adjoining, grooved scales. The striae of neighbouring scales are directed in such a way, that they trace flow-technical sweeping lines from the tip of the muzzle until the end of the tail fluke. This is true for living as well as for fossil forms.

PRINCIPLE:
In the u-shaped ducts formed by adjoining, grooved scales boundary layers develop of their own. The sharp edges of the grooves reduce the interaction between the water flow and the viscous sublayer what finds expression in a reduction of the degree of turbulences. Moreover, the tendency of cross flows is decreased and so the impulse exchange. Therefore, turbulent shear stresses are reduced in the end.

TECHNOLOGY:
For technical flow bodies, a covering with a grooved adhesive film seems to be suitable. By bonding such films to plates at the wall of a water tunnel, their drag resistance could be lowered by approximately 10 percent. At big airplanes (B), the total drag could be diminished to scant 2 and the fuel consumption to 8 percent. These are significant values economically and ecologically. Foils have also disadvantages, e.g. their brittleness (comp. p. 210).

LITERATUR: Reif, W.-E.; Dinkelacker A. (1982): Hydrodynamics of the squamation in fast swimming sharks. N. Jb. Geol. Palaeontol. 164, 184–187. Schweizerbart, Stuttgart. – Bechert, D. W. et al. (1985): The drag reduction of the shark skin. AIAA-paper 85-0546.

Fish slime and "Polyox"

A

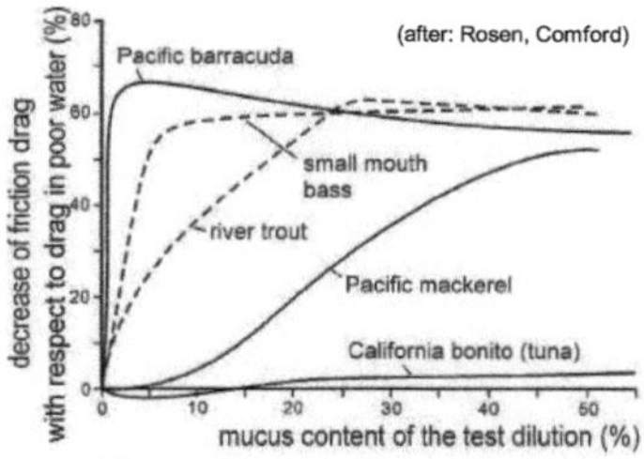

B

Fish slime reduces the flow drag. Analogous substances allow the fire brigade to pump the fire water higher.

BIOLOGY:
Ambush predators among the fish, like for example the pike, *Esox lucius* and the barracuda, *Sphyraena barracuda* (A), must be able to accelerate in a short time extreme rapidly. To diminute the occurring friction drag on that occasion, that reduces the speed possible at a given drive power, they surround their bodies with a substance. This especially dissolves at striking and is added then to the boundary layer flow.

PRINCIPLE:
The phlegm-substance contains long-stretched carbohydrate-molecules that probably roll up in the border-layer and act indeed as flat miniature ball-bearings consequently. Experiments with phlegm-wrapped impact bodies in a water tank produced with 5 percent phlegm of the pacific barracuda an astonishing drag reduction by 65 percent (B). Phlegm of slowly moved fishes was effective only with very high dosages on the other hand.

TECHNOLOGY:
Chemically synthesized polyoxyethylene ("Polyox") works similarly with its long molecule threads. If one hacks these in the mixer, the effect sinks extremely. Since some time, "Polyox"-liquid is added to the fire water tanks of the New Yorker fire brigade what it enables to splash more highly with the same pump-power. Also in oil-pipelines, corresponding additives are used to increase the throughput by reduction of the wall shear stress.

LITERATUR: Rosen, M. W.; Cornford N. E. (1971): Fluid friction of fish slimes. Nature 234, 49–51. – Rechenberg, I. (2001): http://www.bionik.tu-berlin/de/intseit 2/skript/bibu8.pdf. Daraus Zitat: „Bionik, was gar nicht so selten ist, reformiert vages technisches Wissen."

H_2-production by bacterium-alga-symbiosis

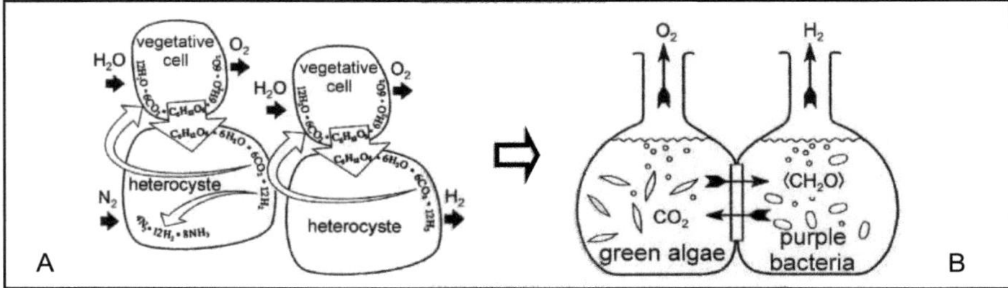

Skilful "artificial" coupling of two natural systems is the basis for a biological hydrogen production.

BIOLOGY:
Blue-"algae" (cyan bacteria), for example *Nostoc muscorum*, are capable to synthesize ammonia. In their normal, vegetative cells normal photosynthesis takes place. Their product, glucose, becomes sucked over membrane ducts into so called heterocysts. There it is fractionized with water to carbon dioxide and hydrogen. The hydrogen is transferred onto nitrogen, where ammonia originates (A, left). If nitrogen is missing, hydrogen-gas is released (A, right).

PRINCIPLE:
The trick of the artificial, biological hydrogen synthesis is to use purple bacteria in the same way as blue-algae use their heterocysts. At absence of N_2 in a compartment of the "biological electrolysis cell" H_2 arises (B, right), when glucose is given to the purple-bacteria. In an other compartment O_2 is produced by green-algae using the decomposition product of glucose (B, left). H_2 and O_2 are led away separately.

TECHNOLOGY:
According to this basic principle various purple-bacteria were positively tested in field-experiments in the Sahara by I. Rechenberg; the artificial biological hydrogen synthesis was proved to work. The total efficiency admittedly is very small (2.5 %), however, large areas can equalize this. With 100 so-called heliomites, hosepipes wound on peaked tapered bars, each of it with 620 m of length, a top-performance of 100 kw would be generated.

LITERATUR: Rosen, M. W.; Cornford N. E. (1971): Fluid friction of fish slimes. Nature 234, 49–51. – Rechenberg, I. (2001): http://www.bionik.tu-berlin/de/intseit 2/skript/bibu8.pdf. Daraus Zitat: „Bionik, was gar nicht so selten ist, reformiert vages technisches Wissen."

Water-production by fog condensation

A

B

At bugs of the Namib, water-drops from fog-swaths condense. This is imitated with big fog-nets.

BIOLOGY:
Over the Namib Desert, in the mornings wide fog-swaths move, however, they never rain down. Darkling beetles, like *Onymacris unguicularis*, situate themselves with lowered head against the swaths (A) and gather the condensed water-droplets, running off the slant of their bodies, with their mouthparts. Also a gecko, *Palmatogecko rangei,* and the sand-viper, *Bitis peringueyi,* as well as the Canaries pine trees with their needles, use the condensation principle.

PRINCIPLE:
Condensation of liquids from the gas-phase occurs when the surface of the body, at which condensation takes place, is relatively cooler. For the condensation of water-droplets from fog-swaths, also electrostatic charge differences may play a role; this is discussed for the 20 cm long needles of the Canaries pine trees, *Pinus canariensis*. Here the condensate drips down in the fine-root-area (comp. p. 189).

TECHNOLOGY:
Through put up big nets (B), this principle of the water production is imitated, for example in Chungungo, Chile, where the villagers almost meet their entire water-demand from "fog catchers" since 1986. 75 large fog nets deliver 11 m^3 water daily. The dripping water is very pure. It is collected in furrows and is led to supply tanks. The technology already was tested until the end of the last millennium in 30 countries.

LITERATUR: Henschel, J. et al. (1998): Namfog: Namibian application of fog collecting systems. Desert research Foundation of Namibia. – Enders, M.; Henschel, J. (2000): Nebel: Wasserquelle in der Wüste. Spektrum der Wissenschaft 2, 38–41.

Package in nature and economy

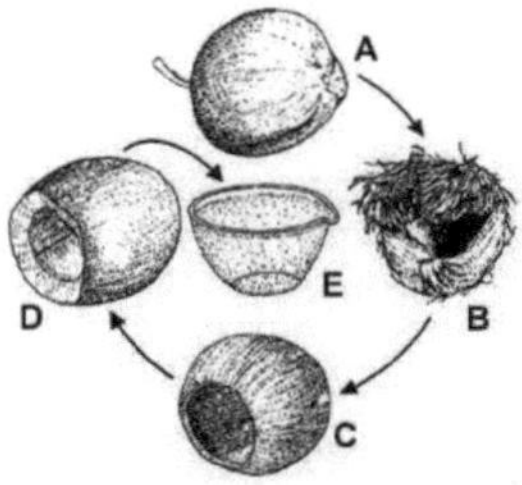

A

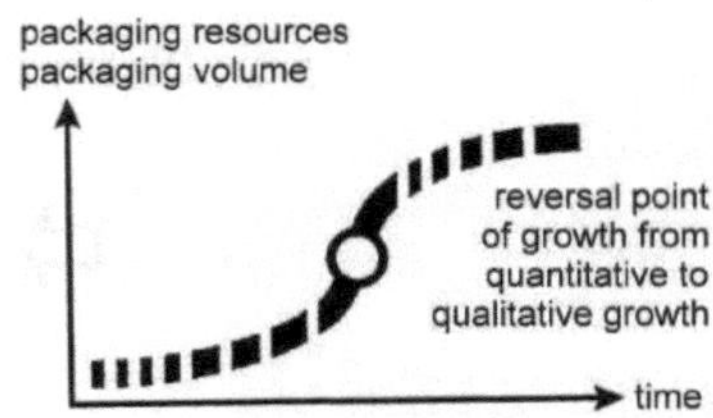

B

Compared with the local and time wise multi functional packages of the nature the economy is unfavourable.

BIOLOGY:
Example coconut (A): This is wrapped by a water rejecting outside cover A. It follows the light, shock proofed fibre-pad that functions as crash protection and, furthermore, let's swim the coconut in the sea-water B. The inside following "stone-shell" forms a pressure proofed capsule C. This surrounds the oil containing mark cover, a high-energy nutrient reserve, D, which finally encases the fruit-water, moisture- and energy donor, E.

PRINCIPLE:
Within fruits, one can name the covers, which in the end surround and protect the important seed-content, as "packages", in the animal kingdom for example eggshells. The individual package covers fulfil locally and, in accordance with the momentary requests, also time wise different tasks. One can speak of multifunction packages. These usually are pressure stable, area- and material saving and totally recyclable.

TECHNOLOGY:
Technical packages frequently are mono-functionally, not sufficiently stable, material-squandering and, above all, they are scarcely or not at all recyclable. They also take on little regard that the task of a package component can change in the course of its lifetime. Finally, the entire package economic system is still oriented towards constant growth instead towards growth that does not damage the environment (B), comp. p. 236, 237.

LITERATUR: Küppers, U. (1995): Bionik der Verpackungen ... In: Nachtigall, W. (Hrsg.): BIONA-report 9, 1–16, Akad. Wiss. Lit. Mainz. Fischer, Stuttgart. – Küppers, U.; Tributsch, H. (2001): Bionik der Verpackung. Verpacktes Leben – Verpackte Technik. Wiley-VCH, Weinheim.

Self-cleaning Lotus-leaf and Lotusan facade paint

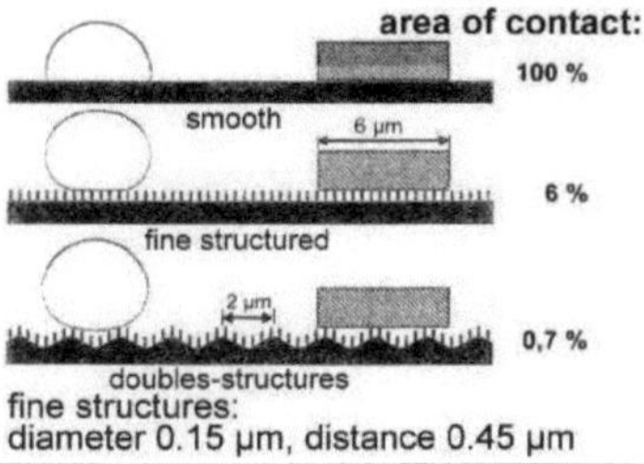

A

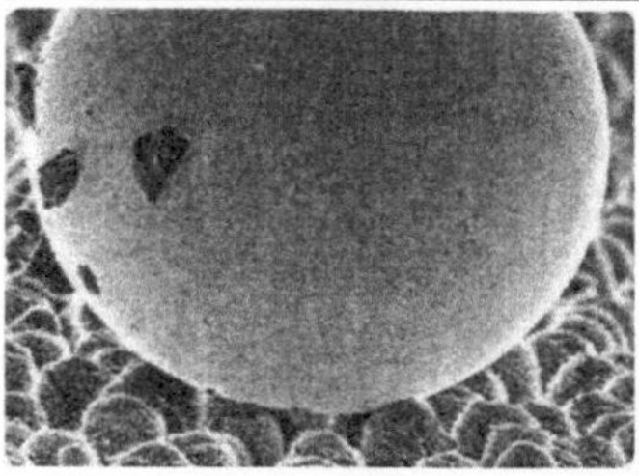
B

Principle: Micro- and nano-structuring of hydrophobic contact surfaces, combined with occasional moistening.

BIOLOGY:
The Indian sacred lily, *Nelumbo nucifera*, always carries clean leaves also after these arose from tough-dirty shore-mud. In the Buddhism, it is regarded as symbol of the purity. Their leaf-surfaces are micro-structured by papillary protrusions of the epidermis-cells and nano-structured because of the cover of these elements by means of small tubes of hydrophobic wax crystalloids (A). Rolling off water drops take fungus spores away.

PRINCIPLE:
Because of their minimal contact surfaces, particles of dirt develop only very low adhesion forces onto these finely studded and very finely structured, hydrophobic surfaces. Water drops inevitably are globular at such surfaces and roll off. Such particles of dirt develop higher adhesion forces to the drop surface, on which they spread broadly. They are caught consequently (B) and taken away with the drop rolling off (comp. p. 204-207).

ABSTRACTION AND APPLICATION:
The patent of the Lotus-effect combines the parameters "finely studded" and "very finely structured " with the parameter "hydrophobic". A first technical application of the principle was the development of the self-cleaning affecting facade paint "Lotusan" of the company Ispo (at that time Dyckerhoff-group, today Sto INC.). In 2008, world-wide approximately 500 000 buildings carried a Lotusan paint. Prerequisite for its effectiveness is occasional raining.

LITERATUR: Barthlott, W.; Neinhuis, C. (1997): Purity of the sacred lotus, or escape from contamination in biological surfaces. Planta 202, 1–8. – Baeyer, H. C. von (2000): The Lotus-Effect. The Sciences (Jan./Feb.), 12–15.

"Modern"

"MODERN"

The Bionics Competent Network BioKoN conceived by J. Hansen, which had its beginning in our Saarbrücker approaches, has an essential share in the further blooming of bionic developments since the later 90er years. The network started with 7 members; over 50 universities, universities of applied sciences and arts, industrial laboratories and companies were unified in its follow-up organization. The number of the jobs and the bionic concepts increased strongly since then, also increased the variety of these researches. Parallel to it, "Biomimetics" is more and more carried out in the area of the Anglo-Saxon language. With it, bionics is worldwide on a good way. The collection of examples for this section essentially begins with the 90s years and is composed of teams of specialists.

MATERIALS

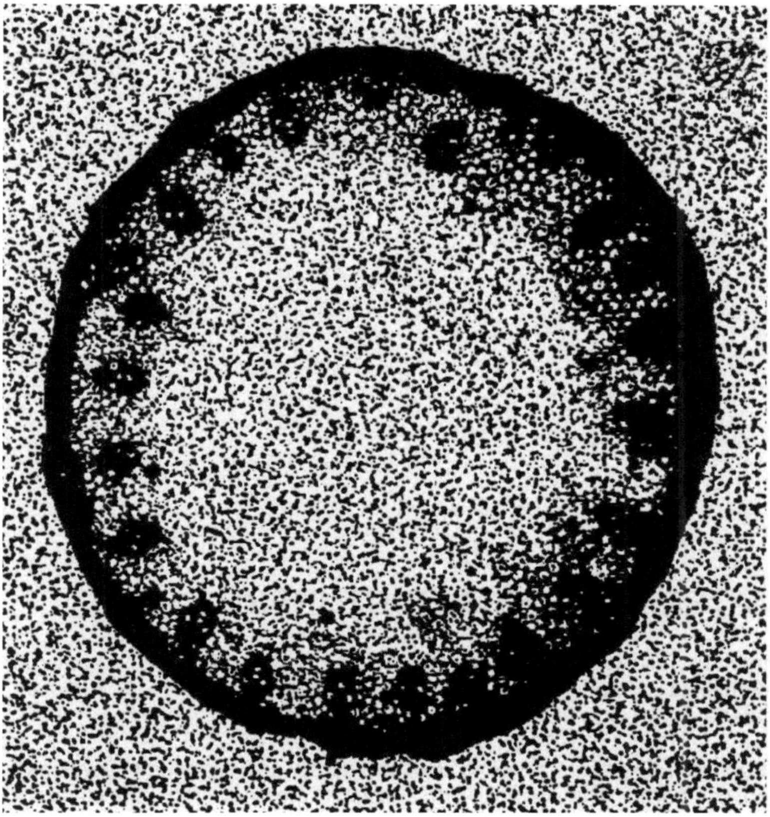

AND STRUKTURES

W. Nachtigall and A. Wisser, *Bionics by Examples*,
DOI 10.1007/978-3-319-05858-0_4

MATERIALS AND STRUCTURES

Materials: Biochemical, bio-physical- and recycling-aspects – material complexes, compound materials – bio-compatible materials and implant materials.

Structures: Elements of biological structures are examined, described and compared. The suitability of biological materials for particular purposes is overviewed. Also unconventional, nature based structures as for example anisotropic compound materials and pneumatic structures or membrane systems spanning over surfaces are examined for their suitability for technical large implementations. Some are addressed in detail in the section "Construction and Acclimatisation". Also induced is the insertion of natural fibres in technical materials as well as bio-plastic. Form generating processes in the biological area offer unconventional models for technical applications.

Self-repairing composite materials

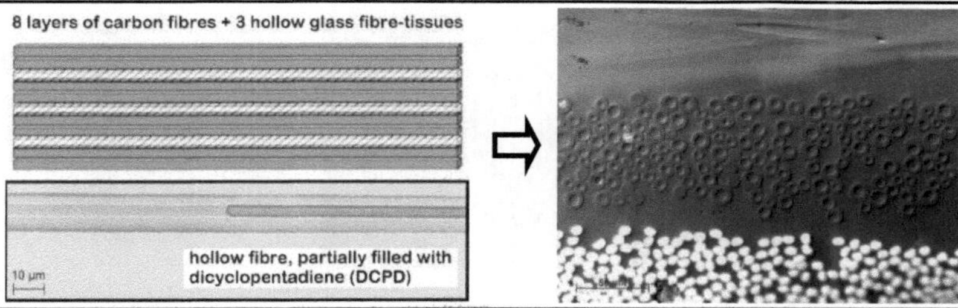

A B

Analogously to self-repairing processes of plants composite materials automatically repair micro-raptures.

BIOLOGY AND TECHNOLOGY:
As on p. 191 explained, intact plants heal cracks by pressing-in a particular tissue under overpressure which cares for closure and subsequent consolidation. The ITCF Denkendorf has developed a system, analogous to the biological processes, which fills automatically raptures in composite materials that frequently suffer from micro-raptures in the polymer matrix under normal load (compare following citations).

SET OF PROBLEMS AND PRINCIPLES OF SOLUTION:
"Recent examinations show that micro-raptures in polymer matrices of fibre-reinforced composites can be immediately closed by incorporated monomer systems. For this purpose, micro-capsules are filled with curable monomers and homogeneously mixed to the entire polymer matrix (A). When micro-raptures appear, the micro-capsules are destroyed locally, and the leaking monomer fluid comes out and starts ... to cross-link."

DETAILS AND REFERENCE TO NATURE:
The cross-linking takes place after the monomer fluid in the micro-raptures had become in contact with a particular catalyst in the polymer-matrix. "The expanding of the micro-raptures up to the complete breakdown of the component can be prevented in this way or at least it is clearly postponed. By this principle, self-repairing mechanisms that each living organism applies when cell tissue is damaged are transferred to composite materials."

LITERATUR: Selbstreparierende Verbundwerkstoffe. Denkendorf News, Sept. 2011, S. 6. (s. Lit. auf S. 191). – Frank, E.; Neumann, J. (2011): Bionische Hohlfasern mit Monomerkern für selbstreparierende Verbundwerkstoffe AiF-Nr. 15777 N/1; Kurzpubl. d. ITCF Denkendorf.

Nano materials, building materials and bionics

A B C

More and more bionic approaches contact the field of nano particles. Nano structures are basis for new key-technologies.

NANO MATERIALS UND INNOVATIVE CONSTRUCTION MATERIALS:
Part of the well known Lotus-effect (p. 65) is based on structured surfaces, i.e. nano structured wax-crystals. In his talk at the congress "Nano-technique and Bionics" (2011) K.-H. Haas emphasizes the multi-functionality of nano based construction materials, in which he also includes Lotus-effect lacquers. Further aspects: Protection against corrosion and alga, wood protection, roof tiles with photo-catalysis, degradation of toxic substance, heat-storage etc..

WHY IS NANO TECHNOLOGY SO INTERESTING?

- Quantum-mechanical behaviour: "New" technical physics by alteration of colour, transparency, hardness, magnetism, el. conductivity (A).
- Increased surfaces: "New" chemistry-processes due to alteration of melting- and boiling-point, chemical reactivity, catalytic efficiency (B).
- Molecular recognition: "New" bio-applications due to combination of self-organization, repair, adaptation and recognition (C).

EFFECTS BY NANO MATERIALS:
Thin nano layers hardly impair the desired substratum-qualities and equip the surfaces with new qualities. Because of their low masses, they are resource-efficient. Furthermore, nano materials can advance into the inside of construction-materials and into smallest pores. Current manufacture and processing-procedures are the sol-gel-process, hybrid-materials, the application of nano filling materials as well as impregnation-procedures.

LITERATUR: Zit. nach Haas, K.-H. (Allianz Nanotechnologie, Fraunhofer-Gesellschaft) (2011): Bautenschutz – Innovative Ansätze durch Nanotechnologie. Kongress Nanotechnik und Bionik für die Bauwirtschaft, Fachforum 1 – Innovative Oberflächen – Bildquelle: BMBF.

Inorganic-organic nano composite in lacquer technique

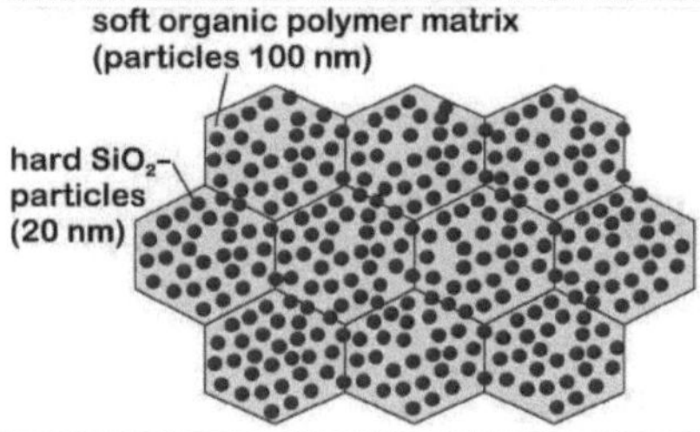

A

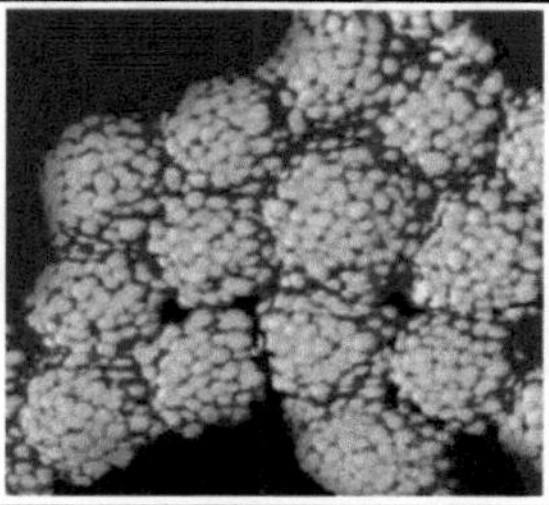

B

The integration of nano based structures in the colour and lacquer technique led already to industrial applications.

BIOLOGY:
Various biological composite structures of inorganic and organic materials show qualities, which neither the one nor the other material presently possesses alone: Hardness and at the same time elasticity in bones, sharpness and at the same time bending stiffness in shark teeth, surface smoothness and at the same time porosity in calcareous shells of snails and mussels. - Bio-analogue nano composites in the field of colours are developed by the BASF.

PRINCIPLE:
Polymer binders require a soft-polymer. However, its surface is not ideal, for example scratch- and dirt-sensitive. In analogy to composite like bio-structures, hard, hydrophilic fine structures are added to the flexible and well adherent soft polymer. By this way structured inorganic organic nano composite polymer binders (A) contain hard SiO_2-partikels (20 nm) in a soft organic polymer-matrix (100 nm).

QUALITIES:
Such nano composites can show exceptional mechanical qualities, again quite accordingly to the natural composites. For example, the E-Module of conventional acrylate decreases strongly with the temperature; the lacquer suffers plastic distortion already with 50° C. The mounting of 40 percent of silica particles (B) sustains the tenacity. Using this technique hydrophobic as well as hydrophilic facades can be fabricated.

LITERATUR: Jahns, E. (BASF SE, Ludwigshafen) (2011): Das Beste aus 2 Welten: Fassadenfarben mit Bindemitteln auf Basis von Nanokompositen. Kongress Nanotechnik und Bionik für die Bauwirtschaft, Fachforum 1 – Innovative Oberflächen.

Artificial spider- and caterpillar-silk

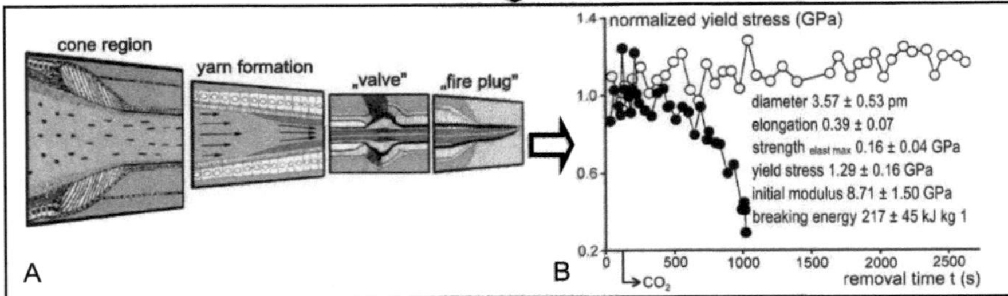

The technical realization of their astonishing potentials is tried for a long time - it is not yet completely successful.

BIOLOGY:
Spider-silk consists of amino acid-ropes with areas of crystalloid structures (beta-leaflets). During crossing the spinning apparatus (A), at first the disorganized fibres are stretched and the leaflets unfold during their solidification and particularly also under the effect of the following burden in different manner. The result is a material with qualities "of which the material scientists can only dream" (p. 58) (B).

NEWER INVESTIGATIONS:
After basic researches, for example of Vollrath and Knight (2001), M. J. Buehler shows with computer-models, how the molecular configuration of the thread changes step-by-step when lengthening up to tearing "after the zipper-principle". P. Fratzl from the MPI of Colloids and Interfaces examines amongst other things the cleft-formation and -prevention, Th. Scheibel / Biomaterials, University Bayreuth, experiments with the molecular components themselves.

REALIZATION:
The "reproduction" works up to over 90 % however the rest asks puzzles that are to be solved only through further basic research. This e.g. would be important for tear resistant and at the same time degradable operation-threads, for neurosurgery, air bags, climbing ropes. G. McKinley, S. Liff and N. Kumar at ISN from the MIT work at a nature-silk analogous material capable of offering high resistance and at the same time high elasticity, suitable for bulletproof vests.

LITERATUR: Reißfestigkeit der Seide entschlüsselt. http;//www.materialica.de/html/_nature_materials _.html. – http://web.mit.edu/newsoffice/2007/nanocomposite.html: Nanocomposite research yields strong and stretchy fibers. – Vollrath, F.; Knight, D. P. (2001): (s. S. 58)

Byssus threads: Abrasion-proof and together flexible

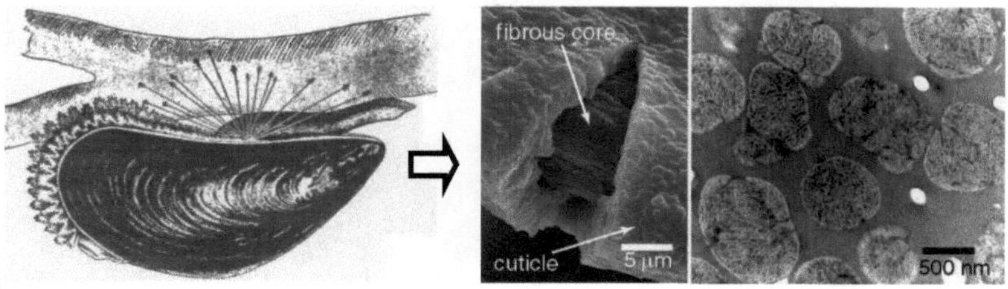

A B

The nature solved the technically significant problem to make structures strongly ductile and nevertheless friction resistant.

BIOLOGY:
These threads developing especially with mussels (*Mytilus edulis*) are very tear-resistant "Byssus threads" (A) and offer a challenge for technical reproduction, because they are flexible until to the double and at the same time at the surface so hard, that they are insensitive against abrasion. Both benefit the mussels, since surfs induce strong traction forces onto their threads and carry fine sand down their surface to corrugate them.

BIOLOGICAL CONSTRUCTION PRINCIPLE:
The Byssus thread ("mussel silk"), composed of high-elastic singles-ropes, is covered with a protection layer, whose "knobbly" surface is attributed to fine granules included in the matrix. These are rich in a high-stable L-DOPA-iron-complex (B), that guarantees through protein-network a high tensile strength and at the same time surface hardness. Consequently, elasticity (matrix) is combined with hardness (inclusions).

TECHNOLOGY:
This "customised connection of metal and protein" should be used for technical polymer protection layers. "Protection coatings are extremely important for the durability of materials and tools. If one looks at the combination of hardness and elasticity, there are only few polymers or compositions that unite these material qualities." At the MPI for of Colloids and Interfaces new concepts are investigated.

LITERATUR: Harrington, M. J., Masic A; Holten-Andersen N; Waite H; Fratzl P.. (2010): Iron-clad fibers: A metal-based biological strategy for hard flexible coatings. Science 4 (Mar.). – http://www.vbio.de/informationen/alle_news/e17162 ?news_id=9015.

Bionics as idea-source for technical insulating materials

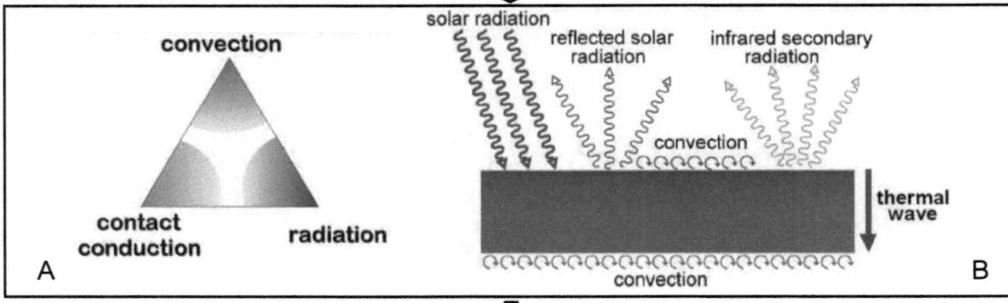

Heat management will be a crucial factor in future. Multiple and essential stimulations can come from bionics.

PHYSICAL WAYS OF HEAT-EXCHANGE:
The "triangle of heat transportation mechanisms" is represented in fig. A, the possibilities of heat transportation as well as of heat exchange of a massive building panel during a clear summer day are shown in fig. B. Apparently, there is a whole series of intervention possibilities into thermo-sensitive building panels, heat insulation, i.e. preventing of heat losses, is only one of them. Natural insulation materials usually are multivalent accordingly.

PRINCIPLES OF NATURE:
T. Stegmaier from the ITV Denkendorf lists the following principles:
- „Insulation
- Adaptive heat-insulation
- Adaptive sun protection
- Heat management
- Refrigeration due to vaporization
- Temperature zoning of the body
- Energy storage
- Hardly heat -, rather chem. storage
- In particular: Function- and system-integration, low-energy".

POSSIBILITIES OF THE REALIZATION:
These principles can be realized in different combination:
- „Insulation (s. cork-oak, *Quercus suber*)
- Insulation with fire inhibition (s. redwood, *Sequoia spec.*)
- Insulation with adaptation (s. penguin, e.g. *Aptenodytes forsteri*)
- Heat-management through reflection (Africa, *Leucadendron argenteum*)
- Transparent thermal insulation (s. polar bear, *Ursus maritimus*)“ and others.

LITERATUR: Zitate nach Stegmaier, T. et al. (2011): Bionik: Ideenquelle und Anregungen für Dämmmaterialien. Kongress Nanotechnik und Bionik – Hightech in der Bauwirtschaft. 18.11.2011, München. – Bildquelle: Blum R., Labor Blum.

Heat insulation with nature-fibres

A

B

The new heat ordinance requires new insulation materials. Today the tendency goes from oil-based to sustainable ones.

INSULATION BASED ON OIL:

The most well-known materials of the "styrofoam" are PUR hard-foams, polystyrene particle foams (EPS) and polystyrene extruder foams (XPS). They unify very low weight and good workability with unbeatable good thermal conductivity coefficients from 0.025–0.045 W $m^{-1}K^{-1}$. Also the smoulder- and fire resistance can be influenced favourably. The dependence on petroleum is inherently adversely; long-term, such materials must be replaced.

INSULATION BASED ON NATURE-FIBERS:

There is a multiplicity of natural materials, that have insulation good properties, alone or in combination with others and with technical support-materials. Hemp (A), flax, cork, sheep-, wood-wool, meadow-grass- and sea-grass-fills, including - exotic - "Neptune-balls" (B) belong to it. The thermal conductivity coefficients are with values of 0.037 till 0.065 W m-1 K-1 not at all so bad (wood-wool till 0.090). Comparison: Vacuum-insulation-panel ≥ 0.004 W m-1 K^{-1}.

USE OF INSULATIONS BASED ON NATURE-FIBERS:

Although their thermal conductivity coefficients are not quite as good as of polystyrene-like materials – that are producible moreover only with high energy input, however – they reach the ones of glassy materials like glass wool, stone wool, foam glass, Perlit (0.035 ÷0.070 W m^{-1} K^{-1}). Insulation materials based on nature-fibres are used as blow-injection-fills and insulating panels. They generally contain additives i.e. boron salt to the fire protection.

LITERATUR: Zitate nach Stegmaier, T. et al. (2011): Bionik: Ideenquelle und Anregungen für Dämmmaterialien. Kongress Nanotechnik und Bionik – Hightech in der Bauwirtschaft. 18.11.2011, München. – Bildquelle: Blum R., Labor Blum.

Nature-fibres in composite materials

A

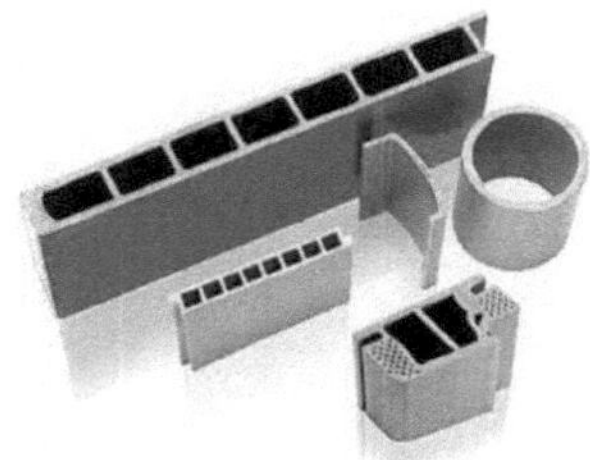

B

As example for this already many-used technology, an ultra-light PVC material with wood fibre filler may be mentioned.

BIOLOGICAL FIBERS IN COMPOSITE MATERIALS:
Wood-fibres (A) are very much tensile; they and others were utilized in various approaches with mechanical advantages as partners in technical composite materials. The previous disadvantage was the high specific mass, which results from the fact that the profile-thickness must be increased. This makes such materials on the one side problematic for use in cars and airplanes, on the other side very expensive and with it non-competitive.

NEW SOLUTION:
The Italian company Friul Filiere SPA developed a new extrusions-technology, with which ultra-light PVC-wood-composite materials can be manufactured for the furniture and building industry (B). They are described as foam fibre composite (FFC). The wood-fibre is dried to process-beginning to less than 2 percent of residual moisture so that it can join ideally with the PVC flowing around.

ADVANTAGES:
The material reaches a low specific mass of 0.6 - 0.7 g cm^{-3} and can be processed like wood. It is regarded recyclable up to 100 percent (PVC?) and so it is ecological. Its physical and mechanical qualities are very good: Stability, moisture resistance, elasticity (which allows to produce also curved parts in the mould casting process), heat- and sound-insulation. It is self deleting in the fire-case. Moreover, it is relatively cheap.

LITERATUR: www.materialica.de/html/pvc.html: Ultraleichtes PVC-Verbundmaterial aus Italien auf dem Deutschen Markt. – http://www.interempresas.net/Plastics/Articles/43806-The-represented-Imvolca-show-their-latest-products-in-the-2010-K.html.

"Different" application of biological components

A   B

Biological components can quite successfully play a "non-biological" role in the technical use.

"TASK CHANGING":
Components of biological systems, for example fluids in cells or interstitial fluids and fibres in membranes, have determined "biological" functions. If such components are applied in the technology, they can bring in qualities that are not in demand in the biological surroundings. A technical "conversion" of this type is not “non-bionic”; the engineer is not tied at the use of biological materials in a "bio-conform" way.

EXAMPLE WATER CLEANING WITH CACTUS SAP:
The somewhat tough juice for example in the tissue of the prickly-pear *Opuntia spec.* (A) serves as water-storage. If one solves such pressed sap in impure water, it - as the American engineer N. Alkantar found - lets coagulate bacteria and debris particles because of their stickiness so that they sink. The overlaying water is pure to 98 percent. Recommendation: Bring to boil impure water with a cactus-disk.

WOOD-FIBERS IN LABORATORY TABLE PLATES:
Wood-fibres are used for stability in the biological environment. Here tensile strength predominates. Through compression of natural fibres together with art-resin and colour-pigments, after a procedure of the Trespa International fully dyed, surface-insensitive laboratory table plates (B) are created. Here, the wood-fibres have the function of an inert filling-material that is claimed only on pressure; the original tensile strength is not used on this occasion.

LITERATUR: Sprado H.-H. et al. (2010): Kaktus, der Bakterien-Killer. PM-Magazin 10, 72. – Sternitzke, I. (2011): Schön und gut. Holzfaser-Labortischplatten aus nachhaltiger Holzwirtschaft. GIT Labor-Fachzeitschrift 7, 456.

Bio-based materials

A

B

S. Peters has structured the already impressing variety of bio-plastics and represented them in their hierarchy.

LIST OF BIO-BASED MATERIALS:
Bio-plastics on basis of polylactic acid – B. o. b. o. polyhydroxybutyric acid – B. o. b. o thermoplastic starch – B. o. b. o. cellulose – B. o. b. o. vegetable oils - B. o. b. o. lignin – B. o. b. o. algae – B. o. animal basis – acrylic glass o. b. o. sugar – natural rubber – wood polymer materials – coconut wood systems – bamboo – thermal tempered natural woods – thermo hydro mechanically condensed wood – cork polymer materials – almond polymer materials.

CONTINUATION AND APPRECIATION:
Algae based materials – fungi based materials (A), – nature fibre composite – linoleum – bark cloth materials – corn spindle plates (B).
Peter's composition marks in well-pictorial abridgment the type of material, its qualities, uses as well as processing-steps and gives application-examples. Also with "conventional" materials", one learns a great deal; about some one has even not yet heard.

UNCONVENTIONAL EXAMPLES:
Bio-plastics of animal origin: Chitin → Chitosan, usable for example for fat bonding (sewage-works), in medicine or for toothpaste.
Almond polymer materials: Ground almond peels in degradable resin matrix → employable in the furniture industry or as coating material.
Fungi based materials: Naturally growing fungi can cause a solid bonding system between vegetable waste-materials → replacement of styrofoam.

LITERATUR: Peters, S. (2011): Materialrevolution. Nachhaltige und multifunktionelle Materialien für Design und Architektur. Birkhäuser, Basel. Zum Thema: Abschnitt „Biobasierte Materialien“, S. 30–59.

Bio-based plastics - no contradiction in itself

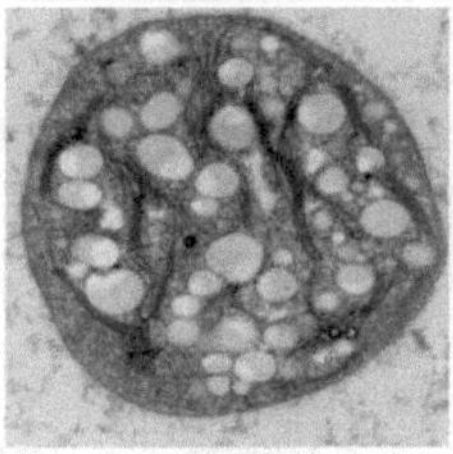

chloroplast with PHP-inclusions

In 2010, 2.6·10⁶ t of petroleum-based plastics were produced. Could bio-based plastics be able to accomplish compensation?

PROBLEM:
Plastics are to be produced and processed favourably. They proved to be useful in a wide range. However, they are i.e. not biodegradable and have to become replaced in the after-petroleum-age. Bio-based plastics, which are investigated since several years, could represent a lasting and climate-friendly alternative. They are used already today in various products, in the year 2010 for example already 10^6 t.

BASIC MATERIALS:
At present, as basic-material e.g. starch from corn, wheat, potatoes, Tapioca is investigated. It is changed chemically into thermo-plastic polyesters or Polyurethane. This is also possible biotechnologically via plants (A) or genetic engineered changed bacteria (*Escherichia coli*), that store such granulated spherules into their membrane. Potatoes with a gene from cyanobacteria for example can produce biodegradable polyaspartate etc.

PRODUCTS:
For products (B), the raw materials chitin (from insects and crayfish), cellulose (from wood), gelatine (from algae), casein (from milk; s. p. 95) and others are already used today. There are already coke-bottles, watering cans, ski-shoes, computer-keyboards, packages for bio-noodles, form-peels for meat, packages for hot foods and so forth. Bio-based plastics, however, are not necessarily biological-degradable.

LITERATUR: Fachkongress Biokunststoffe (2010): Von der Nudelverpackung bis zum einbruchssicheren Fenster. www.vbio.de/informationen – BMBF DVD (2008): Biokunststoffe für die Industrie. www.youtube.com/watch?feature=player_embedded&v=LYVnPaUxwyw.

Resource efficiency - increase by bio-plastics

A

innovation for flatware

	2001	2008
injection mould	PS-tool	tool for Biograde C9550
avaiability of tool	ca. 80 %	> 98 %
reject rate	2–10%	< 0,1 %
injection pressure	2 000–2 500 bar	1 700 bar
wall thickness (fork)	1.0–1.5 mm	0.7–1.2 mm
moulding weight (fork)	5.5 g	4.2 g
cycle times	17 s	6–8 s

B

Bio-plastics will only become generally accepted if their processing is more resource efficient than this from petroleum-

TECHNOLOGY:
Plastics very frequently are processed with the method of injection moulding. For this, corresponding tools and procedure-details, partially very expensive - especially at the fork- and spoon-production, are necessary (A). An important task for the commercial acceptance of bio-plastic-materials is to adjust the technical parameters of the material to the detailed procedure so, that it becomes as resource-efficient as possible.

EXAMPLE:
The material "Biograde", one from Fraunhofer UMSCHAU together with the FKuR plastic Ltds developed bio-plastics on cellulose-basis, is suitable to injection moulding. During this procedure, its material-technical qualities can be tuned optimally with respect to both the design of the casting-products and of the injection moulding tools and, finally, with reference to the efficiency of the manufacturing process.

TECHNICS:
T. Wodke and C. Michel of Fraunhofer UMSCHAU have shown which improvements were possible during the transition of a conventional plastic to Biograde C 9550 with respect to the increase of the resource efficiency within 7 years (B). The tool-availability could be increased close to the technical limit; all other technically and financially significant parameters could very clearly be lowered.

LITERATUR: Ressourceneffizienz beim Spritzgießen mit Biokunststoffen. Fraunhofer UMSICHT. Jahresberichte 2009/2010, S. 40 f. Kategorie „Nachwachsende Rohstoffe".

Biodegradable bio-plastics

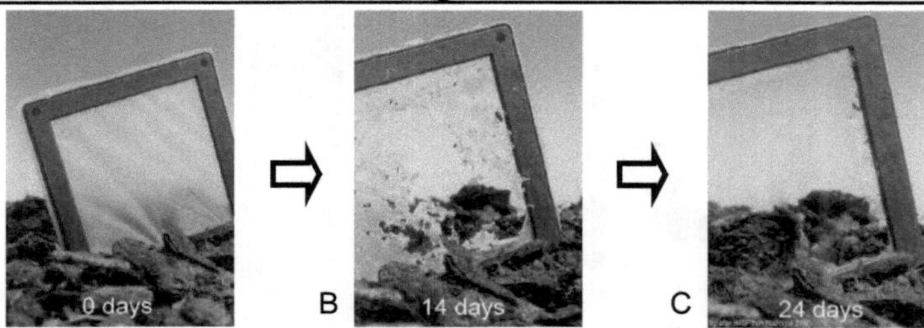

A B C

Bio-plastics are not necessary biodegradable; however, this would be of greatest importance for a changeover.

PROBLEM:
As quoted on p. 81, a changeover from petroleum-based to bio-based plastics is of highest meaning for the future. There are already examples for the latter, however, only a few are decomposable, that means biodegradable (A–C). The VBio says to the myriads of conventional not bearing and tear proof plastic bags: "Exactly this stability and durability make the conventional bags to an always larger growing environmental problem ... ".

BIOLOGY:
In Australia, plastic-packages amount 25 percent of the waste. Chitin structures that are present in insect, crab and spider carapaces, are disintegrated on the other hand on the compost works. In the food-industry, crabs and shrimp carapaces accumulate in large numbers that, in contrast to also useable grain-starch, can not be used for food-purposes furthermore. Wood garbage (cellulose) and poly-lactic acid are added.

REALIZATION:
M. Nysperos of a food-company and E. Palombo as well as the postgraduates S. Chattopadhyay and C. Way of the Swinburne University in Melbourne developed structure-formulas for solid, though degradable bio-plastics. They demonstrated the degradability at first for chemical compounds of polylactic acid and lignocellulose on basis of a grain-starch-polymer. In the near future, the latter should be replaced by a "waste-polymer".

LITERATUR: Vbio-Newsletter (2010): Biologisch abbaubares Plastik durch Schalentiere. – Müller, R. J. (2001): Biologisch abbaubare Kunststoffe. BIUZ 30(4), 218–225 – Hingewiesen sei auch auf rasch sich entwickelnde Ansätze der BASF Ludwigshafen, s. Abb. A–C.

Bio-plastics leave the exotic status gradually

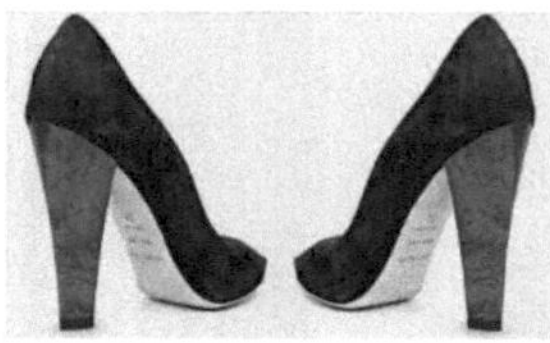

A

B

In the year 2009, 240 10^6 t of oil-based plastics were produced worldwide, in Germany 17 10^6 t – bio-plastics percentage < 1.

REGARDED STILL TODAY AS "EXOTICS", BIO-PLASTICS WILL INCREASE THEIR PERCENTAGE. EXAMPLES (after Dohnke) 2011):

- Agar Agar (alga-material) → one-way-cup (design-office New York)
- Agriplast (plastic from grass-cut) → cups, plates
- Arboform (lignin based) → pumps (A), techn. cabinets (B) (Tecnaro / Ilsfeld)
- EcoCradle (hard-foams on fungus-culture-basis) → packages (Ecovative)
- Ingeo (material based on corn starch) → shapes, packages (NatureWorks)

UPLIFTING:
Oil-based plastics have two big advantages. They are on the one hand in their qualities and with brand names known and anchored with the user. On the other hand, they are available for nearly any application-purposes. Bio-based plastics have difficulties there. They can keep up with respect to quality and use. They are not yet available as one likes. However, they lie in the trend of an increasing eco-awareness.

VIEW AND SOME INTERNET-ADDRESSES:
Oil-based plastics admittedly are market-dominating at the moment, but if nature-based plastics experience social and political acceptance, however, they won't remain exotics.
AgriPlast: www.biowert.de - *Arboform*: www.tecnaro.de - *EcoCradle*: www.ecovativedesign.com - *Ingeo*: www.natureworkslic.com - Kay Dohnke: www.dohnke.de - Dr. Sascha Peters: www.haute-innovation.com.

LITERATUR: Dohnke, K. (2011): Sauber gemacht! DB *mobil* 5,62–66. Abdruck: Dohnke, K. (2011): Zurück zur Natur. Kunststoffe 8, 36–38. – Peters, S. (2011): Materialrevolution. Nachhaltige und multifunktionale Materialien für Design und Architektur. Birkhäuser, Basel.

Bio-plastics → from nature, for sustainability

A

B

Form and injection moulding processes as well as sandwich-hybrid-constructions for bio-based nature-fibre-composites.

NATURE FIBRE REINFORCED PLASTICS (NFP):
Materials are called NFP's, when for stability-reinforcement natural fibres of flax, hemp, kenaf, sisal or abaca are incorporated. In order to be able to produce modern materials containing nature-fibres, one requires binders for their production like e.g. polypropylenes (PP). Or one replaces the synthetic polymers by bio polymers. To do this, animal and vegetable adhesives, glues and resins can be used.

ADVANTAGES OF NFP'S:
For the production of nature fibre reinforced composite materials the good qualities of e.g. wood-fibre and plastic are used in order to create wood plastic composites (WPC), whose quality-potential unites the advantages from the world of both raw materials. The advantage over full-wood-products is the free plasticity, higher moisture resistance and, interconnected with it, better weather resistance of the nature-fibre-composite.

NON-POLLUTING:
Sustainable by application of native plants; up to 100 percent CO_2-neutral; 100 percent biodegradable (NF-PLA); no application of health-endangering materials.
SPECIAL MATERIAL-QUALITIES:
Low weight; very impact resistant; tough; high shock absorption; extremely light (lighter than GFK and MDF); surface ideal for individual outside-decor.

LITERATUR: Informationsstand 2012: jakob-winter.com bzw. www.naturfaserverbundwerkstoffe.de/de. – S. auch www.wpc-kongress.de – www.nachwachsende-rohstoffe.info und www.nova-institut.de/bio.

STYLING

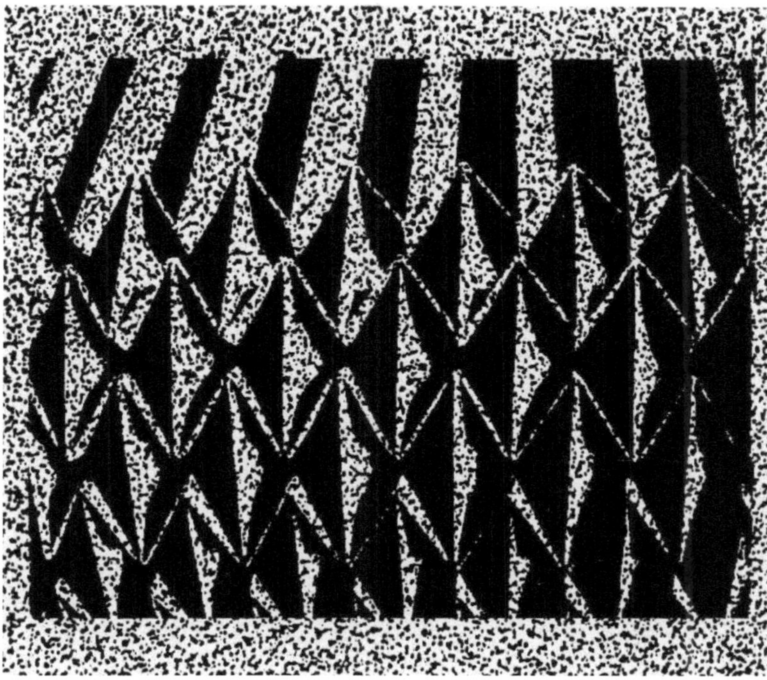

AND DESIGN

W. Nachtigall and A. Wisser, *Bionics by Examples*,
DOI 10.1007/978-3-319-05858-0_5

STYLING AND DESIGN

Not-functional and functional design – bionic-design.

Within a creature, the "outer form" together with the "inner-life" always forms a functional total. A "bionic-design", which earns the name, must correspond this functional aspect if it doesn't want to slip off into pure form-similarity.

Is there a "bio-design" or a "biological design?"

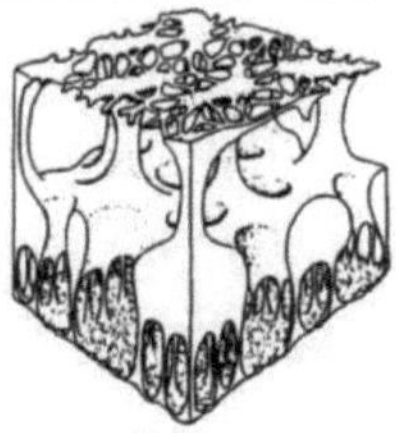

A B

Merely the latter concept can be referred to the constructive design of plants and animals.

NATURE AS DESIGN-MODEL:
The inflationary application of the prefix "bio" admittedly has already tradition; however a "bio-design" is not suitable for classification characteristics. With the designers it is as with the architects. One can get stimulation for his creative action, where one wants, and one can combine stimulation from the nature with others as one wants. Design is not pretended by something but originates creatively in a brain.

AND "BIOLOGICAL DESIGN?"
Nearer considerations (s. section: To the concept of "Biological Design" in W. N. 2005) show that this concept is virtually congruent with the biological term "construction-morphology." What counts, is the constructive forming of creatures (A, B). One can call this "designed" linguistically and can find the connection with it about the technical design-concept, if it is clear that one doesn't see any personified designer behind it.

CATALOGS FOR "BIOLOGICAL DESIGN":
The concept in the just defined sense also is applicable to example-collections that present the functional form-canon of the nature in picture and text. A fullness of stimulation for the designer, construction engineer and architect can arise from it, at which he doesn't approach so easily otherwise. The book of B. Hill is to a greater extent addressed to the designer. That of W. N., with its 1771 figures and function-oriented texts, is a rich source for everyone.

LITERATUR: Hill, B. (1999): Naturorientierte Lösungsfindung-Entwickeln und Konstruieren nach biologischen Vorbildern. expert-Verlag, Renningen - Nachtigall, W. (2005) – Biologisches Design. Systematischer Katalog für Bionisches Gestalten. Springer. U.M.v. A. Wisser.

Formal and functional design and inspiration

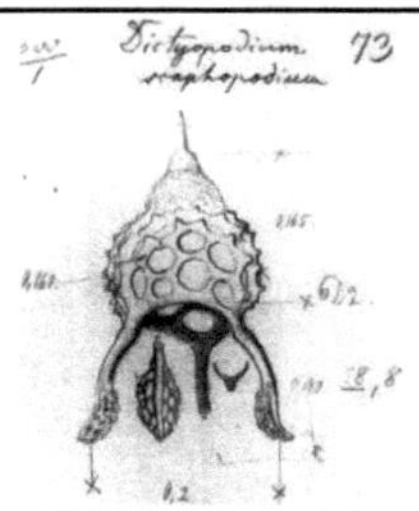

A

B

In the design-area, both stimulation-possibilities are given; however, only a functional approach is "bionic."

FORMAL APPROACH:
From the publications of Ernst Haeckel, widespread about the turn of the century, the architect René Binet got stimulations for his constructions during the world's fair 1900 in Paris. The formation of the monumental entrance to the exhibition area (B) became well known. Its form is reminiscent of a species of radiolarians, *Dictyopodium scaphopodium*, and a spherical structure with 3-times step-wised top that seems to stand on 3 legs (A).

CRITICISM:
Against a "zoomorphism" of this type, that reappears also in other architectural designs and particularly in the jewellery-design (belt-buckles), one cannot bring up any criticism from the function-point of view. Similar to the widespread "phytomorphic" art nouveau ornamental nature-forms have given stimulations for transfers of all types, from naturalistic after-creations to strongly reduced abstractions.

FUNCTIONAL ASPECT:
As in the construction-bionics and in the architecture-area ("biomorphism") the sentence is applicable also to the design, that pure form-similarity cannot be called "bionic." This would disagree with the definition of bionics and lead quickly into formal or esoteric spheres. Real bionic transfer often is inconspicuous. So, for example abstractions of vegetable sclerenchyma-formations for stiff technical tubes are recognizably only in the cross-section (p. 113).

LITERATUR: Haeckel, E. (1904): Kunstformen der Natur. Bibliographisches Institut Leipzig, Wien. Neudruck der Tafeln im gleichbenannten Buch (1998), Prestel, München/New York. – Nachtigall, W. (2005): Bau-Bionik. Natur-Analogien-Technik. Springer, Berlin.

Biomorphic and bionic architecture

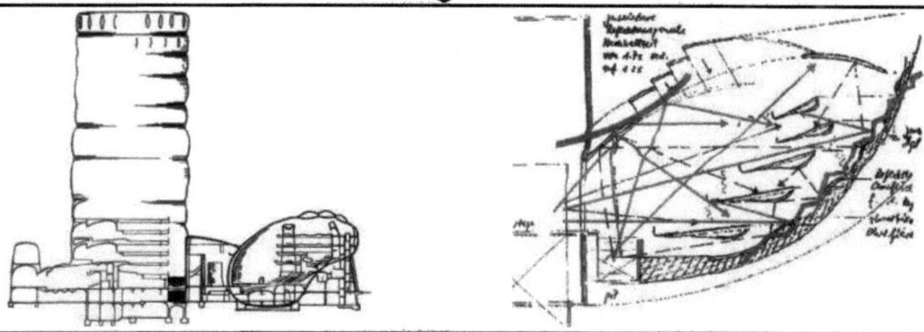

A B

Only external organically appearing architecture is not bionical. Its is differently, if such a form is functionally embossed.

BIOMORPH ARCHITECTURE:
As a prototype of biomorphic architecture Friedrich Kieslers "endless house" can be named, that was inspired from contiguous cells or cellular forms. Today, the 1959-model is exhibited in the Museum of Modern Art. The visionary drafted also a theatre whose stage-construction has the form of an intestine, and whose auditorium the form of a stomach (A); both forms are not functional for such a building.

BIONIC ARCHITECTURE:
As a competition-design for the new opera in Oslo (2001), the architect-office F. Birke / Stuttgart in cooperation with G. Rummel and W. Nachtigall / Saarbrücken has proposed a special frame shape for the auditorium (B). This admittedly looks at a first glance like the Kiesler-form, however, it is both in the optical and in the acoustic field functional: Visitors should at the same time watch and listen as well as possible from all places.

BIOLOGICAL MODELS FOR THE OSLO-DESIGN:
For the optical field, the reverse optical path of the eye of the marine snail *Haliotis* and the further developed mammal-eye with its "equitable" light-sensitive cells were the inspiration. For the acoustic field, the ear-mussel of the bat *Rhinolophus mehelyi* was taken as model. It possesses a "non-linear folding" which leads the sound to the eardrum from all directions ideally (reversion: eardrum ⇔ stage).

LITERATUR: Kiesler, F. (Hrsg.: Museen der Stadt Wien) (1997): Notes on Architecture as Sculpture, 1966, S. 140. – Kiesler, F. (1966): Inside the endless house. S. 136. – Birke, P. (2001): Bionische Strategie ... Oper in Oslo. BIONA-report 15, 52–57.

Diatoms and the design of shells and mats

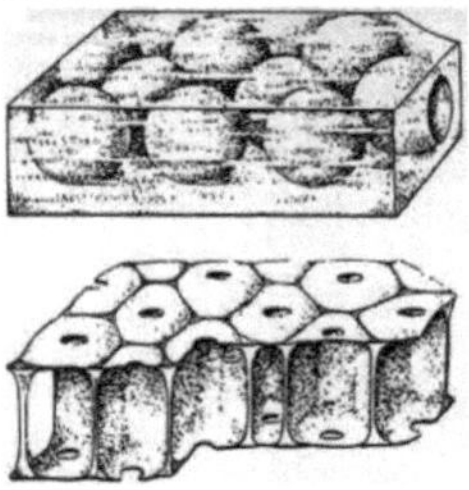
A

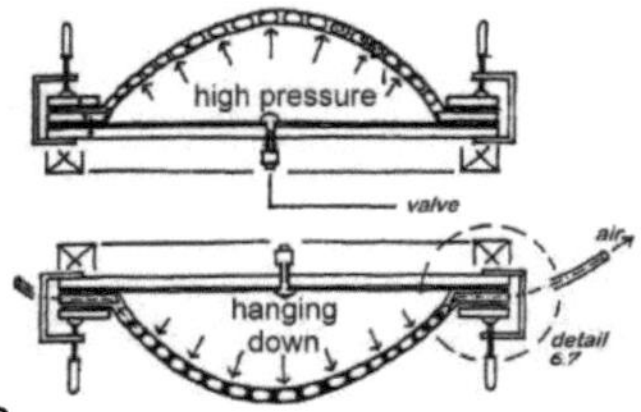

B

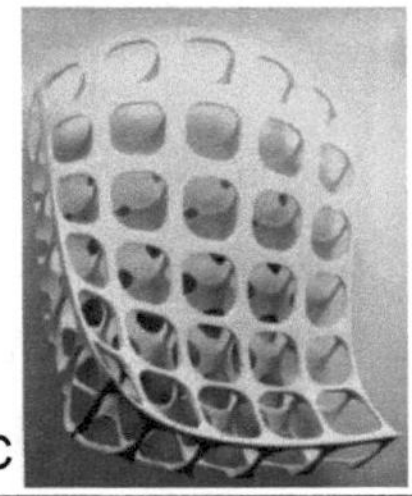
C

The "fat-droplet development hypothesis" for diatoms led to the conception of versatile usable shells and mats.

BIOLOGY:
After a hypothesis of G. Helmcke from the fifties of the last century, diatoms form their filigree skeletons through accumulation of fat-droplets on the surface of the still soft body, which flatten them approximated hexagonally, and following filling of the resulting gaps with liquid SiO_2 (A). This idea has given stimulation - (Noser) 1985 (B) and also recently again (Garin 2006) (C) - for the design of technical shells and mats.

NOSERS SHELLS:
Football-blisters were pressed between chipboards, and the gaps poured out with plaster of Paris or polyester. Forms of little boxes with 8 holes (due to the former blister) resulted. Also shells with cross-section-forms of chain lines were tried (B, $\leq$ 5 m of diameter). After curing and turning inside out, they were self-supporting and could be used, e.g. spanned with foils, as pool covering, for example.

GARINS MATS:
After similar pilot tests with pressed blisters, structures resulted according to figure C. After digital parameterisation and computer-operated modification series of similar spatial structures could be manufactured using the rapid prototyping technique. These could be tested for their loading capacity. Increased, they can for example be used as flexible mats or for the modulation of air-currents.

LITERATUR: Helmcke, G. (1959): Form und Funktion der Diatomeenschalen. Beitr. Naturk. Nieders. 12, 110–114. – Noser, T. (1985): Natur als Baumeister. HS Künste, Berlin – Garin, G. S. (2006): Porous mats. In: Hensel, N.; Menges, A. (Hrsg..): Morpho-Ecologies. AA London.

Zoomorphism and symbolism

A

B

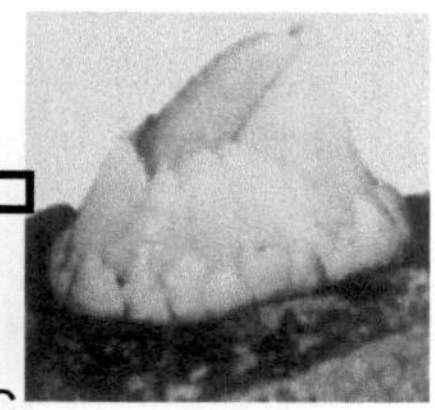
C

The form-language of architecture occasionally uses animal-models in symbolic sense - ok, but this is not bionic.

Jean-Jacques LEQUIEU, Michael SORKIN:
Early examples are zoomorphic designs of the French architect Jean-Jacques Lequieu, 1757 -1826, for example, in form of a bull (A). This "architecture parlante" should symbolize strength and dignity. Michael Sorkin's design of a "jellyfish hotel" (2010) for the Chinese city Tianjin plays with form-similarities, however, no functional assignment is shown in the division in "head" and "tail".

SANTIAGO CALATRAVA:
The vibrant ease of some designs and realized constructions of the Spanish architect frequently results from the metamorphosis of artistic impressions of animals. So, the serial supporting structures of the 1986-89 built 9. October-bridge in Valencia go back on the sketch of a horse in lateral view. The street runs along over the back-region and a wide pedestrian-way over the head-region (horse as symbol of "load-carrying?").

JØRN UTZON:
The opera-house in Sidney (design Jørn Utzon, 1957 -1973) virtually became the symbol of the modern Australia. One says, that it (B) reminds of together-standing mussels; zoologists rather think of barnacles. *Waikalasma boucheti* (C) is regarded as model; however, this was discovered not before 1996 ("as if"-inspiration). Petra Gruber (2011) described such bio- and anthropomorphisms in detail in her book about architecture of life and buildings.

LITERATUR: Gruber, P. (2011): Biomimetics in architecture. Architecture of life and buildings. Springer, Berlin. – Nachtigall, W. (2003): Bau-Bionik. Natur – Analogien – Technik. Springer, Berlin. – Nachtigall, W. (1997): Vorbild Natur. Springer, Berlin.

A new design-principle for super-hydrophobic surfaces

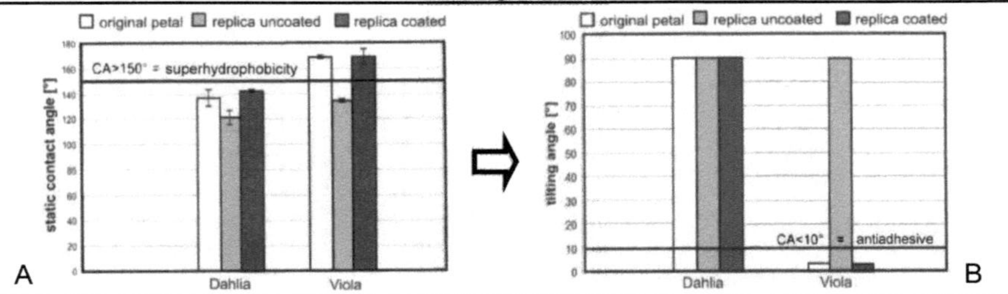

Super hydrophobia was tested on replicas using flat wax-layers over finest folding instead of spatial wax-crystalloids.

BIOLOGY AND THE TRANSFER-PROBLEM:
The Lotus leaf wins its super-hydrophobic qualities (angles of contact > 150°) (comp. p. 65) through micro knops and their coating with nano wax crystalloids. Technical replicas reproducing also finest knob-like details can be manufactured; however, the transfer of the wax-crystalloids doesn't succeed. For the production of technical surfaces, it therefore would be important to produce super-hydrophobia different to the latter biological prototype.

PRINCIPLE:
A.J. Schulte et al., analyzing hydrophobic and super-hydrophobic petals, have ascertained, that water-drops have a globular form as expected, but usually do not unwind from inclined surfaces ("petal-effect"). On the wild violet *Viola tricolor*, however, they unroll as on the Lotus-leaf. This type possesses not only especially long cell-end-vaults, but replaces also the wax-crystalloids through a wax-film over finest cuticle folding.

TECHNOLOGY:
Such a folding (approximately 260 nm) can be cast - without wax-film - using a particular procedure. Technical surfaces manufactured after it are, in the comparison for example with a dahlia-leaf, super hydrophobic, after coated (A) with a hydrophobic fluoride polymer (angle of contact 106.5°). Also the "petal effect" is reduced dramatically (B). With it, a new basis is established for self-cleaning surfaces.

LITERATUR: Schulte, A. J. et al. (2011): Hierarchically structured super-hydrophobic flowers with low hysteresis of the wild pansy (*Viola tricolor*) – new design principles for biomimetic materials. Beilstein J. Nanotechnol. 2, 228–236.

Eco-Design: Clothing from milk?

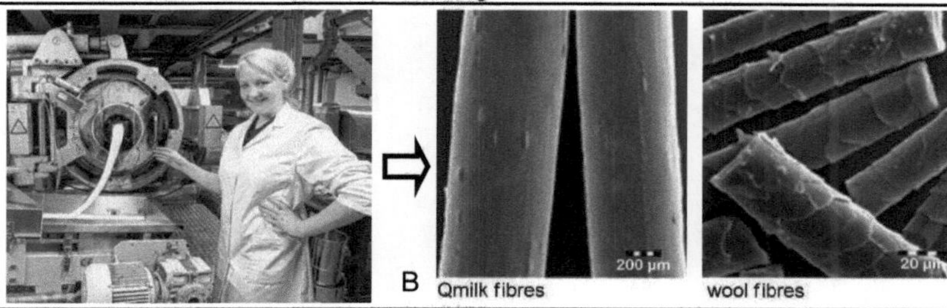

A B Qmilk fibres wool fibres C

A new biogenic fibre on basis of not drinkable milk is anti-allergic and for many textiles persistently producible.

MANUFACTURE:
From milk, not usable for food, technical casein is produced. This is dried, pulverized and then scrambled in a heavy duty mixer under renewed addition of water and other natural materials to a liquid paste. Pressed by strainers or nozzles (A) milk-fibres (brand-name: Qmilk) result, to whose manufacture only 2 kg of water for 1 kg of material is required (synthetic material fibres: up to 25 000!). So this is really a water saving procedure.

QUALITIES:
The Qmilk fibre (B) is thinner than a wool fibre (C), anti-allergic and, since smooth, skin-gentle. For clothing, viscose- and elastane-fibres can be woven in up to 30 percent, for example. Up to a point, it also could replace the increasingly expensive "bio"-cotton. Since it doesn't show any pesticide-residues, it is regarded as anti-allergic. In addition, it should behave temperature-balancing, is well wearable and may prevent skin aging.

FIELDS OF USE:
Besides the clothing industry, the fibre developed by A. Domaske (A) could find application in the area of home-textiles, approximately for blankets, mattresses (surface and filling) and sleepwear, where it could demonstrate its moisture-balancing and bactericidal qualities. Potential uses of refined fibres may be found in the car-industry as well as in form of chemical free textiles in the medicine-technology.

LITERATUR: Domaske, A. (2011): Qmilch; www.qmilk.eu/index2.html – Sticht, C. (2011): Werden wir bald alle Kleidung aus Milch tragen? Hamburger Abendblatt. Rubrik Öko-Mode. 29.9.

CONSTRUCTIONS

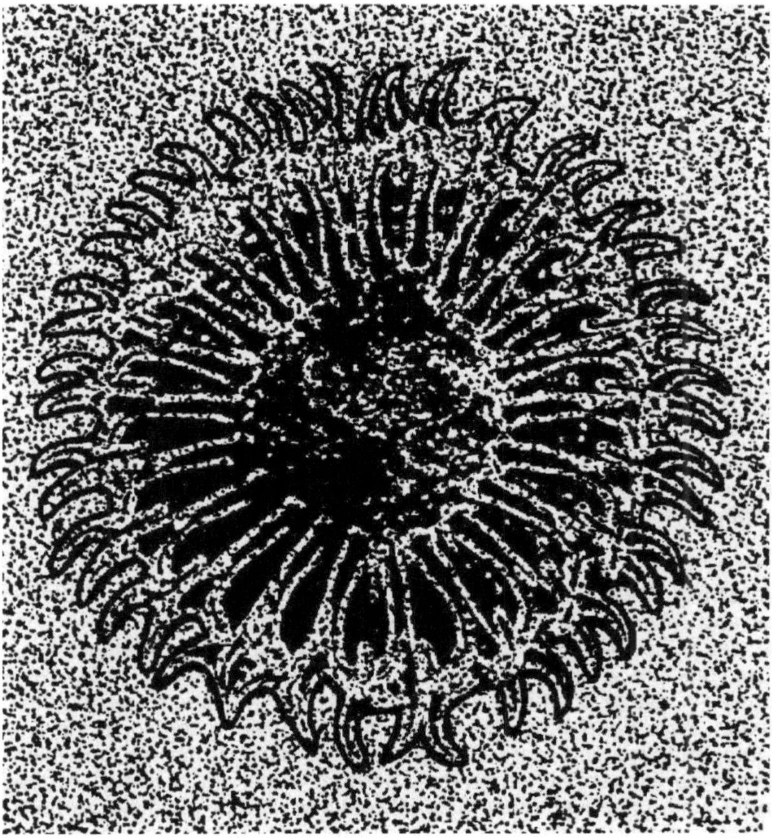

AND EQUIPMENT

W. Nachtigall and A. Wisser, *Bionics by Examples*,
DOI 10.1007/978-3-319-05858-0_6

CONSTRUCTIONS AND EQUIPMENT

Construction-morphology and functional anatomy – General and applied biomechanics – Constructions – Devices – Mechatronic – Micro-systems – Nano-systems.

Construction-elements and mechanisms from the fields of the biological and technical world are analyzed and compared. The interaction between the construction-elements and the complete working constructions are worked out. Stunning similarities are found, so for example in pump-constructions (saliva-pumps in insects, vertebrate-hearts, technical pumps). However, compared with technology, the animate world caused more strongly integrative constructions, in which the single elements often have to perform a majority of tasks. At the same time, unconventional material features like for example partially different elasticity play a role. With reference to technical constructions, a better integrative adjustment of single components for multipurpose tasks is significant. The bionic of devices occupies the development of useable complete constructions according to models from nature. Particularly in the fields of the connection and bearing technology, of pumps and conveyor technology, of hydraulics, pneumatics and flow-mechanics diverse stimulating encouragements can be found.

Turbine optimization according to the giant shark

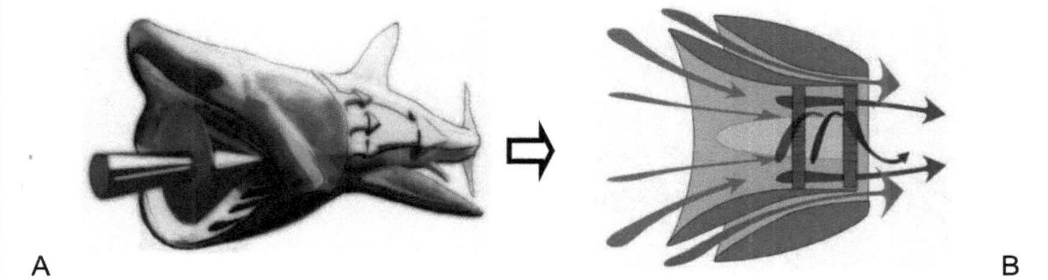

Giant sharks are purely passive plankton filter feeders. Water-turbines improve their efficiency due to this principle by 40 %.

BIOLOGY:
In contrast for example to whale-sharks, which support actively the reception of their plankton-prey through suction and gulps, the up to 14 m long and 4 t heavy giant-shark *Cetorhinus maximus*, swimming at speeds of approximately 4 km h^{-1}, filters out with at least 1000 m^3 water per hour purely passively, alone by "skilful" utilization of the pressure circumstances within the flow (A). Thereby, plankton carried along gets caught with the gill.

PRINCIPLE:
The question is why the water doesn't take the way of the lowest pressure drag, this means, why it does not stream completely around the head, instead of flowing through the drag generating gills. The solution lies in the head form and the transition from head to torso in the region of the gills. Through the design, a low pressure is created there, that sucks the water, otherwise backing up itself, from the drag generating gill-basket.

TECHNOLOGY:
Exactly this problem also faces water turbines with docked generators for electric power that stand in housings the free flow. The water tends to flow about them instead of flowing through them. Anthony Reale, an industry designer at the college for Creative Studies Detroit, solved this well known problem in a bionic-analogous manner with a doubly converged inlet nozzle, an entrance in an opening, so to say (B).

LITERATUR: http://www.designboom.com/weblog/cat/8/view/12858/anthony-reale-strait-power.html v. 01.18.11.– Grzimeks Tierleben Fische 1, Lizenzausgabe (1980). dtv , München. – S. auch Lb. Techn. Strömungslehre und Strömungsmaschinen.

Wind-concentrator "Berwian"

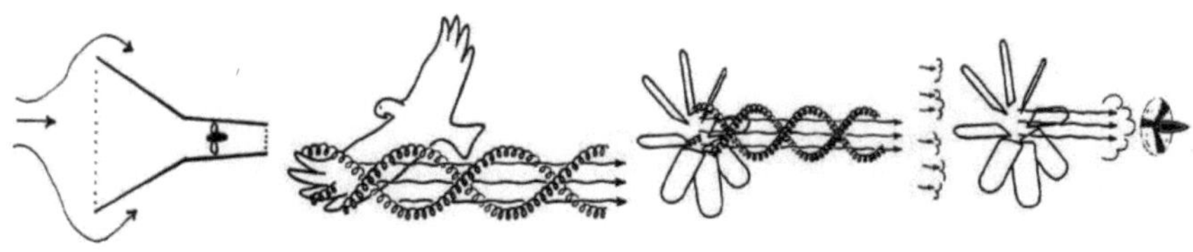

With a bionic vortex coil trick, something like an effective "Nuremberg funnel for the wind" succeeds.

PROBLEM:
This, itself already classic approach is inserted here, because it is suited to the preceding page: flow always proceeds over regions of low drag. - A faster running wind turbine emits higher output. If one placed it into the inside of a funnel (A), however, it could not run faster since the wind then would stream essentially around the funnel edge. The funnel doesn't work as a "Wind concentrator".

BIOLOGY:
I. Rechenberg studied the vortex production by the fingered primaries of large terrestrial birds (storks). Each free primary is flown for itself and generates, independent from the others, a small vortex spiral. These spiral vortexes surround each other forming a vortex coil (B). In their inside, a higher wind speed prevails in comparison to the outside flow: Consequently, such a system works as a "wind concentrator".

TECHNOLOGY:
The author has placed wing profiles in radial assembly. Thus, they form an annular "stator". Usable are the vortices that peel away from the wing inside edges (C). A small, high-speed wind turbine is placed into this central vortex coil (D). Now, it can draw out a higher output because of the higher wind-speed. Since this output is proportionally in the cube of the wind-speed, the system works with increase factors up to 10.

LITERATUR: Rechenberg, I. (1984): Berwian konzentriert den Wind. Sonnenenergie 2, 6–10. – Abb. aus Faltprospekt TU Berlin (1984).

Bionic shovel profiles for an axial ventilator

A 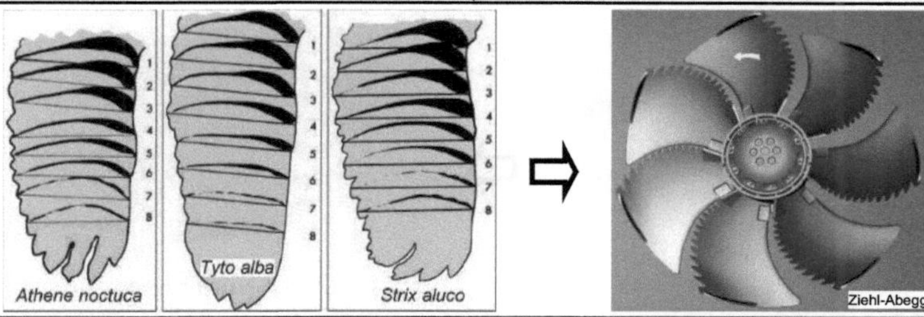 B

Continuing examinations of owl-wings – development of more efficient shovel profiles for ventilators.

BIOLOGY, FOR EXAMPLE WING PROFILES OF THE OWL:
Examinations of wings (A) of different owl-types show that the basal wing-sections, 1 – 3, are profiled, thick and strong cambered. On the other hand, the middle profiles, 4 – 6, are less admittedly profiled, more thinly and cambered more weakly. The last ones do not reach up lift coefficients c_A of the first named, however, show lower drag coefficients c_W. Therefore they can be suitable as models for more efficient fan-profiles.

PRINCIPLE:
On the search for higher efficiencies and lower sound development with fan shovels, the middle profiles and the edge structures of owl-wings proved to be productive. They reach with good c_A- and c_W-values at fan-typical angles of attack favourable glide ratios without higher motive power. Under these conditions also only small flow separation takes place and with it less noise develops.

TECHNOLOGY:
The new FE2owlet-ventilator of Ziehl-Abegg possesses on the one hand protracted profiled shovels (B), on the other hand, uses at their rear-edge "owl-wing"-structures. In comparison with earlier used, not bionically optimized shovels at a flow rate of 3000 $m^3 h^{-1}$, the power density increases about q = 16 % and the static fan efficiency increases about η_{stat} = 5 %. Conveniently the noise is reduced about 6 dB (!).

LITERATUR: Nachtigall, W. (1975): Vogelflügel und Gleitflug. Eine Einführung in die aerodynamische Betrachtungsweise des Flügels. J. Ornithol. 116(1), 1–38. – Ziehl-Abegg AG (Künzelsau): Bionik & Energieeffizienz. Roadshow 2008.

Wind turbine blades with "butterfly scales"

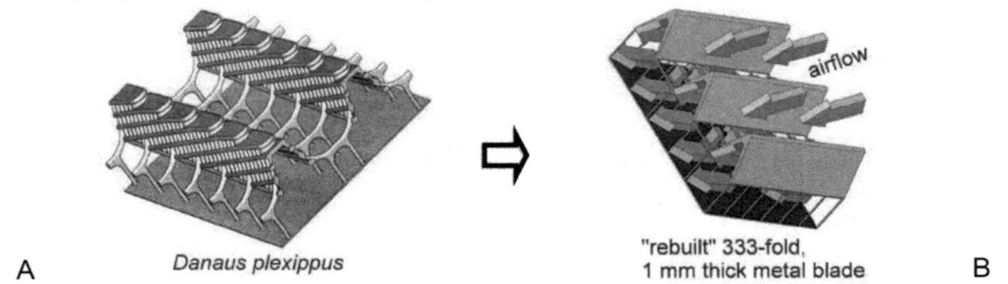

Butterfly scales increase lift and stiffness of their wings - can this also be applied to large scale wings?

BIOLOGY:
The scales on some insect-wings, preferably butterfly-wings (A), are a typical example for multi-functional effects, that in the animate world one and the same morphological formation can have. Here for example: Increase of the lift and the stiffness, production of iridescent colours, temperature-balance. Flow effects are already known since a longer time (Nachtigall 1965); however, it is yet not completely clear how they work.

PRINCIPLE:
In the technical transfer of biological effects one must first desist from the multi-functionality and look at a selected effect in more detail. Here it was the aerodynamic effect. At the same, it is to be considered that the large engineering design is working in completely different characteristic ranges (here: Reynolds number Re, percentage of virtual mass, etc.). This may limit the transferability, but not in this case.

TECHNOLOGY:
I. Kovalev from the Israeli Kinneret College has reconstructed technically analogously the scales of *Danaus plexippus* (A) (B) and fixed the resulting device on a NACA 230 profile oscillating in a wind tunnel. In comparison with the smooth profile, the "scaled" one reached a 1.15-times higher lift at $Re = 2 \cdot 10^5$, a 1.14-times higher virtual mass and with it a 1.1-times lower frequency as well as also a 1.37-times reduced damping factor.

LITERATUR: Kovalev, I. (2010): From butterfly to wind turbine. Wind Engineering 34(4), 351–350. – Nachtigall, W. (1965): Die aerodynamische Funktion der Schmetterlingsschuppen. Naturwissenschaften 52(9), 216 f.

More efficient rotor blades with "back whale edges"

A

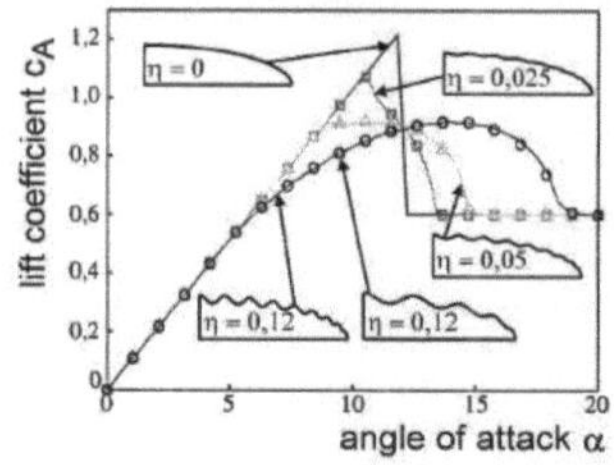

B

The optimization of wind rotor blades is of big practical importance. Fins of back whales have given stimulation.

BIOLOGY:
Back whales of the species *Megaptera novaeangliae* belong to the biggest, fast swimming high sea's mammals with flow adapted bodies. Their mass can reach up to 30 t. Their good motive force efficiency and mobility are caused by the enormous, "hump shaped" knot-rows, especially on the leading edge of their pectoral fins (A), that play a role above all with the precise and very quick control (Woodward et al. 2006).

PRINCIPLE:
After works of Johari et al. (2007), E. V. Nierop et al. (2008), using models of pectoral fins with $Re = 5 \cdot 10^5$, have recorded in a wind tunnel among other things the characteristic curves c_A (α) (c_A lift coefficient, α angle of attack) of fins with smooth and "hump shaped" leading edges (B). The flow separates from the smooth profile after it has reached a $c_A \approx \alpha$ 1.2 at $\alpha \approx 12$ °; consequently, the lift suddenly breaks down ("stall"). With humps, the graphs are more even.

TECHNOLOGY:
Bionically adapted profile leading edges are at the best, if a hump-period corresponds approximately to 12 percent of the profile length ($\eta = 0.12$). Here, "only" a maximum of $c_A \approx 0.85$ is reached, however, this is postponed to $\alpha \approx$ 15 °: "stall resistant profile (delayed stall)." So rotor blades can work at higher α and thus increase the performance (Prototypes: Company "Whale power"; Toronto; efficiency increase of fans up to 20 percent).

LITERATUR: Nierop, E. A. von; Alben, S.; Brenner, M. P. (2008): Physical Review Letters 100, 1–4, 054502. – Johari, H. et al. (2007): AIAA J. 45, 2634.– www.whalepower.com – Woodward, B. L.; Winn, J. P.; Fish, F. E. (2006): J. Morphol. 267, 1284.

Oscillating fan according to fanning bees

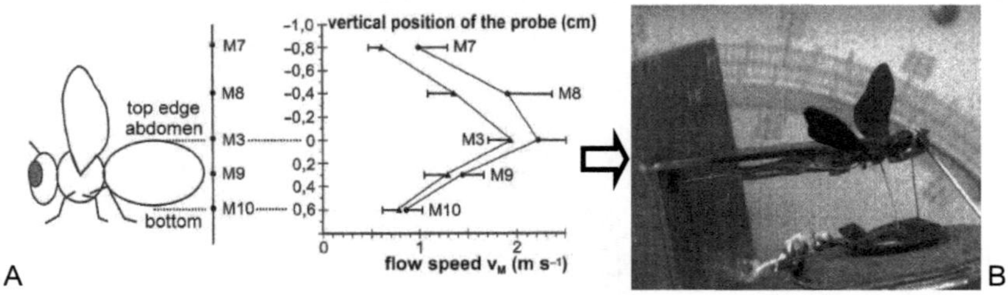

Almost silent oscillating fans using the principle of fanning bees could replace disturbingly buzzing computer-fans.

BIOLOGY:
When "fanning" in front of the entrance, honey bees, who let swing their wings at vertical plane of rotation like a stand propeller, generate a powerful airflow with a cross-section that corresponds approximately to the "wing-disk." Here, the speed-distribution is somewhat parabolic with tops over 2 m s^{-1} (A). The air beam is strongly directional. Probably, the bees save energy fanning at their resonance frequency.

PRINCIPLE OF THE REALIZATION:
An "artificial bee" was built (B) that moves its wings in a wind tunnel similar to the honey bee when fanning. (This kinematics corresponds approximately to the one of the “Sterzeln”. On that occasion, the air stream drives away chemical messengers.) At the " artificial bee", R. Spillner and M. Junge carried out smoke thread flow performances. H. Scharstein also succeeded in power measurements at artificially moved original insect wings.

TECHNOLOGY:
The wings of the " artificial bee" swing optimally at 60 s^{-1}. This corresponds to the resonance case, with which the energy input is minimized, the amplitude and the total degree of effectiveness is maximized. The tuning of the oscillation and rotation is separately adjustable. From the system "technical bee", an effective, quietly oscillating fan of extremely low energy needs may be developed for refrigeration of computer-components.

LITERATUR: Spillner, R. (2000): „Mechanische Biene“. WO2000EP02164. Priorität DE 19991010731. IPC F04D33/00, 2000-09-14. – Scharstein, H. (1998): Kräfte- und Leistungsbilanz bei der künstlichen Schlagbewegung einzelner Insektenflügel. BIONA-report 12, 257–270.

Oscillating fan according to bird wings

A

B

Often, a principle is differently realized in the nature: Stimulation of diverse technical implementations.

BIOLOGY:
Every wing in the animal kingdom moves in such a way, that oscillation and rotation are connected phase-favourably. With it, as has been shown several times in this book, lift and thrust are generated in an optimal way. If the animal sits like the fanning honey bees (comp. p. 104) such a "stand oscillator" on the other hand generates an airflow. The bees induce a powerful, narrowly bounded flow in order to ventilate their hive.

PRINCIPLE:
This oscillation coupling principle is used e.g. also by birds (A) and bats, however, not for the purpose of inducing airflow in the stand. However, it doesn't speak anything against transforming the principle. So, slowly oscillating, big artificial wings are conceivable, that - positioned in a room - ventilate this without draft but effectively. If water is evaporated on that occasion, they also cool the room. And they look quite elegantly anyway.

TECHNOLOGY:
R. Spillner of the bionic motion PLC, Koblenz, together with Phoenix Design, Stuttgart, has brought to maturity such a concept (B). The wings made of a white glossy material are 1.6 m high. They work pair wise. They can be remotely controlled in 5 steps and generate with 18–30 oscillations per minute air-flows up to 0.45 m s^{-1}, that man does not yet feel as draft. They ventilate remarkable 200–230 m^3 h^{-1} on that occasion.

LITERATUR: Spillner, R. (2000): „Mechanische Biene". WO2000EP02164. Priorität DE 19991010731. IPC F04D33/00, 2000-09-14. – Scharstein, H. (1998): Kräfte- und Leistungsbilanz bei der künstlichen Schlagbewegung einzelner Insektenflügel. BIONA-report 12, 257–270.

Partially harmonic-linear oscillating wing generator

A

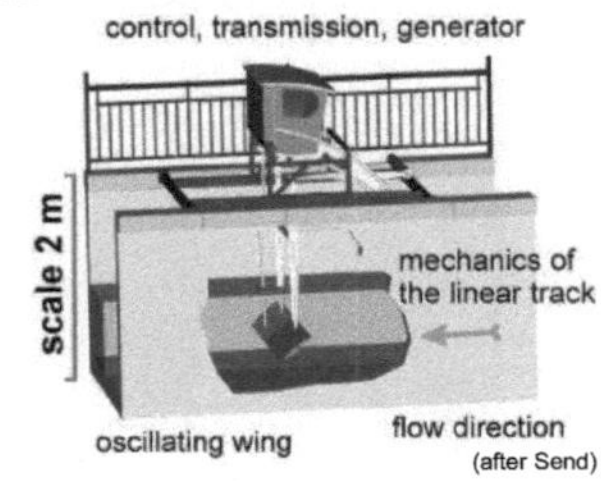

B

Driven according to the wing movements of locusts, this generator draws a large power out of a flow.

BIOLOGY:
At the active flight of locusts, the wings (A) move partially harmoniously, e.g. with longer approximately linear sections in the middle of the down- and upstroke. There are, as it is frequently in the case in insect flight, oscillating and rotating components. Between these, there is a phase shift. The phase angle is adjusted so, that an as big as possible output percentage of the muscle motor can be used to induce a flow.

PRINCIPLE:
Through shifting of the phase angle between down- and upstroke-component the energy transport may be reversed. In this case, the oscillating system draws power out of the flow. This appears as well at fluttering of a flag or at the feared wing flutter of an airplane that can lead to a wing break if not damped immediately. A new generator can use this kind of energy transfer.

TECHNOLOGY:
In the oscillating wing generator of W. Send, a rectangular wing blade adapted to the cross section of a river and fastened on bars is moved up and down by the flow (beating oscillation). This is induced by a mechanism that provides an additional rotary oscillation, and adjusts a suitable phase angle. The linear sections preferably should be prolonged (B). A tandem version with 2 wings is developed at the time.

LITERATUR: Send, W. (2006): Nach dem ...: Der Hubflügelgenerator. In: BioKoN (Hrsg.) Innov.-motor Natur, 176–186. – Birnbaum, W. (1924): Das ebene Problem des schlagenden Flügels. Z. Angew. Math. Mech., 277–292. – Wortmann, M.; Zarnack, W. (1993): Wing ..., J. Exp. Biol. 182, 57–69.

Peristalsis of a gut as a model for micro-pumps

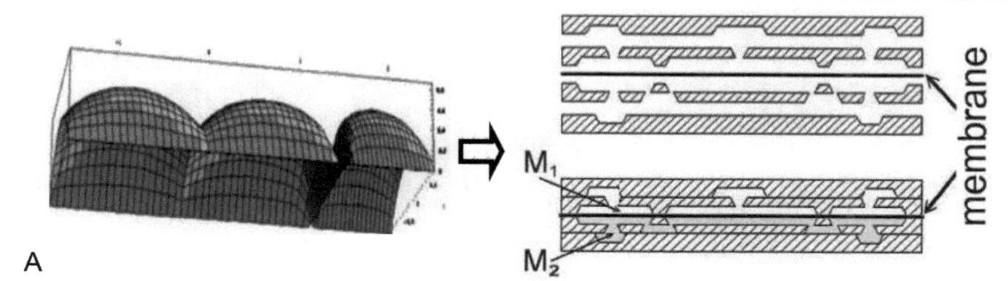

Failure sensitive micro pumps for medicament dosing systems etc. become replaced by peristalsis-systems.

OPERATIONAL PROBLEMS IN CONVENTIONAL PUMPS:
"Bubble-jet" systems, piezoelectric stimulated oscillating membrane pumps, fine piston-cylinder-pumps and other dynamic pumps generate high drags and become blocked very easily. In systems dosing medicaments as well as in the pharmacological and biotechnical research, however, tough solutions prevail. Often also suspensions, namely salves, pastes, polymer-meltdown, even dough and mash must be transported frequently.

PRINCIPLE OF THE INTESTINE-PERISTALSIS:
By ring- and longitudinal-muscles, a distortion-wave of the intestine-wall is induced. At intestine segment, somewhat apart the cross-section is first reduced and then, with a certain phase-postponement, the segment lying in direction of the substratum-transportation is shortened (A). The transportation is not due to pressure-differences, but because of differences of the flow resistance between narrowed and extended intestine parts.

TECHNOLOGY:
In contrast to the spatial intestine-system, the analogous micro pump, designed by M. Kallenbach, is carried out planar. A ductile membrane separates the conduct of the transported medium M_2 from the conduct of a drive-medium (M_1; compressed air) (B). A rhythm generator shifts the pressure in the latter chambers so, that the medium M_2 is sucked in and is pressed out periodically.

LITERATUR: Kallenbach, M.; Smella, E.; Holdenried, J. (2000): Electrochemically driven polypyrrole bilayers for moving and positioning bulk micromachined silicon plates. In: J. Micro Elektromech. Syst., S. 373–383 – Patent Nr. P 19724240,5; Fa. Höfer & Bechtel, Hanau.

The "Fin Ray Effect®" and its technical utilization

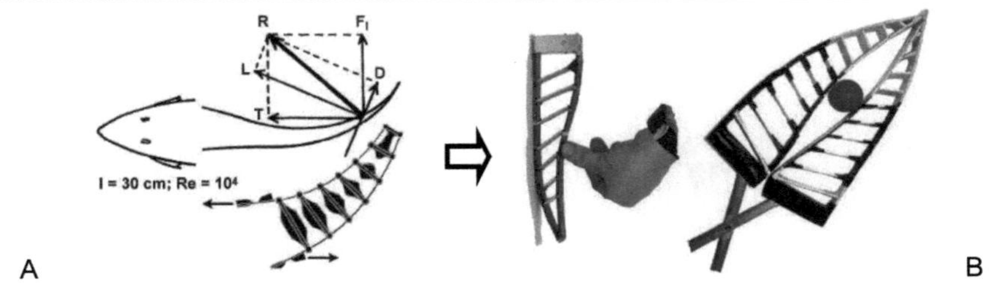

The construction-principle of fish fins, that "counter-bend" automatically, leads to self fitting gripper arms.

BIOLOGY:
When a trout moves the tail fluke to and fro, it bends under the combined effect of its own stiffness and the water pressure, so that beside other components a lateral force F_l and thrust T originate (A). An inherent automatism ("Fin Ray Effect®") counteracts an excessive bending: Oblique connections of fine bone-braces between the basal and the inner membrane generate a counter bending.

PRINCIPLE:
The fine bone-braces between the covering surfaces of the tail fin are diagonally oriented. If the surfaces are shifted against each other during a fin beat, these oblique configurations cause the trailing edge's bending in direction of the beat. The water pressure normally overcomes this curvature-tendency, so that a functional bending results according to fig. A. The "Fin Ray Effect®" could be a protection against over-bending.

TECHNOLOGY:
The described mechanism causes, that a contact-surface, constructed in such a way, snuggles up automatically to a touched object, a gripping-arm for example around a round-tube (B). Various utilization of the mechanism is conceivable, for example well sticking operation-silverware or automatically pressing chair backs. The effect already was applied to the construction of the free floating Festo "Aqua Ray".

LITERATUR: Nachtigall, W.; Mitarb. von A. Wisser (2006): Ökophysik. Springer, Berlin. – Fin Ray Effect®, Evologics GmbH, Berlin (2007): Zit. zu Aqua_ray: www.festo.com/cms/en_corp/9786.htm

Bionic grippers for micro-robotics

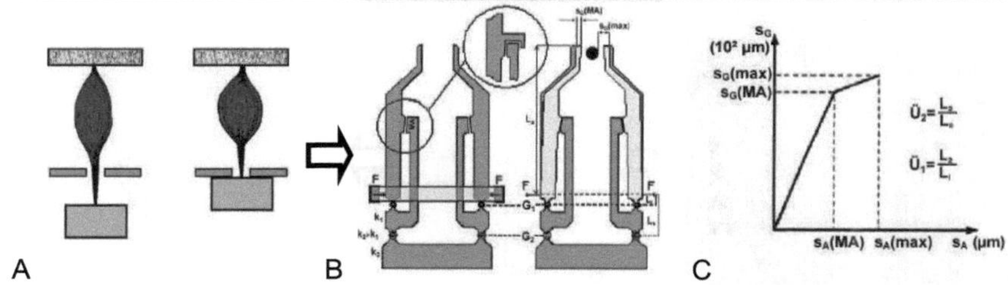

A B C

The basic structure of a gripper can be fitted for the requirements of the micro-technology with bionic design-criterions.

TECHNICAL REQUESTS AND BIOLOGY:
For micro-purposes, for example the positioning of grasped objects in bioengineering-processes, developments of macro-robotics can not simply be reduced in size except the "reduction-tolerant" characteristics of the skeleton-muscle. Accordingly, models can rather be small biological objects, as mandibles of arthropods, which combine the grip-precision with a certain softness ("compliance").

DEVELOPMENT:
The gripper conceived by R. Salim (comp. p. 57) was developed further by R. Keoschkerian. Different variations of micro-glass-grippers that are based on interlocking gripping could be realized with it, e.g. "quick and powerful grippers with changeable kinematics and two-stage transmissions." They contain, in accordance with the "end-stop-contraction" of the skeleton-muscle (A) several end-stop-elements for the gradual change of the stiffness (B).

TECHNOLOGY:
With this gripper-system, a stepped characteristic for the actor-travel-handling-range with transition of a transmission ratio to another could be realized (C). It marks, as mentioned, the transition of an isotonic contraction to an isometric one, in accordance with the end-stop of the skeleton-muscle. With it, the actor can quickly and well-directed grip at first and then powerfully fasten (immobilising) the object.

LITERATUR: Salim R. (1995): Mikrotechnischer Silizium-Greifer für die Mikromontage, Symposium für Mikrosystemtechnik, S. 105–111. – Salim, R. et al. (1997): Micro grippers created in microstructurable glass. Microsystem Technologies. Bd. 4, S. 32–34.

Moisture driven mechanics of the Tamarisk type

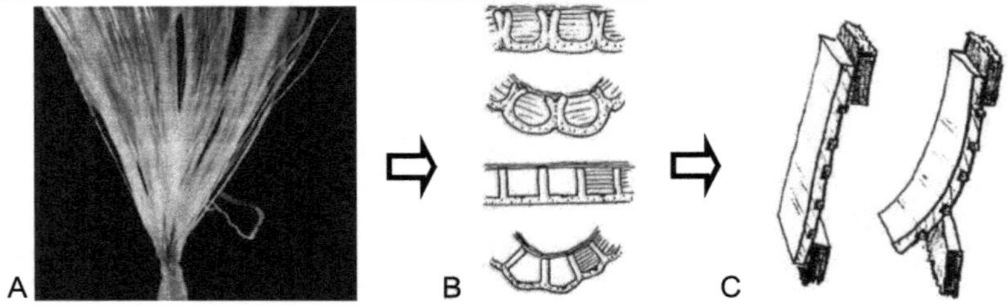

In bimetallic-operated ventilation-flaps of greenhouses, the process variable "moisture" is adjusted only indirectly.

BIOLOGY:
The spores (dispersal bodies) of the Tamarisk, *Tamarix gallica*, carry a pappus (mop of hair) of hygroscopic hairs (A). If the little parachute is driven over ground-near moist air-layers, the hairs are folded up very quickly. The spores sink more quickly, then land and can take roots well there. With about 35 percent of relative humidity ("dry air") the settling speed amounts to 9.8 cm s^{-1}, with 70 percent, however, already to 15.3 cm s^{-1}.

PRINCIPLE:
The hairs of the pappus consist of consecutively laying hygroscopic elements (B) at the basis. One can abstract them as cubbyholes with three solid sides and one side spanned by an extremely fine membrane. The single bending of the single elements is added up, so that strong total bending occurs. Therefore the pappus hairs move on a wide bow, and the entire pappus system responds fast. That is the basis for the automatic landing.

TECHNOLOGY:
Hygroscopic activated bending, adjust and switch facilities are hardly used (e.g. for weather-houses) in technology until now. Bimetal-moved ventilation-flaps for greenhouses open with higher temperatures, not with higher humidity. Therefore the Tamarix principle could lead to a direct humidity-regulation, e.g. over hinge-loose opening covers (C). Also, among others, humidity switches and cheap simple hygrometers are conceivable.

LITERATUR: Nachtigall, W. (2011): Biomechanik von Flugsamen. Teil 6: Diasporen der Tamariske mit rasch reagierenden hygroskopischen Pappushaaren; Funktionsmorphologie – Ausbreitung – Bionik. Mikrokosmos 100(2), 86–91.

Indirect evaporation cooling according to the skin

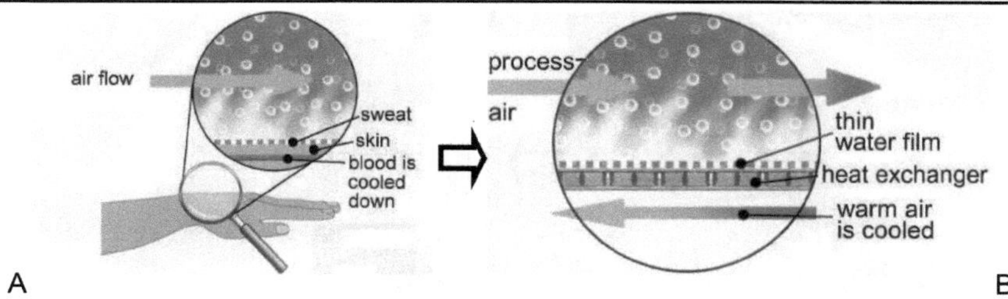

Skin cooling by sweat-evaporation and counter current processes were models for novel technical cooling-systems.

BIOLOGY:
The mammal skin generally possesses sweat-glands. The thin sweat film on the skin evaporates, leads away approximately 2.4 kJ of heat per gram and contributes so effectively to cooling (A).
The counter current process is used for the heat recovery (dolphin fin, duck leg and tuna head) as well as for the increase of O_2-uptake/CO_2-output many times by animals in fish gills and other exchangers.

PRINCIPLE OF THE KAMPMANN KLIMANAUTEN:
The air-conditioner possesses a wavy heat exchanger of large surface with a particular hydrophilic surface coating. Consequently, an applied water-film remains on the surface and can be easily evaporated.
Gill-like rib slits provide good heat transfer. An optimal evaporation is made possible because one third of the pre-cooled air is led back as process-air by the counter current over the heat exchanger.

EFFECTIVENESS:
Especially due to the use of evaporation to cool the inflowing air (until about 14° C under room-air-temperature), the patented Klimanaut technology works with good efficiencies that consequently make compact implementations of the appliance possible. On a typical "hot afternoon" with 30° C outside-air-temperature, it generates internal process-air of 27° C and delivers inflowing air of 18° C. Volume-streams of 400–6000 m^3/h are possible.

LITERATUR:
Kampmann GmbH, Friedrich Ebert Str. 128–130, 49811 Lingen (Ems); www.kampmann.de

Fold awnings and clamp mechanisms

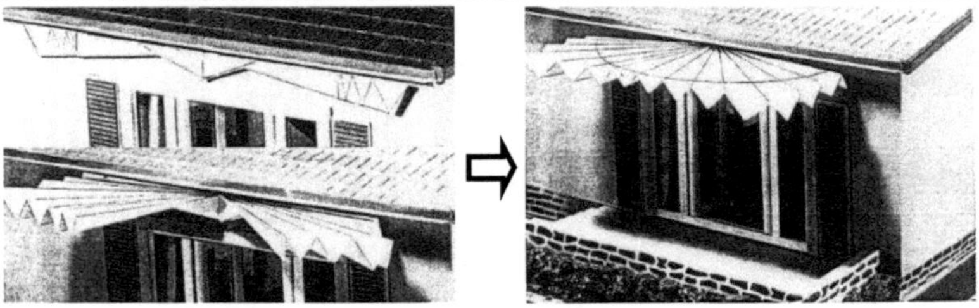

A B

The manner, how insects interlock and fold up their wings, leads to concepts for technical fold constructions.

BIOLOGY:
Locusts fold up their wings before storage in longitudinal folds and tilt them space-saving over the abdomen. The locusts use a membrane (wing-spread) braced between supporting structures (wing-venations) and a special joint, in which, beside the wing-beat muscles, the wing folding is done by a particular "folding muscle". In flies, one admittedly does not know any wing-folding, but a lot of material-saving warping.

PRINCIPLES:
Principles that one abstracts from nature forms can be combined. The two pack-in-principles "longitudinal-fold-formation" and "tilt of the fold-products", as they are realized by locusts in their wing-mechanics, can lead to a technical product that works quite accordingly: Analogy also with respect to the type of combination. For the fly-wing, the following principle of construction counts: "Stability with preferably few and with it light bearing veins."

TECHNOLOGY:
The airplane-constructor H. Hertel published 1963 a book "Biology and techniques" that already contains in its characteristic subtitle "structure - form - movement" a classification of bionics. In it he describes the development of a fold awning (A, B) after the principle of the locust-wing. In 1996, a light self stabilizing clamping system for surfboard-sails related to the wing of the blue fly was proposed by A. Wisser.

LITERATUR: Hertel, H. (1963): Biologie und Technik. Struktur – Form – Bewegung. Krausskopf, Mainz. – Wisser, A. (1996): Vorgespannte, faltbare Insektenflügel als Vorbild für technische Segel. In: Nachtigall, W., Wisser, A. (Hrsg.): BIONA-report 10, 181-182. Fischer, Stuttgart.

A technical "plant-stalk"

Many plant-stalks of high slenderness combine high stability with flexibility: Models for "technical stalks."

BIOLOGY:
With slenderness (length : mean diameter) of more than 400 : 1, for example, grain stalks are very notably. The whistle-grass, *Molinia coerulea*, is even more slender. Also larger types like reed, vibration-absorbing reed grass and bamboo combine buckling resistance with elasticity. The winter horsetail, *Equisetum hyemale*, shows a peripheral layer of "double-T-beams" that include cavities between themselves (A): a lightweight-construction.

PRINCIPLE:
Plant-stalks gain their stiffness by the fact, that high-tensile ropes (sclerenchyma) are embedded in a pressure-proof matrix (parenchyma). Therefore, they are analogue the steel concrete, in which the high-tensile steel rein-forcement lies in a pressure-proof cement matrix, which, after curing, keeps the tensile steels at distance. Researchers of the university of Freiburg and the ITV Denkendorf converted this and other principles into "technical plant stalks".

TECHNOLOGY:
The "technical plant stalk" (C) is a bionic fibre-composite-material, in which fibre bundles of many tensile threads are woven into a network (B), that is filled with a synthetic resin matrix. After Speck et al. not less then 6 biological prototypes have been the force behind: Biological lightweight → sandwich structure, optimized fibre orientation and gradients in the wall of the stalk → macro-gradient, binding fibre matrix → micro-gradient etc.

LITERATUR: Milwich, M. et al. (2007): Der technische Pflanzenhalm: Ein bionisches Schmaltextil. Melliand Textilberichte – Band- und Flechtindustrie 44(2), 33–38. – Speck, T. et al. (Hrsg.) (2011): Bionik in Botanischen Gärten. Verb. Bot. Gärt. Publ. Bot. Garten Freiburg.

Self-sharpening incisors like cutting tools

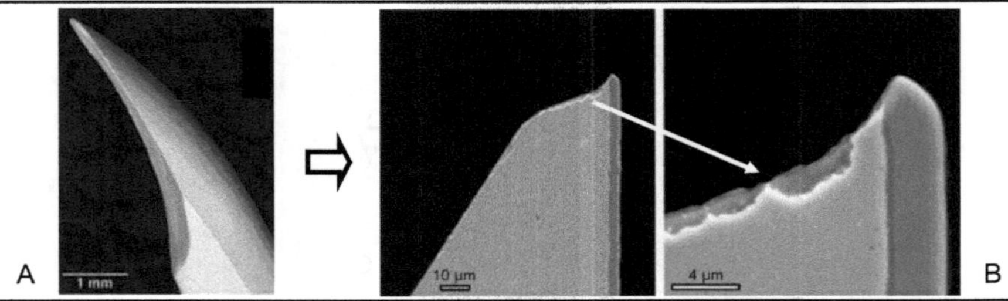

Self-sharpening according to the principles of enamel and dentine for industrial cutting tools.

BIOLOGY:
Teeth of the herbivores and specifically these of rodents (A) are exposed to extreme abrasive stresses. During evolution, denture-types developed, which use the abrasiveness of food for self sharpening of the teeth. The soft-elastic dentin is not enclosed by the hard enamel completely, so that it can rub off purposefully. The micro-structure of the enamel guarantees high hard- and toughness, so that a filigree, sharp edge can be durably formed at the tooth-top without breaking.

PRINCIPLE:
The principle of the layer construction of a tooth of mammals especially that of rodents was converted to functional technical layers. The hierarchical micro-structure of the enamel was reproduced by a ceramic multilayer-nanocomposite-architecture (B). The layer of only 4 µm shows excellent toughness and very high hardness. The steel-substratum was adapted so, that the edge becomes self-sharpened through abrasive loads.

TECHNOLOGY:
Rodentics®-knifes can be used in cutting procedures during which fine abrasive particles (cutting of paper, wood, plastics) have a wear out effect. In the start-up phase, the blade sharpens autonomously. Cutting forces remain permanently low. Through the high hardness of the layer, the knife hardly wears out. The energy and material-efficiency of the cutting process is improved significantly.

LITERATUR: Rechberger, M. (2012): Selbstschärfende Schneidwerkzeuge für abrasive Schnittgüter – eine bionische Entwicklung. Dissertation Universität Duisburg-Essen.

Molecular nano motors

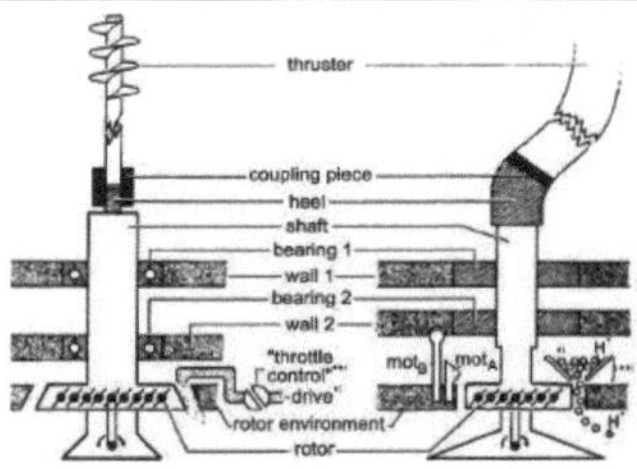

A

DNA-Rotaxane

B

Nano-motors of the type of the bacteria flagellum drive could interact with biological substrates, for example with proteins.

BIOLOGY:
The proton-driven mechanism of the bacterium flagellum has a shaft, attached by two ball-bearings in the two bacterium membranes that can rotate freely (A, left) (the only example in the animate world). By this shaft the attached flagellum is set to helical rotations. Frequencies up to 300 Hz are reached. An analogous micro-technology is discussed (A, right). The vision is, that mobile molecular machines may circulate in the bloodstream.

ROTAXANE-AXES:
Since some time, researchers succeeded to produce molecular ball-bearings and other elements under utilization of self-formation processes. The problem was and is the freely revolving, mechanically stable axis. A solution is offered by the Rotaxanes ("wheel-axes") of the organic chemistry. They consist of a long-stretched axis and a mobile ring threaded over it. At the terminals the axis carries "stoppers" in form of two crossed rings.

ON THE WAY TO MINI-ROBOTS:
The organic-chemical Rotaxanes are too small. D. Ackermann et al. with M. Famulok of the LIMES-Institute of the University Bonn were successful to produce larger molecular-systems (B) under application of DNA-modules. Their nucleotide doubles rope also provides higher mechanical stability. The system can be coupled with other molecular elements. Further research concerns the double bearings according to the type of the bacterium-drive.

LITERATUR: Nanomotoren mit DNA (2011). GIT Labor-Fachzeitschrift 03, 142 – Ackermann, D. et al. (2010): A double stranden DNA rotaxane. Nat. Nanotechnol. 5, 436–442 – Zur Bakteriengeißel s. Ref. in: Nachtigall, W. (2005): Bionik, S. 125–126, Springer, Berlin.

BUILDING

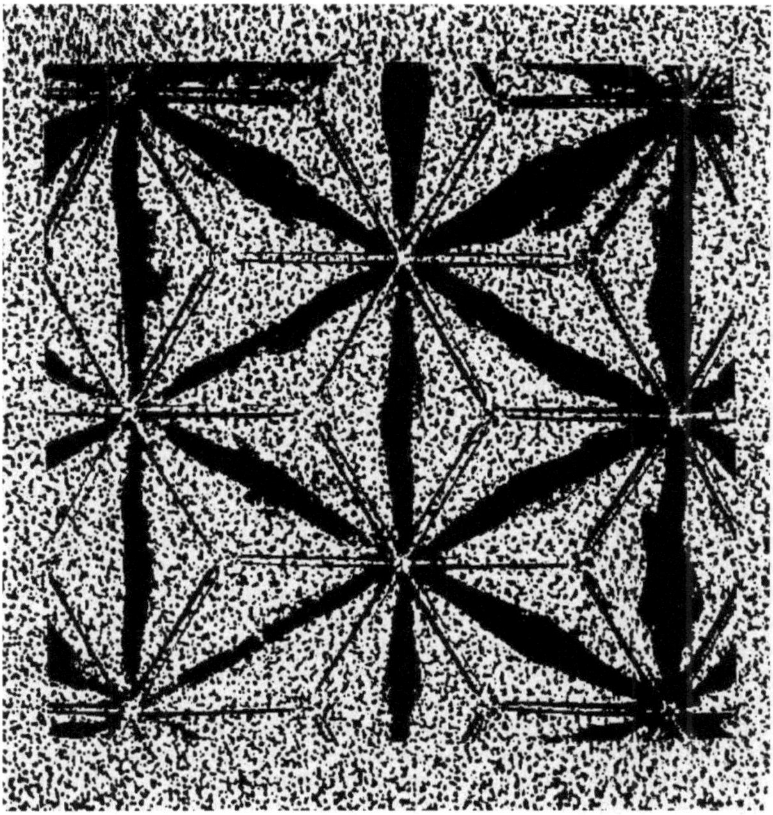

AND CLIMATIZATION

W. Nachtigall and A. Wisser, *Bionics by Examples*,
DOI 10.1007/978-3-319-05858-0_7

BUILDING AND CLIMATIZATION

Animal-constructions – Buildings and Constructions – Climatisation, energy-optimization.

"Natural" building on the one hand means return to traditional construction-materials that are used also in biology (for example clay-materials with their biologically interesting construction characteristics). On the other hand, one gains ideas from the study of biological lightweight-constructions for temporary technical lightweight-constructions. Stimulation can come along, for example, from rope-constructions (cobwebs), membrane and peel-constructions (biological peels and tanks), protective covers, that allow gas-exchange (egg shells), floor-constructions, integration of suspended units, variable constructions, constructions with better recyclable materials than the technology knows until now, ideal area coverage (leaf superimpositions) and surface-utilizations (honeycomb-principle). Adjustments of individual residential-elements relative to the total area have to be taken into account, in their alignment for example to sun and wind, in analogy to leaf superimpositions and bloom constructions and other more. Passive ventilation, cooling and heating are essential points of view. The study of natural constructions as well as the analysis of so-called primitive constructions for example in central-America and North-Africa can lead to unconventional designs and facilities. Alone, the ideal alignment to sun and wind, roof-forms, niches in the earth, ideal basement and air-flow from the cool soil to the summer-warm rooms, air circulation according to the termite-constructions with gas-exchange under application of porous materials could save 80 percent of (electric) energy for the summery cooling and 40-60 percent of the energy for the winter-heating. Symbiotic integration of plants into the living-landscape can be used for the improvement of the partial pressure of oxygen and for food supplies.

What does bionics give to the architect?

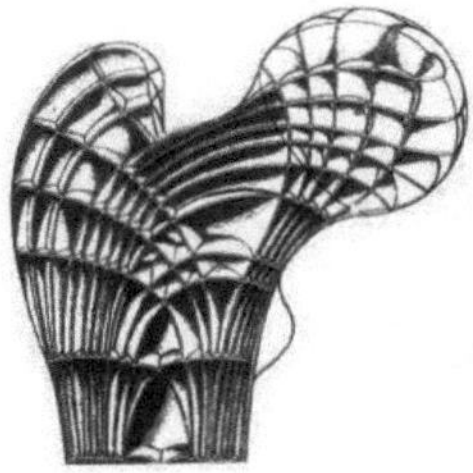

A

B

"Copying the nature is completely useless." How can this stunning sentence from a Bauwelt-interview be understood?

NOT TO BE COPIED:
In the interview between the biologist Werner Nachtigall and the architect Göran Pohl – both Bionics convinced – an appropriate manner of view of the architect on the nature was discussed. Can this view give stimulation for architectural designing? There was agreement that the nature can not be copied. Coppying nature leads at most to a "biomorphic", not functional architecture of superficial takeover of aspects from the form canon from nature.

BIONIK AS ONE OF MANY TOOLS:
The key-sentence of G.P. was: "As an architect, I should use in the development of my buildings everything available for me, also the nature. Bionics is one of many different creativity tools". At a bionic proceeding, it is not a matter of finding throughout nature-inspired approaches, but one must "also" look at natural role models, not just exclude them. One cannot see "closeness to nature" at the building.

BIONICS AS FUNCTION-ANALOGY:
The look on the nature can help to get ideas, that one gets not or not so fast otherwise. However, the basis is not superficial, but functional similarity (A, B, comp. p. 23). In the bionic proceeding, it therefore is about the "abstraction of general principles from the biology and their technically adequate realization" (W.N.): Example: termite-analogous ventilation-system, put in with success e.g. in the West Gate-Building, Harare (comp. p. 50).

LITERATUR: Schultz, B. (2011): „Die Natur zu kopieren ist völlig sinnlos". Was bringt Bionik dem Architekten? ... Interview mit W. Nachtigall und G. Pohl, Bauwelt 102, Aug. 2011, 14-17.

Material masses as thermal storage

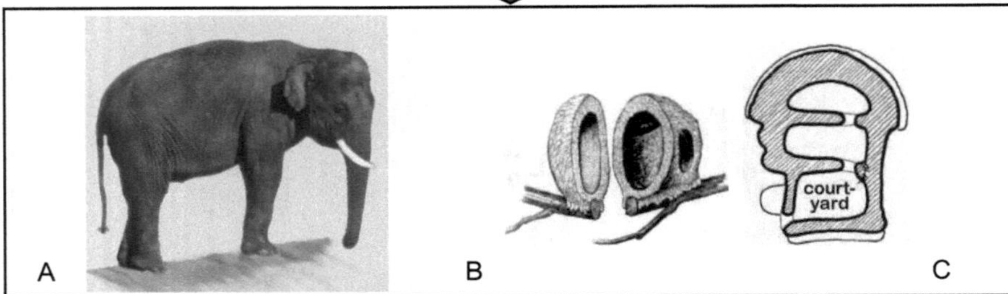

The heat tolerance of elephants is based on their thermal reserve; similarly work certain low- and high-tech constructions.

BIOLOGY:
The Asian elephant *Elephas maximus* (A) tolerates the high day temperatures (30–35° C) of the surroundings because his body-tissue works as thermal storage. During the day, heat is stored instead of that it is passed on to sensitive inner organs; therefore these don't come under heat-stress. At night the body temperature sinks and so heat is given off. Camels and other desert mammals behave similarly here.

THERMAL BUILDING-BALANCE, LOW-TECH:
The Pueblos in North America use amongst others (C) since time immemorial the corresponding thermal qualities of their adobe construction materials (fibre-reinforced clay). Amongst others, they adjust material alignment and wall-thicknesses so, that the day heat is stored in the material at first and arrives indoors during the cool night. So, the interior is tempered over 24 h pleasantly. The clay nest of the ovenbird *Furnarius* behaves similarly (B).

THERMAL BUILDING-BALANCE, HIGH-TECH:
With the Fraunhofer and the BASF, a system was developed how the thermal storage qualities of construction materials can be adjusted auto-regulative in similar manner. Plaster construction plates become added, for example, micro-encapsulated paraffins as a phase-change-material. These work as cold storage. "Without energy-supply, they cool buildings ... and prevent, that areas heat up over 21–26° C in the summer."

LITERATUR: Bullinger, H.J. (2011): Mit Hightech in Zukunft bauen ... Kongr. Nanotech. Bionik. Messe München - Weissenböck, N., Arnold, W, .Ruf, T. (2011): Taking the heat thermoregulation in Asian elephants under different climatic conditions. J. Comp. Physiol. B; 22.09 [Epub]

A skyscraper-concept according to a model of a trunk

A 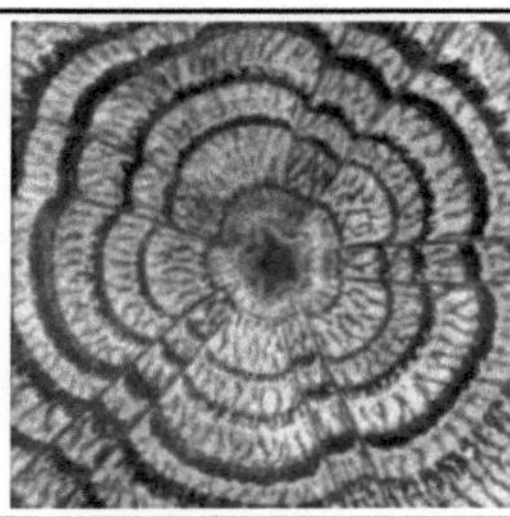 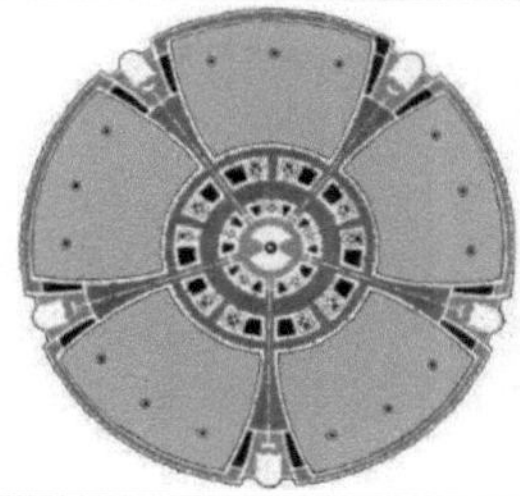 B

Skyscrapers are today often not focused on the people. A bionic concept tries to change this.

TRIGGERS AND FUTURE REQUESTS:
"The 11th September 2011 with his terrible events left mature in me long-dormant ideas on how to revolutionize skyscraper construction in all its functions ... especially making skyscrapers more humane and safer" (D. Oligmüller). - Requests: Consideration of human behaviours especially in case of danger; maximum security; orientation to the daylight; visual contact with the outside world.

TREE TRUNK MODEL:
"The tree withstands the dynamic burdens during storm and with it the side forces. By means of his mark rays (A), he provides the individual "annual ring circles" (comparable with room forming walls) with nutrients and dowels them simultaneously in an intelligent connection of functions". The principle of the mark rays gave rise to the basic idea for the development of new floor plans according to bionic principles (B).

TECHNOLOGY:
The predominant part of the fundamental walls is built in a certain "cell-structure" that connects high loading capacity and high fire protection with low weight. The snail-shaped wound opposite stairwell (analogously the "mark tube") forms the centre of the core. From it diagonal opening-ways (analogously the mark rays) lead to the outside facade from it and finish in outside stairways there: in panic, the human being runs towards the light.

LITERATUR: Oligmüller, D. (2003): How to scratch skies. Wie .... Eine zeitgemäße Hochhausstruktur im Spannungsfeld vielfältiger bionischer Aspekte. Zusammengefasst aus: Wisser, A. (Hrsg): Biona-report 15, 254-262 u. Nachtigall, W. (2003): Bau-Bionik, 1. Aufl.,pp. 162-168. Springer.

Flexible membrane structures inspired by nature

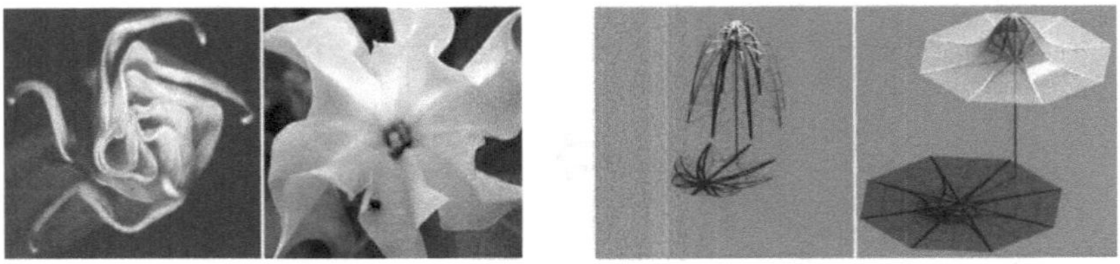

A B

Variable constructions are easily adapted to fast weather changes and are energy-optimized for themselves.

BIOLOGY:
Petals are narrowly packed in the bud and when blooming, unfold to remarkable size, for example the bloom of the thorn apple, *Datura spec.* (A). Some blooms can open and close several times through growth- or osmosis-processes, mostly in the day-night-rhythm, for example the blossoms of the winter aconite, Er*anthis hyemalis*. Also hygroscopic covering in rain occurs, for example with the silver-thistle, *Carlina acaulis*.

PRINCIPLE:
Petals can be reinforced at the edge or at the basis; the opening-closing movements originate from there. Whilst the beam-structures are moved, the fine spread, that is intensified by traits in bigger blossoms normally, follows passively after them. Also technical umbrellas, used in the garden and for covering sports arenas and strolling areas, are convertible constructions, using this principle (B).

TECHNOLOGY:
In a joint research project of the universities of Stuttgart and Freiburg and the ITV Denkendorf, the potential of such constructions for energy optimized constructions is examined. Basic parameters are: "Large elastic distortion – adaptive stiffness – fibre-reinforced composites with big breaking elongation – self tightening membrane materials." Expected are impulses for building industry regarding sustainability, energetic and consumption of resources.

LITERATUR: Verbundprojekt zur Bekanntmachung «BIONA»: Biegsame Flächentragwerke auf der Grundlage bionischer Prinzipien. Projektpartner Institut für Tragekonstruktionen, Uni Stuttgart, Plant Biomechanics Group, Uni Freiburg, Inst. f. textile Verfahrenstechnik Denkendorf.

Biological pneus and application of the pneu principle

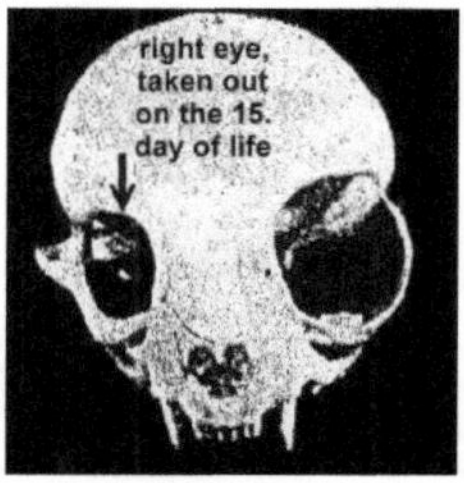

A

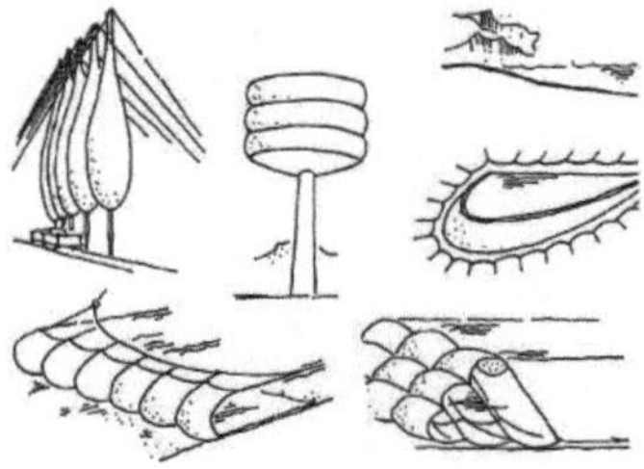

B

"The pneu" represents the universal construction principle of life. In technology, it leads to many diverse applications.

BIOLOGY:
The three parameters of pneus – non-rigid membrane, inside and outside medium, pressure difference between both media – and their interplay to achieve pressure stabilization within organisms were already known in the early 19. century. However, only in the 70-er years of this 20. century, the term "pneu" was introduced by the architect Frei Otto. He also showed that "the pneu" represents "the" universal biological construction principle.

UNIVERSAL VALIDITY OF THE PNEU PRINCIPLE:
No biological supporting structures were found until now, that would not have been built "pneumatically" at least. This is true even for the skull capsule of a mammal, and the eye-bubble that is contained in this capsule is a "pneu in a pneu". This is shown by a smaller eye-socket in the skull if, through operation, the counter pressure of one optic vesicle lacks (A). It doesn't depend on the type of the media; a filled potato-sack is also a "pneu".

TECHNICAL REALIZATION:
The trinomial pneu principle, abstracted from nature, was especially in the school (s. IL report) of Frei Otto converted technically in a very diverse manner or was at least proposed for technical realization in schematic diagrams (B). To these belong for example light pneumatic frameworks as e.g. air domes, variable pneus as for example sluice constructions or pneumatic, car enwrapping garages and various similar blueprints.

LITERATUR: IL-Bericht 12 (1975): Wandelbare Pneus. Stuttgart. - Otto, F. (1978): Der Pneu - Bauprinzip des Lebens. Bild d. Wiss. 10, 124-135. - Nachtigall, W. (1985/86): Der Pneu-Begriff in der Botanik des 19. und 20. Jahrhunderts. Jb. Wiss.kolleg Berlin, 313-327.

Pressure-stabilization: From pneu to tensairity

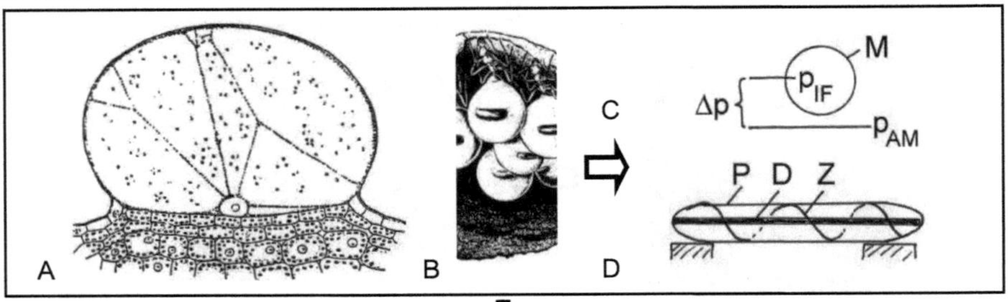

As well as the classic "pneu"-principle also the newer "tensairity"-principle use pressure-stabilized membranes.

BIOLOGY:
Form stabilization by the inner pressure bigger than the outside pressure is a well known biological principle of form maintenance and stability production that is realized in each vegetable parenchyma cell. The big "water cells" of the soda-plant *Mesembryanthemum crystallinum* (A) are notably, for example. In the animal kingdom, the "honey-pots" of tropical ants with "blown-up" abdominals (B) belong approximately to this group, too.

PRINCIPLE "PNEU":
A system of three morphological elements is defined as "pneu", namely a non rigid membrane M, an inner filling IF and an outside medium AM. Between IF and AM there must be a pressure difference Δp (C). For $p_{IF} > p_{AM}$ the membrane is blown-up, until its membrane stress at each point equals the pressure. For $p_{IF} < p_{AM}$ the membrane shrinks and then must be supported internally.

PRINCIPLE "TENSAIRITY":
A composite construction, as shown in fig. D, is called "tensairity." In such a construction the inner pressure of a sausage shaped pneu P ($p_i \approx 100$ Pa) containing a built-in pressure rod D must cooperate with a stabilizing, spiral-shaped reinforcement Z wreathed around. Such constructions can be executed as lightweight structures and yet on a larger scale. With long-stretched hose balloons temporary bridges could be built, for example (p. 125, fig. B).

LITERATUR: Schaur, E., Bach, K. (Hrsg.) (1976): Pneus in Natur und Technik. IL 9, - Nachtigall, W. (1979): Pneus - Beispiele aus der Zoologie. In Schaur/Bach, s.o., 182-191 - Luchsinger, R.H. et al. (2004): The new structural concept Tensairity. Basic principles. s.u.

Analysis and application of the tensairity principle

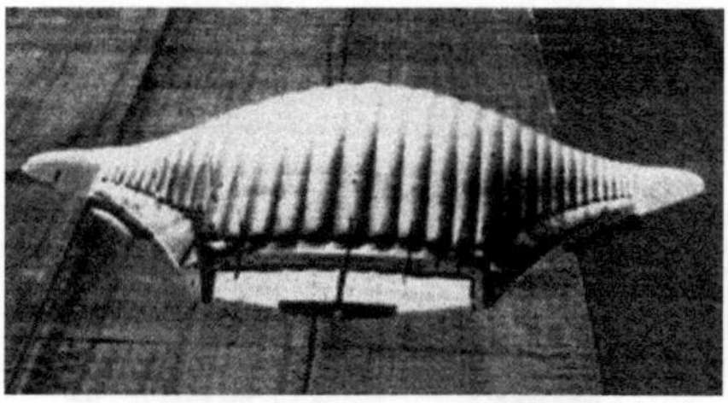

A

B

C

In retrospect, it was apparent that the 3 construction elements of the tensairity-principle occur also in the micro area of biology.

BIOLOGY:
The tensairity principle was actually not developed exactly after microscopic model although it's most important basis, the pressure stiffened membrane, is biological commonplace. Rather, it was apparent just in retrospect, as mentions that the three construction principles of tensairity constructions occur in an approached analogous manner also in the biology. Such technical structures let themselves consequently develop in comparison.

TECHNOLOGY:
The foldable membrane cover, the roll-able reinforcement tape and a possibly pluggable or screw-able pressure rod could be transported space-saving and with a pressure blower quickly could be brought into function. Large demonstrators like the "Stingray" (A) or the tensairity-emergency bridge (B) were realised by the Swiss companies "Airlight Ltd." in cooperation with "prospective concepts ag".

APPLICATION:
Tensairity constructions can be very helpful especially for emergency and temporary bridges, as well as for short time usable roofs (C), supporting columns, advertising columns, show pillars and other temporary constructions, that are to be positioned quickly. The inherent disadvantage of such structures, their sensitivity to stitch damages, can be moderated by a bionic inner coating with self repair effect (p. 127).

LITERATUR: Luchsinger, R.H., Pedretti, M., Reinhard, A. (2004): Pressure induced stability: From pneumatic structures to Tensairity. J. Bionic Engineering 1, 141-148. - Rampf, M. Speck, O., Speck, T., Luchsinger, R. (2011): Self-Repairing Membranes ..., J. Bionic Engineering 8, 242-250.

Tensairity: Air pressure as stabilizer in large building

A

B

With a first large-roof covering, this "structurally new", but "biologically ancient" principle has been introduced well.

BIOLOGY and TECHNOLOGY:
The a posteriori found agreement with biological principles was discussed on the preceding side. In contrast to conventional pneumatic modules, in which the air filling has a supporting function, the air pressure possesses just a stabilizing function in tensairity modules. That is a different principle, compared for example with air supported halls (Known biology analogue: "water bubbles" of *Mesembryanthemum crystallinum*, comp. p. 124, fig. A).

PRINCIPLE:
In contrast to the last-named example, tensairity structures don't abstain on pressure and tensile load bearers, however these can be shaped more lightly because of the overpressure in the interconnected pneumatic elements that stabilize the upper flange against buckling. So this can be more cost-efficient. "The tensairity bearers ... are light, but nevertheless strong and durable. Moreover, they allow a large form variety ...".

DEMONSTRATION IN FULL-SCALE:
The architect's office Luscher/Lausanne and the company Airlight/Blasca have covered the widening of a parking garage with 70 off-street parkings at the railway station Montreux (A) with a tensairity roof. The roof-foil bearers span 28 m support-free (!) and consist of a pressure stressed curved upper belt (RHS-steel profile of only 10x20x0,5 cm) and a tensile stressed lower girdle separated by a spindle pneu with an internal pressure of only 125 mbar (B).

LITERATUR: Pedretti, M., Luscher, R. (2007): Tensairity-Patent - Eine pneumatische Tenso-Struktur. Stahlbau 26 (5), 314-319. Ernst & Sohn, Berlin. - Pedretti, A. et al. (2007): The new structural concept Tensairity ... In: Zingoni, A. (ed) (2004): Progr. Struct. Eng. ... London.

Bionic self-repair of pneumatic systems

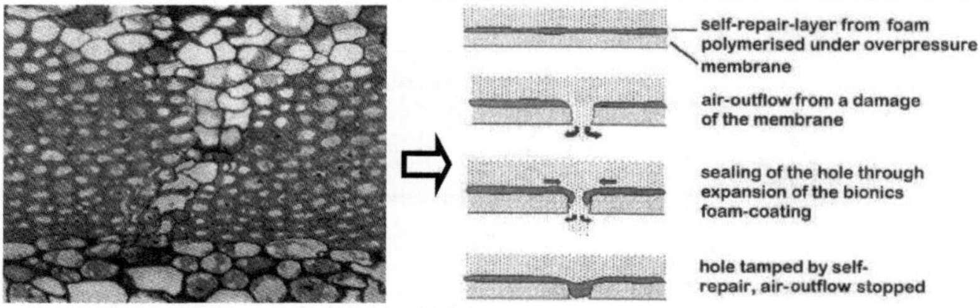

A B

Plants provide closure through inserting histological elements, driven by inner pressure, that close the sore.

BIOLOGY:
In lianas and, especially, in herbaceous plants, whose cells stand under a relatively high, form-stabilizing inner-pressure, stitch- and cut-injuries are closed relatively quickly in the way that parenchyma tissue moves in consequence of this pressure into the wound. It closes the opening there, finally stops the rip-continuation and lignifies. So, the rip or stitch-canal can finally be sealed permanently.

BIOLOGICAL PARAMETERS:
The quick wound closure (A) takes place straight bio-physically, without additional biochemical synthesis-processes. During this process, the form of the parenchyma cells changes from "round" to "irregular." Thus the relation circumference to surface increases, in the liana *Aristolochia macrophylla* from 160 to 230 mm^{-1}. Simultaneously, the cell wall thickness depending on elongation falls off e.g. from 1.2 to 0.9 µm whilst the cell-volume remains constant.

REALIZATION:
Technical realization takes place by applying a bionic layer on the inside of a pneu or a tensairity structure. This consists of a foam, that hardens (B) by overpressure. The time, in which the pressure has fallen off up to the instability of the pneumatic structure (with a 2,6-mm-nail characteristically approximately 2 minutes), can be increased by this method about up to 3 scales (in the example this means some hours).

LITERATUR: Busch, S. et al. (2010): Morphological aspects of self-repair of lesions ... Proceedings of the Royal Society London B, 277: 2113-2120. - Rampf, M. et al. (2012): Structural and mechanical properties of flexible polyurethane foams ... Journal of Cellular Plastics 48: 53-69.

Textile based transparent insulation

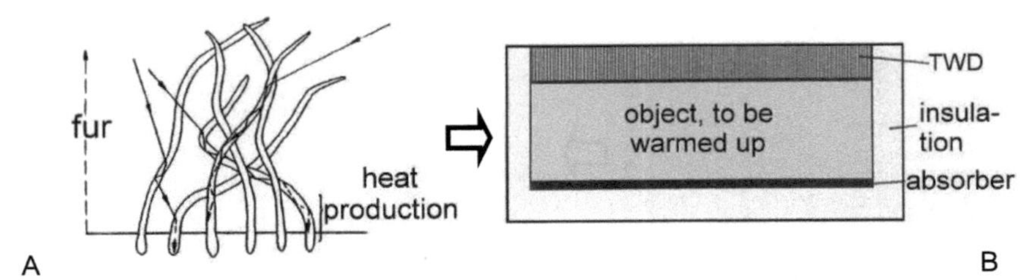

According to the principles of the polar bear fur a fibre based, transportable technical collector has been developed.

REALIZATION:
At the ITV Denkendorf, the polar bear fur principle (A; comp. p. 51) was converted technically by specifically coated distance elements. These are weaved after textile type ("distance fabric") and were transparent or coloured. The result is a large scale producible product, that passes on sunlight and is heat insulating because of its air-inclusions and prevention of convection by compartmentalisation: a transparent insulation TWD (B).

FUNCTIONAL CONSTRUCTION:
Light-leading polymers were used, provided with a particular silicone coating. These are translucent in the visible light, whereas UV-radiation is not let pass. Another coating reduces heat-loss by IR-emission. A third (outside-) coating finally reduces contamination and thereby achieves a high transparency. The textile based product is flexible, break-proofed and transportable.

CHARACTERISTICS:
The following parameters were reached: Thickness 5–60 mm, area related density 1.2–2.0 kg m^{-2}, translucence 80–95 percent, heat conductivity only 0.08–0.09 W K^{-1} m^{-1}, coefficient of heat transmission 2,2–3.0 W K^{-1} m^{-2}. These parameters are differently adjustable by alterations of the distance fabric and the coating parameters. Production advantages opposite to this product of other TWD materials are draw-ability and curvature-adaptation.

LITERATUR: Stegmaier, T., Linke, M., Stefanikis, J. (2006): Bionisch inspirierte flexible transparente Wärmedämmung für solarthermische Anwendungen. Bionik-Industriekongreß 2006, 72-79, Biokon e.V., Berlin.

Flexible "polar bear cover" on textile-basis

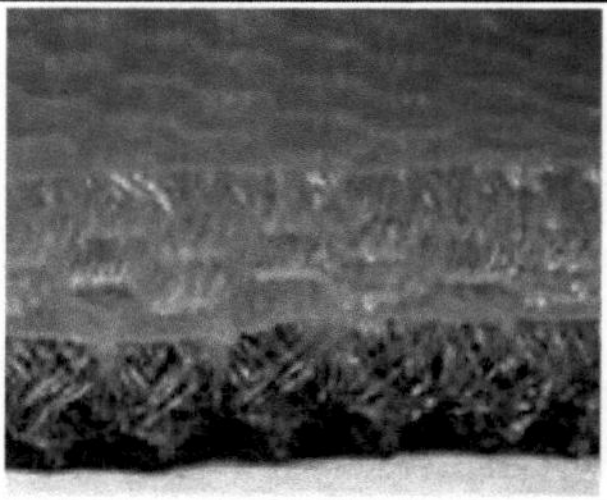

A

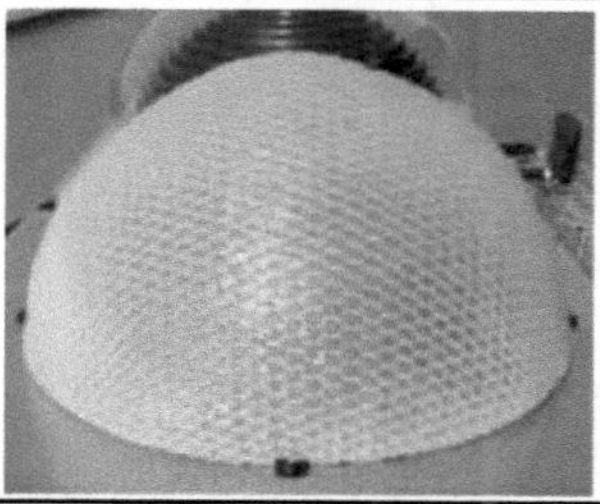

B

A flexible fabric, that mimics the polar bear principle of insulation, avoids the disadvantages of stiff components.

TRANSLUCIDE HEAT INSULATION MATERIAL:
The principle of this structure analogous the polar-bear fur (comp. p. 51), called "Transparent insulation material" (TIM), has led to box-shaped elements, that are put in as parts of building walls. Here they work well, however no spatially bent building-parts can be equipped with it. This disadvantage avoids a textile based and therefore high flexible TIM-material, developed by the ITV Denkendorf.

PRINCIPLE:
The textile upper side (A) is transparently coated and allows passage of light- and heat-radiations. Below it a distance fabric is located, that contains many air-pads for heat insulation. However, this does not work according to the principle of the polar bear-hair. On the other hand, the textile wearing a black membrane, which heats up and can radiate heat to structures lying flat on it, or on a medium such as air or water that flows around them.

APPLICATION:
It offers itself, as mentioned, especially for spatially bent surfaces, for example globe sun collectors for the production of hot water that can be isolated with this flexible as well as light "light-catching" special tissue (B). This envelope should increase also the efficiency of such collectors. A quite important use is, according to statements of the institute, also in the field of installations for seawater desalination.

LITERATUR: Umweltministerium Baden-Württemberg (Hrsg.) (2009): PatenteNatur-NaturPatente. Was Bionik der Umwelt bringt. Publ. des Umweltministeriums, Stuttgart. - Patentantrag des Instituts für Textil- und Verfahrenstechnik ITV, Denkendorf.

Butterfly's scales and light reaction facade

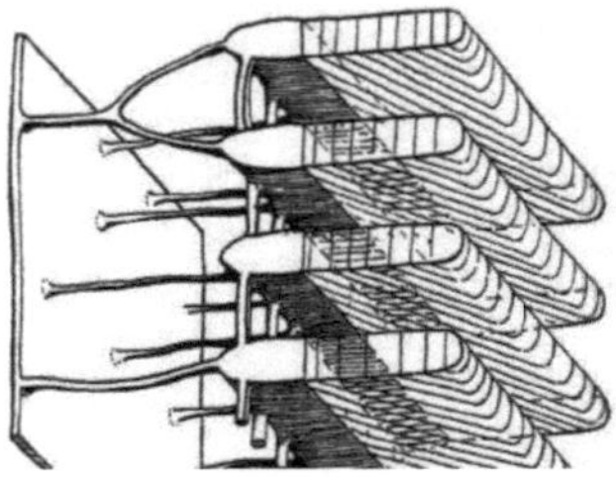

The principle of changing iridescent hues of the common morph leads to new facades for thermal regulation of buildings.

BIOLOGY:
Like most butterflies, also the great, very well examined common morph, *Morpho amathonte*, possesses scales, that generate iridescent hues (structure colours, no pigment colours) because of their strictly directional ordered, sub-microscopic small leaf like structures (A). These let the butterfly wings flash colourfully brightly. They work as warning colours, scare dress and as signals that play a very essential role for partner-finding.

PRINCIPLE:
Colours of thin platelets originate through their light interference. All the beams of white light, diagonally arising, are partially reflected at the upper side of a little platelet, to the other part broken to the underside and reflected again from there. The two parallel reflected rays interfere. This causes partial colour obliteration, so that the rest-colour appears. This is defined by the direction of arrival of the light beam.

ABSTRACTION AND TECHNICAL REALIZATION:
Scale-like shaped micro-elements (B) could stain building facades in an adjustable structure-colour. This could change strongly according to the momentary sun-stand. After this method, also the visible light could be allowed to pass fully, and only the UV- or the IR-radiation could be filtered out. In this way, the facades could contribute to the thermal building control and consequently to saving energy.

LITERATUR: Schmitz, H., Tributsch, H. (1994): Die Eigenschaften von Schmetterlingsflügeln als Solarabsorber. Verh. Deutsch. Zool. Ges. 87,112. - Braun, D.H. (2004): Bionic inspired building skins. Fortschr.-Ber. VDI 15/249, 3-11; 002.

Hingeless bionic shading of facades

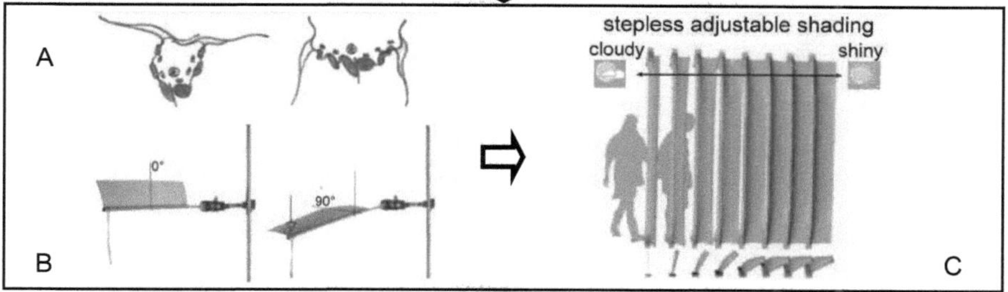

With skilful transformation of the torsion buckling of the bird of paradise flowers, a hingeless folding mechanic results.

BIOLOGY:
The bloom of the bird of paradise flower, *Strelitzia reginae*, is pollinated by birds. It offers a kind of "perch" to them, formed by connate petals. The weight of a resting bird bends an inherent mechanics in such a way that "simultaneously in a horizontal movement a flat lamella is clapped to the outside" and releases the way to the pollen. This mechanism works without hinges and in lab-experiments up to 6000 times.

MECHANICAL PRINCIPLE:
Bilateral ordered strips, interconnected by tender tissue, are covered by a flat lamina. The weight distorts strips and lamina so that the bloom opens (A). Physically, one speaks of a torsion buckling. Also a more simple flat reproduction (foil-strips at a flexible plastic-rod) bends itself under load joint freely to the side away (B), basis for a bio-inspired variable construction.

REALIZATION:
In cooperation between the botanical garden of the university Freiburg (T. Speck, T. Masselter, S. Poppinga) with researchers of the Institute of Building Structures and Structural Design (ITKE) of the University Stuttgart, the Institute for Textile and Process Engineering in Denkendorf and the firm Clauss-Markisen, a joint-free variable construction "Flectofin" was developed for the architecture, that appears employable above all to the facade shading (C).

LITERATUR: Lienhard, J. et al. (2011): Flectofin: a nature based hinge-less flapping mechanism. Bioinspiration & Biomimetics. Vol. 6, Nr. 4. - Knippers, J. et al. (2011): Gelenkloser, stufenlos verformbarer Klappmechanismus. Eur. Pat. Office. Filing: 10013852.8.

Natural construction principles: View of a civil engineer

heterogenity

anisotropy

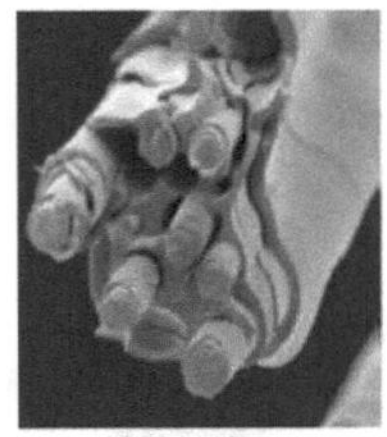

hierarchy

Since Frei Otto's SFB 230, "natural constructions" are taken into account by civil engineers and architects.

NATURAL CONSTRUCTION PRINCIPLES (J. KNIPPERS, STUTTG.):
"*Heterogeneity*: Physical and chemical properties are locally adjusted.
Anisotropy: Physical and chemical qualities are focused.
Hierarchy: Hierarchical construction over several levels.
Redundancy: Differently formed and interconnected similar elements.
Multi functionality: Stability, but also transportation of nutrients and chemical information as well as for energy reception."

BASIC QUESTIONS OF THE TRANSITION TO TECHNOLOGY:

"*Transferability*: Is transferability from nature to technology possible?

Topic: Which ones of natural principles can be used technically?

Methodology: How can such principles be applied in engineering? What does this mean for the architectural design?"

THREE-STEP-WAY:
The way, propagated by the author, corresponds exactly to bionic action:
"*Analysis*: Morphological analysis and quantitative characterization of the form-function connection.
Abstraction: Digital recording of the geometry and numerical simulation of the mechanical qualities.
Transfer: Constructive realization and implementation architecture."

LITERATUR: Knippers, J. (2011): Wandelbarer Leichtbau in der Architektur. Kongress Nanotechnik und Bionik - Hightech in der Bauwirtschaft. 18.1.2111, München - Nachtigall, W. (2010): Bionik als Wissenschaft. Erkennen → Abstrahieren → Umsetzen. Springer, Berlin etc.

The future of bionics in the architecture

W.N.  G.P.

An architect and a biologist commented in an interview: Bionics will be embedded "in a self evident way".

ARCHITECT (G. POHL):
"Already fascinating realizations were won from the understanding of natural structures that now slowly penetrate into the material world and into our facade world and will ultimately turn upside down also production processes. With modern computer- and production-technology, a bionic inspired technology, that can lead to material saving and reduction of energy consumption, is realisable. I believe that we stand just at the beginning."

BIOLOGIST (W. NACHTIGALL):
"The same will be the case in cars. The car in 30, 40 years will certainly look like our present to day cars, but has practically nothing in common with these. Each of the possibly 3000 individual parts will be changed and optimized. I believe that it is similar in the field of constructions. Maybe the buildings of the future will look like a "norm-building" of today, but – whether bionic or not bionic – they will be certainly different in every detail."

FORM, FUNCTION AND BIONICS:
"The nature has its own forms. It is up to the architects to decide, whether they mimic these ("biomorphic architecture") or not. Anyway, bionics is based exclusively on function analogy. If I have a form, in which a function is disguised, and the function only can run out, if exactly this form represents the cover, then, the form is bionic. If, however, I form a reproduction, only because it is so pretty, this doesn't have to do anything with bionics." (W. N.)

LITERATUR: Schultz. B. (2011): „Die Natur zu kopieren ist völlig sinnlos". Was bringt Bionik dem Architekten? ... Interview mit W. Nachtigall und G. Pohl, Bauwelt 102, Aug. 2011, 14-17.

ROBOTICS

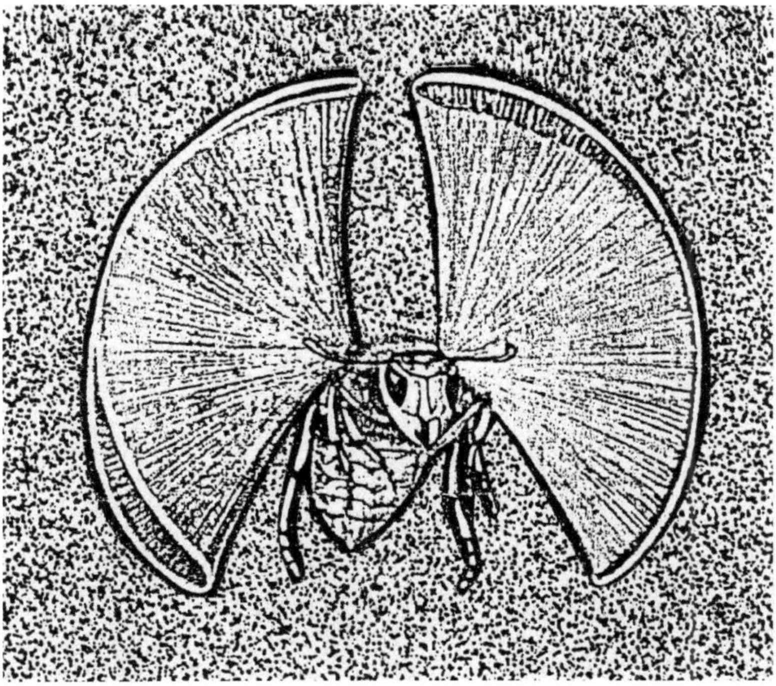

AND LOCOMOTION

W. Nachtigall and A. Wisser, *Bionics by Examples*,
DOI 10.1007/978-3-319-05858-0_8

ROBOTICS AND LOCOMOTION

Movement and locomotion in biology – Robotics – Locomotion on solid substratum – Locomotion in fluids.

Running, swimming, flying are the main forms of locomotion in the animal kingdom. Fluid mechanical interesting interactions between locomotors and the surrounding medium are found in the area of lower and middle Reynolds-numbers (microorganisms, insects) as well as in the region of very high Reynolds-numbers that reach the Re-area of airliners (whales). Questions of flow adaptation of moved bodies, the drive mechanism of locomotory organs and their flow mechanical efficiencies are in the foreground. Also questions of function morphological formation for example of wings can give interesting stimulation, for example the surface roughness of bird wings because due to the roughness of the plumage. Here positive boundary layer effects result in certain cases.

"BigDog" - a biologically inspired running robot

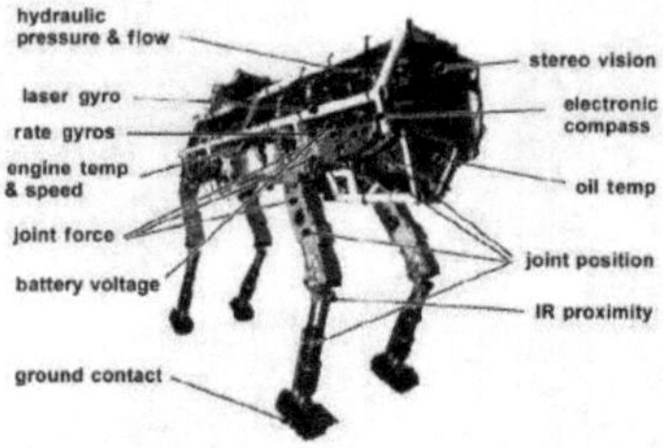

A

B

Four-legged running machines are developed in diverse approaches at present. "BigDog" is a far thriven example.

BIOLOGY:
Dogs run extremely auto-stable especially with faster gait. Their modes of leg movement and the type of the auto-mechanical, central nervous and peripheral control of their singles legs as well as the control-specific interactions of each leg in relation to the others are well-known. At Boston Dynamics, Cambridge, Mass., they were the starting point for the development of a large, bio-inspired auto stable running machine (A).

ABSTRACTION:
The parameters of the above-mentioned characteristics were abstracted. Movement possibilities (degrees of freedom) of the leg-limbs in form of two active hip-joints, an active knee-joint as well as a passive linear compliance in the lower leg were used. Positional and positioning reflexes, loading and unloading cycles, possibilities of leg communication as well as a "virtual leg scheme" were used for the coordination control of the real legs.

REALIZATION:
This already far developed running machine is 1 m long, 1 m high and 0.3 m wide. It weighs 90 kg and carries 50 kg of payload. Still, auto stable jumps over a 1 m high fence, running up-hill with 45° (100 percent) dispositions and running speeds of 5 m s^{-1} are planned. The machine can put in already different running styles, among them trot (B) and gallop (horse). Stability problems, present still to be solved, are related to rising and sitting down.

LITERATUR: Buehler, M. et al. (2006): Biologically inspired robots at Boston Dynamics. In: Bionik-Industriekongress Innovationsmotor Natur, BIOKON, Berlin 2006. S. 40–48.

Bionic elephant-trunk handling-assistant

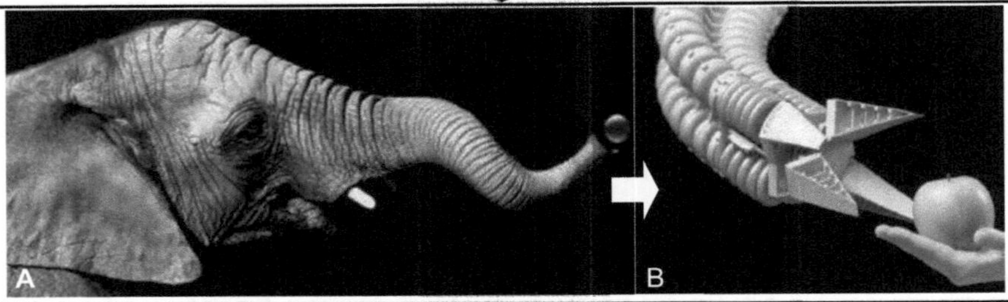

According to an elephant trunk, a "bionic handling-assistant" for human technology interaction was developed.

BIOLOGY AND TECHNOLOGY:
The elephant trunk (A) can be moved into every direction and can be held in every position. It can pick up objects extraordinarily "soft" with its end section. "It is flexible, transfers high strengths and serves as precise gripping-tool. Approximately 40000 muscles, interwoven to frets, make the trunk especially movable and pliable. Pneumatics and mechatronics are used as basic principles for the technological realization of these natural principles."

PARADIGM CHANGE BY BIONICS:
"The bionic handling assistant of Festo (B) was developed in cooperation with the Fraunhofer institute for production technology and automation. It is the goal of the Bionic Learning Network to produce new carriers of technology, through application of the bionics, to say the carryover of bionic principles to technology." This Festo quotation shows that today it is not just about particular "imitation" of nature, but about creating novel designs.

CONCEPTION AND APPLICATION:
Festo entitles the following fields for the conception and application of the handling-assistant (German future award 2010): Learning, to combine many and different cross sectional technologies – intelligent controlling integrated in a solution package – making handling systems move freely – smooth gripping – lightweight construction and generative production (Polyamid) – safe human machine interaction – suitable for household and industry.

LITERATUR: Festo AG & Co.KG (2011): Robotino ® XT. Mobiles Lernen flexibel erweitert. Firmenschrift. Darmstadt – FESTO AG & Co KG (2010): Bionischer Handling-Assistent. Flexibel und nachgiebig bewegen. Firmenschrift. Darmstadt.

Robots as geriatric nurses

A

Fig. modified after the source FOCUS 45/11

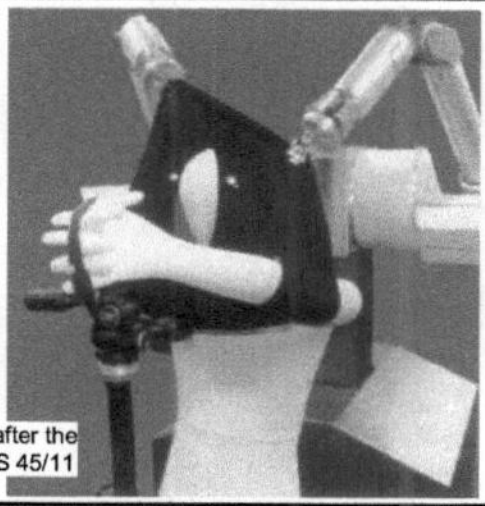

B

For simpler tasks in the area of the old people's care, especially built and programmed robots have a big future.

PROBLEM:
With the increasing life expectancy, also the need for care and with it the future chances for the profession of the old people's nurse increase. But human services are expensive – in the longer term may often be extortionate – and not available for every patient 24 hours a day. Here personally associated care robots (A) could care take a number of simple care tasks that are required around the clock.

EXAMPLE: ROBOTS AS DRESSING SERVICE:
The prototype of a "dressing robot" developed in Japan by the NAIST (Nara Institutes of Science and Technology) can pull "a doll a T-shirt over the arms from the back, for example" (B; Focus). The necessary sequences of his two 7-jointed gripper arms were previously practiced in a human. Robots should also be helpful with lifting, raising, fetching objects and delivering food.

FURTHER DEVELOPMENT:
Especially in the dense populated and strongly over aged Japan one gives big future chances to such developments for old people's homes and clinics. However, above all they should find also their way to old people in private one-person household, also in Europe and North America. Toyota established already such a line of business. Prototypes of such robot systems are produced at present also by Panasonic and others.

LITERATUR: www.innovations-report.de/html/berichte/interdisziplinaere_forschung/bericht-56887.html. – Fleschner, F. (2011): Roboter als Ankleidehilfe. Focus 45/11, 133.

The first, fit to drive bionic car

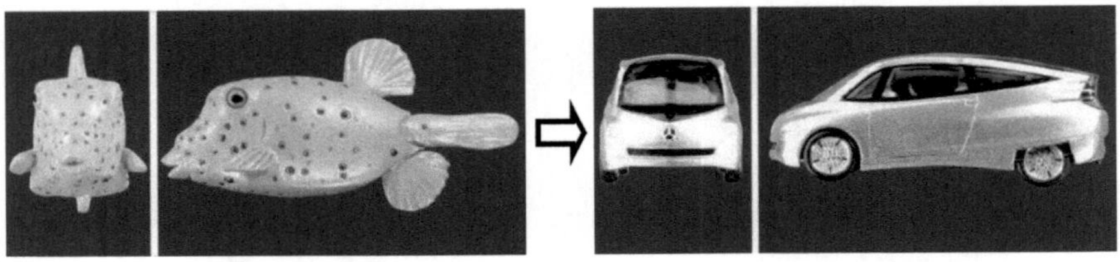

Stimulated by flow adapted nature forms, a four-seater with a drag coefficient $c_d = 0.19$ was developed.

BIOLOGY:
The bodies of biological swimmers and fliers are often characterized by very low flow drag. Natural objects, used as prototypes for technology, should be sufficiently big and should move, so, that the Reynolds numbers are not too different in biology and technology. The gentoo penguin (length 0.75 m) swims with $Re \leq 10^6$, for example, and his frontal area drag coefficient is as low as $c_d = 0.07$. House martin: $c_d \approx 0.1$.

PROCEDURE:
The department for predevelopment (A. Jambor) of Mercedes-Benz was in search of natural models for a novel car, whose form should provide the smallest possible drag and thus the lowest possible fuel consumption without loss of interior size and comfort. The chosen design prototype of the Mercedes-Benz researchers led by D. Gürtler, was that of a boxfish".

TECHNOLOGY:
Measurements of the c_d were made at casts from the boxfish *Ostracion meleagris* (A) at first; they brought a c_d of 0.06. Thereupon, bigger models were manufactured, whose contours were as similar as possible to the frontal plane of the boxfish, at least. A form was found (B) in wind experiments on a big-model for a driveable vehicle, whose c_d of 0.19 was unusually low; the series car has a c_d of not smaller than 0.27.

LITERATUR: Nachtigall, W. (o. J.): Bericht über schnelle Schwimmer u. Flieger geringen Widerstands (unpubl.). – DaimlerChrysler AG (Hrsg.) (2000): Die Geschichte einer Leidenschaft. Festschrift. – Jambor, A. (2006): Bionik und ... im Automobilbau. Industrie-Kongress Bionik, S. 188–189.

Locomotion according to swimming rays

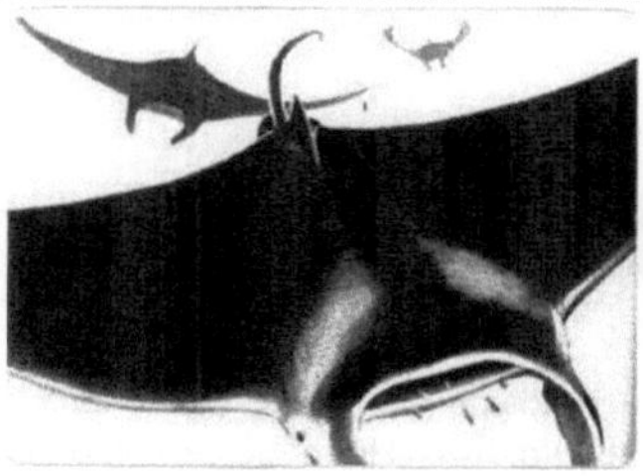

A

B

The technical realization of swimming with undulating fin seams led to an artificial ray: "Aqua Ray".

BIOLOGY:
Due to the wing like processes of their lateral fin seams, rays "fly" under water (A), generating thrust in nearly every beating phase in a flow-favourable way by combining undulatory beating movements and torsions of these processes – without the propeller-typical reliefs of vortex rows. The giant oceanic manta ray, *Manta birostris*, is especially remarkable because of its size and the slow beating movements.

PRINCIPLE:
On the one hand, the research and the technical realization of this principle of undulating fin seams serve to understand the physics of this drive-form. On the other hand, one can develop demonstrators that can move freely under water. These allow to change systematically the kinematic parameters of the fin seam drive and further to optimize flow-mechanics of this efficient drive-form. So “Technical Biology” and “Bionics” are pursued.

TECHNOLOGY:
The propulsion and control unit of the "Aqua Ray" (B; span width = 96 cm) from the company Festo functions in combination with the Fin Ray Effect® (comp. p. 108). The "wings" consist of an elastic material; the skin is built of polyamide and elastomers. The elastic qualities are coordinated so, that the undulations and the optimal angles of attack of the sections are automated. The appliance can be used for probe holders in marine research.

LITERATUR: Festo-Firmenprospekt (2008): „Aqua Ray“. Wasserhydraulisch betriebener Manta-Rochen mit Schlagflügelantrieb. Denkendorf.

Body undulations for an airship-drive

A

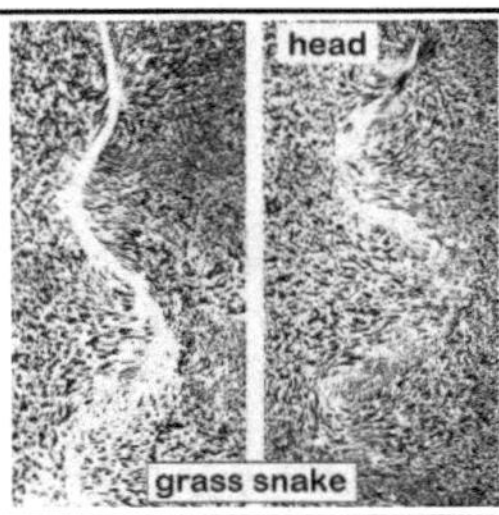

prototype according to S. Michel
photo. by H. Hertel B

Elongated swimmers use body undulations as a principle for locomotion. Airships probably could do this, too.

BIOLOGY:
Most fishes doe not have stiff bodies. Instead, their body is moved in form of transversal oscillations as, for example, is shown by the shark-catfish. So, the tail fin is supplied by a "preformed flow." But also the body itself generates a part of the thrust. During its thrust-generation, sections lying farther behind profit of the flow preformation by sections farther in front. In extreme, this is shown by long-stretched winding swimmers as eels or snakes (A).

PRINCIPLE:
Preformation of flow means for the tail fin more favourable flow-conditions, better conditions of angles of attack and more effective vortex shedding by high-low-pressure-zones periodically running back. Hereby the tail fin can work with higher efficiency that improves the power balance. But also without a tail fin thrust can be generated in this way if the body can swing transversally as winding swimmers show.

TECHNOLOGY:
The two principles "preformation of flow" and "additional thrust-production by active-flexible bodies" should be transferred according a concept-sketch to flexible airship-bodies (B). Quickly working technical actors, that can execute movements with high amplitude, are required for this. Possibly, this is suitable for solar driven, slow flying smaller airships that must handle solar-energy efficiently.

LITERATUR: Anonymus (2007): Forellenantrieb für Luftschiffe. Geo 04/07, 199. – Kunz, L. (1988): Kinematische und strömungsdynamische Untersuchungen beim Haiwels (*Pangasius sutchi*). Diplomarbeit; Math. Nat. Fak. Universität des Saarlandes (unpubl.).

Studies about a bionic megaliner in future

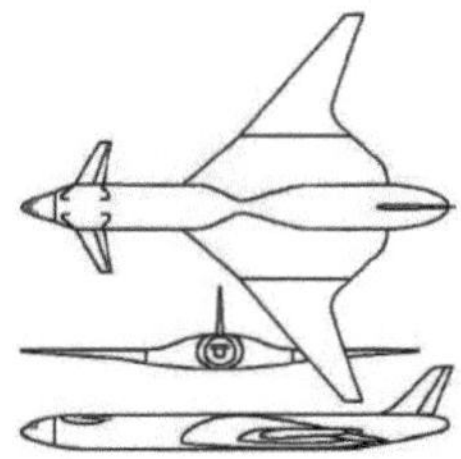

A

B

In the development of future large airplanes for several 100 passengers (A, B) there are numbers of bionic approaches.

BIOLOGY:
Besides new technological developments, also the following biological results are discussed: Streamlined body-cross-sections of big whales – very favourable drag coefficients of penguin bodies – contours, harmonic geometries and smooth surfaces by penguins – electro kinetic effects of living cell-structures – influence of the heat transfer in bird-plumages by variable electrostatic fields.

MECHANICAL PRINCIPLES:
Whale-bodies with rather broad instead of high-elliptical body-cross-section are well flow adapted. Penguin bodies show extremely low pressure drag coefficients, smaller than 0.07; abstracted rotation-body even coefficients of 0.03. Differently big penguins are marked through harmoniously tuned body geometries. Their surface drag coefficient is with $c_{d\ s} \approx$ 0.002 extremely low.

ELECTROSTATIC PRINCIPLES AND TECHNOLOGY:
By the membrane-voltage, electric field-strengths are generated up to 10^7 V m^{-1}, which can oscillate with high frequencies. Coherent electro-acoustic surface-waves could contribute to the boundary-layer-stabilization of biological bodies in a flow. Analogously, the formation of electrical or Helmholtz-bilayers could stabilize technical flows. To go further, also electro dynamic pulse engines become conceivable.

LITERATUR: Göksel, B. (1998): Studien zu einem bionischen Megaliner zukünftiger Generation. Dt. Luft- u. Raumfahrtkongr. Bremen, DGLR-Jb., Bd. I, S. 771–785. – Bannasch, R. (1996): Widerstandsarme Strömungskörper ... Biona-rep. 10, 151–167.

Bastard wings and pre-wings

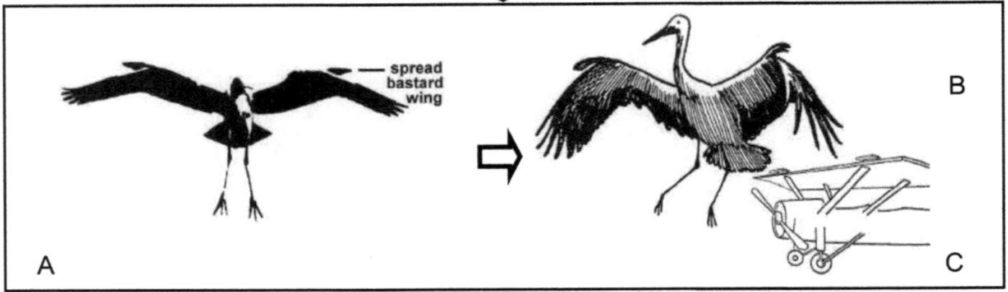

The observation of the effect of spread bastard wings (alula) in birds led to the development of lift increasing pre-wings.

BIOLOGY:
Birds, for example herons, often extend the bastard wings during landing (A, B) or during a curve flight. The bastard wings are small clusters of feather that normally are integrated into the wing-front-contour. This was already noticed by Leonardo da Vinci. But first in the 60ies of the 20. century, the proof succeeded, that for example sparrows increase their lift by positively influencing the flow over the upper side of its wings at high angles of attack.

PRINCIPLE:
Through the slit in an extended bastard wing and the front wing-contour, additional air is led to the upper side. So, the flow over the upper side is accelerated; it is supplied with extra energy. Consequently it can better overcome the increase of pressure that appears with high angles of attack. Therefore the boundary layer does not or not completely separates. Through this effect on the other hand, the wing can generate high lift.

TECHNOLOGY:
That high-lift is generated by extending the bastard wing, was assumed hypothetically already in the 20er years. According to this observation, independently from each other, Lachmann in Germany and Handley-Page in England have designed pre-wings for airplanes. Those were small and fixed at first and were installed at the wing there, where the bastard wings sit by birds (C). Later, they were long-stretched and retractable.

LITERATUR: Nachtigall, W.; Kempf, B. (1971): Vergl. Untersungen zur flugbiologischen Funktion des Daumenfittichs (*Alula spuria*) bei Vögeln. I. Der Daumenfittich als Hochauftriebserzeuger. Z. Vgl. Physiol. 71, 326–341. – Dubs, F. (1990): Aerodynamik ... Birkhäuser, Basel.

Ends of wings, spread like fingers, influence the flow

A

B

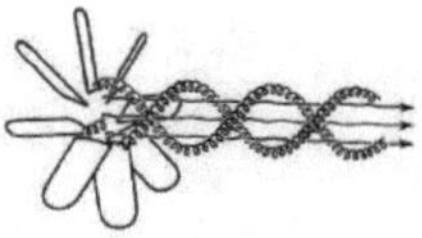

C

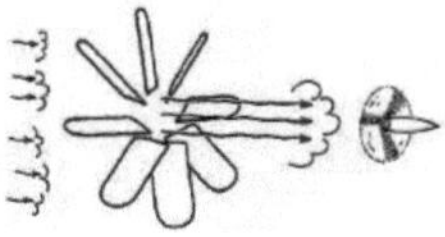

D

The free hand swings of large over land gliders led to the reduction of the induced drag and to vortex coils.

BIOLOGY:
Eagles, vultures and storks, but also smaller, partially gliding-capable griffins like buzzards, have free primary feathers. These are – at least by storks that are well examined relating to this – spread automatically because of their typical bearing, so that they are flown round individually. These single flows interact in an optimal manner. So, the induced drag that is connected with lift production is reduced.

PRINCIPLE:
The induced drag is a consequence of the flow around the ends of the wings, that appears because of the pressure difference between the wing-underside (high pressure) and the wing-upper side (low pressure). Together with other drag components it adds up to the total drag that should be as small as possible. The cascade of the free primary feathers generates a vortex spool that delivers a kind of thrust contribution and reduces the total drag consequently.

TECHNOLOGY:
A cascade of end winglets at a technical wing (A, B), for example, makes a landing at higher angles of attack possible and reduces consequently the landing distance. When the winglets are positioned in a circular manner, a vortex-spool results with high flow-speed in its centre ("wind concentrator") (C). A small turbine placed in this centre (D) runs faster than in the free wind and therefore provides more power ("Berwian", comp. p. 100).

LITERATUR: Rechenberg, I. (1984): Berwian konzentriert den Wind. Sonnenenergie 2, 6–10. – Rechenberg, I. (1994): Evolutionsstrategie 94. Frommann-Holzboog, Stuttgart. – „Winggrid". www.bionik.tu-berlin.de/institut/skript/B1-08Fo8.ppt

Ornithopters – bird-like wing-beating aircrafts

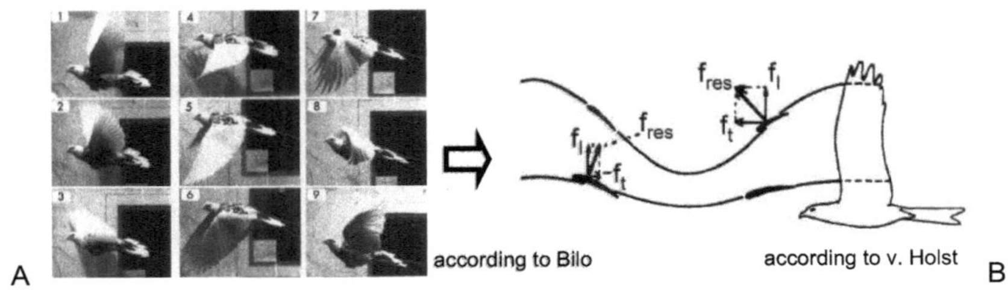

Lippisch (comp. p. 148) and v. Holst have already executed tests in the 30er years. Today, these are highly topical, again.

BIOLOGY:

Large birds move their wings from behind-above to front-below and back again (A), approximately in form of a sine-shaped beating oscillation with $z(t) = h_1$ cos $2\pi ft) \omega t$ (z = vertical distance). Moreover they tilt them, also approximately sine-shaped, in form of a rotary oscillation $\alpha(t) = \alpha_0 + \alpha_1$ cos $(2\pi ft) + \varphi$ (α = angle of attack) around their longitudinal-axis (further fine attitudes are not looked at here). These two components of oscillations are in a fixed phase relationship.

PRINCIPLE:

As consequence of this two coupled components of oscillations, there result favourable, that means positive and relatively small angles of attack over wide parts of a beating-period. As further consequence, favourable aerodynamic forces result, that means big lift forces f_l vertically to the beating direction and relatively small drag forces f_d in beating direction. Finally, their resultant f_{res} is partitioned into thrust f_t and lift f_l (B) (comp. p. 148).

TECHNOLOGY:

In the area of models the beating wing principle works, as already mentioned above, quite well. Early big implementations regularly went wrong, because the moved masses were not controllable and the materials were inappropriate. In the 70s, effective beating wing drives could be realized in unmanned small airplanes (drones) with spans of approximately 3 m. Besides micro air vehicles (MAVs, p. 151), such drones should profit from the beating wing drive.

LITERATUR: Nachtigall, W. (1983): Warum die Vögel fliegen. Rasch u. Röhrig, Hamburg. – Holst, E. v. (1947): Über „Künstliche Vögel“ ... J. Ornithol. 91, 406–447. – De Laurier, J. D.; Harris, J. N. (1993): A study of mechanical flapping-wing flight. Aero J. 97, 277–286.

Measurements in wing-beating "artificial birds"

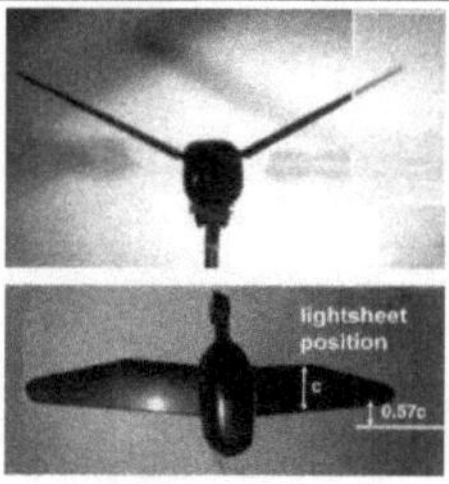

A

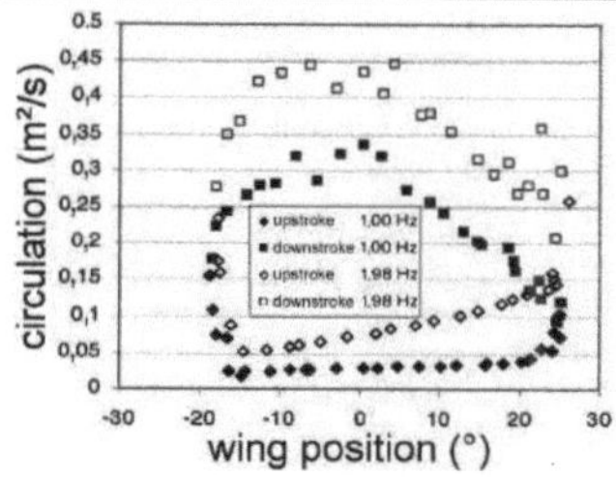

B

Bio-physical flight phenomena, that are not measurable at the original, can be studied using suitable models.

TECHNOLOGY FOLLOWING BIOLOGICAL PROTOTYPES:
An artificial bird, resembling the ringlet-goose, was put into a wind-tunnel of the TU Darmstadt, so that it moves under conditions, that are as realistic as possible for a ringlet-goose: span 1.1 m, wing-profile Wortman Fx 60-126, defined wing-flexibility, angle of attack -2° to 12°, flow speed 12 m/s, Reynolds-number 8×10^4 to $1{,}3 \times 10^5$, amplitude 20° to 55°, frequency 0 to 2.2 s^{-1}, reduced frequency 0.03 to 0.27.

PRINCIPLE:
The stationary and non-stationary air-forces were measured using this model (A). Formations of vortexes were registered simultaneously to the beat by laser anemometry, from which the air-forces were re-calculated. These two approaches to air-force-evaluation were compared. The sense of the action was to get information of the aerodynamic force generation at beating flight, which one cannot attain by direct measurements from a living object.

DATA AND RE-TRANSFER:
As example, the graphic plot of the distribution of the circulation over the wing-positions during down- and upstroke is demonstrated (B) for the beating frequencies of 1 and 2 s^{-1}. Parameters of this application were: Flexible wing, amplitude 44°, angle of attack 2°, flow velocity 12 ms^{-1}. By doubling the frequency, the maximal circulation in the middle of the downstroke increases from 0.35 to 0.45 m^2/s. Such data are important for MAVs (comp. p. 151).

LITERATUR: Tropea, C.; Hubel, T. (2006): Vorbild Vogel: Instationäre Effekte in der Aerodynamik. Industriekongress Bionik, 141–173, Biokon, Berlin.

A 2 m model, based on the birds' beating-flight

A B

A "technical bird" is based on obligatorily phase-coupled beating- and rotation-oscillations of the wings.

BIOLOGY:
With biological organs for propulsion, two phase coupled components of oscillations are the rule. Within fishes (p. 16 and 38) one speaks of bending and rotation oscillations, within insects (p. 42, 43) and birds (p. 146) of beating and rotation-oscillations. These are obligatorily coupled up to a certain degree, within fishes by the elasticity of the fin, within insects by sclerites of their wing joint, within birds by eccentric muscles.

PRINCIPLE:
This *passive* coupling takes care, that a favourable phase-relationship between the two components of oscillations results mechanically, which automates the drive. However, the phase angle can be changed in limits by *active* influences (muscle-contractions). Within insects and birds, this provides a fast adaptation of the drive to the momentary requirements of the lift- and thrust-production.

TECHNOLOGY:
While in technical toy birds only the beating oscillations are motor-driven (rotation oscillations "automatically"), the company FESTO (project-leaders M. Fischer) has, re-created according to the silver seagull (A), built the "SmartBird" (B) in cooperation with the aerodynamics W. Send, in which suitable rotation oscillations of the "hand-" and "arm-wing" were realised through mechanical permanent coupling. So, an amazingly "naturally" appearing flight results.

Send W. (2010): Tragflächen als Biege/Torsionsantrieb. Können wir noch von der Natur lernen? Kolloquium Luftverkehr, TU Darmstadt – Festo AG & Co KG (2010): SmartBird. Aerodynamischer Leichtbau mit aktiver Torsion. Firmenschrift. New (2013): "Artificial dragonfly (Festo).

Human flight with beating wings succeeded first time

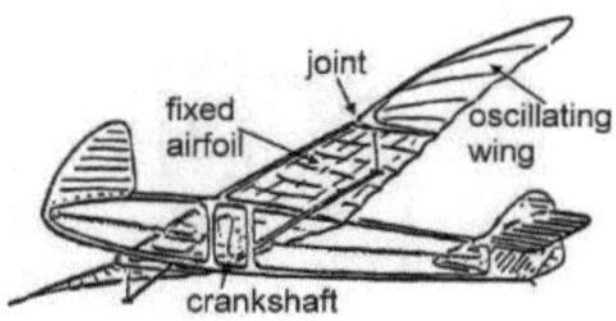

A

B

The old mankind-dream, to be able to fly like a bird, became true – admittedly not according to Leonardo's ideas.

BIOLOGY:
In contrast to small-birds, in which all wing-parts are strongly shift and twist during beating, large birds like eagles, vultures and condors move their arm-wings with relatively low amplitudes. Their angles of attack do not change very much, and the flow essentially generates lift. Contrary, the external lying hand-wings are strongly adjusted during down- and upstroke and essentially generate thrust.

PRINCIPLE:
The principle of the distribution of lift and thrust, typical for the large birds, has advantages. For example, the two wing-parts can be optimized on their own, related to their respective main-function. In contrast to Leonardo da Vinci's idea (comp. p. 3), this functional separation has already been proved in beating-wing flight-models by A. Lippisch since 1940 (A): He had connected beating "hand-wings" (rubber-motor) to stiff "arm-wings."

TECHNOLOGY:
At the 2.8.2010, the first man powered beating-wing airplane has itself held in the air "from own strength". According to the concept of the "Gossamer Condor" (McCready, 23.8.1977, German museum, Munich), students built the super-light airplane "Snowbird" with 32 m of span (B). It was held by the powerful pilot T. Reichert, TU Toronto, in air over 145 m and 19,3 s ($v \approx 27$ km h^{-1}). Students of 3 universities were involved.

LITERATUR: Reichert, T. (2010): Human-powered ornithopter becomes first ever to achieve sustained flight. www.engineering.utoronto.ca (22.9.2010). – Lippisch, A. (1940): Schwingenflug. Der Segelflieger 11, 11 f. – Leonardo da Vinci (1505): Sul volo degli ucelli. Florenz.

Basic researches for micro-flight-objects

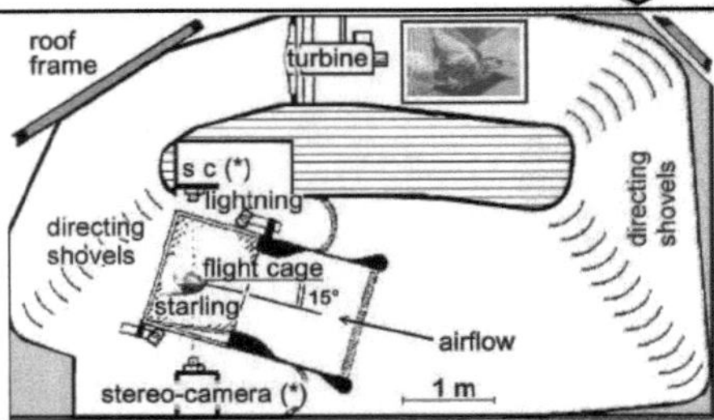

A

B

In a technical-biological-joint project of the DFG know-how of three technical universities flows together.

BIOLOGY:
Veil-owls and lanner falcon, whose flight is examined in detail again, should serve as models for new micro-flight-objects. Veil-owls are held and trained for flight-tests at the RWTH Aachen. Until now, first experiments outside from wind-tunnels were made in cooperation with members of the Universität der Bundeswehr/Neubiberg, C. Kähler a. A. Friedl, and the DLR Göttingen, R. Konrath a. T. Wolf. From 2012, falcons fly in a wind-tunnel in Munich (B).

AIM OF THE BASIC RESEARCHES:
After basic realizations about the bird-flight, e.g. from D. Bilo (1971, comp. p.146) further research with up to date technical measuring possibilities is inevitable for further applications. With the measuring-system, developed at the Universität der Bundeswehr, the profile-geometry and the movement-form can precisely be determined in order to make continuing numerical examinations of the speed and pressure-fields possible.

MEASURING-TECHNOLOGY:
Until now, in Germany larger wind tunnels specifically for the bird flight were built in Saarbrucken (A) and Seewiesen. The tunnel in Neubiberg is also suitable for examinations of the glide- and beating-flight. For a time wise and spatially high-definition analysis of the bird-wings modern digital picture-evaluation-methods are available. But, procedures for data reconstruction under exploitation of different observation perspectives were made earlier in Saarbrucken at starlings and domestic pigeons (comp. p. 42).

LITERATUR: http://www.vbio.de/informationen/alle_news/e17162?news_id-11009 Schleiereulen als Vorbild für Mikro-Flugobjekte. – Bilo, D. (1971): Flugbiophysik v. Kleinvögeln. Z. Vgl. Physiol. 71, 382–454. – Nachtigall, W. (1998): Starlings … in wind tunnels. J. Av. Biol. 29, 478–484.

Parameters of wing-beating micro-air-vehicles (MAVs)

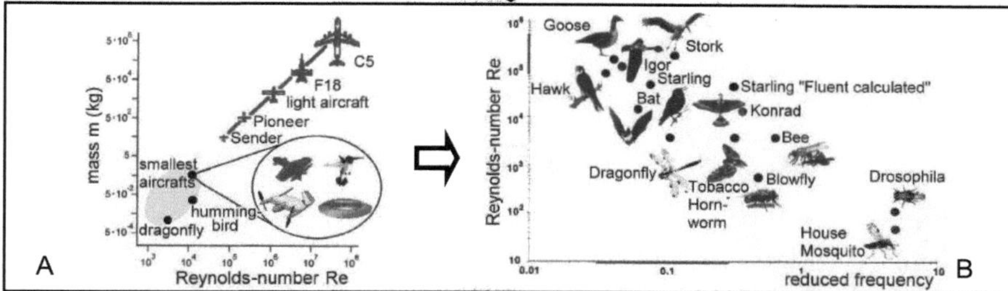

The most important "dimensionless" comparison-parameters are the Reynolds number Re and the reduced frequency k.

REYNOLDS-NUMBER:
Biological flyers move according to their size (mass) with different speeds. For comparisons, one can make the speed v (m s^{-1}) "dimensionless" referencing to a benchmark ν/l (m^2 s^{-1}/ m = m s^{-1}) with the dimension of a speed. That means, one defines the Reynolds number Re = $v \cdot l/\nu$ (v flow velocity, l length, ν kinematic viscosity (for air = 1,51 10^{-5} m^2/s^{-1} at 20° C). The Reynolds-number is the basic parameter for animal swimming and flying.

REDUCED FREQUENCY:
Biological flyers also move their wings with different beating-frequency f (s^{-1}). In order to be able to compare different sizes, one can make the frequency "dimensionless" referencing to a benchmark v/l π ((m/s)/ m = s^{-1}) with the dimension of a frequency. That means, one defines the reduced frequency k = f l π/v (s^{-1}). Here, f is the frequency (s^{-1}), l a length, e.g. the wing-depth (m) and v the flow velocity (m/s).

APPLICATIONS:
For the MAVs built until now , masses of 5 to 50 g and Reynolds-numbers from 10^4 to 10^5 are typical, contrary to masses of 500 tons and Reynolds-numbers near 10^8 of large airplanes (A). Between the years 1996 and 2002, the average masses of MAVs descent from 75 to less than 20 g. For MAVs, reduced frequencies between 0.03 and 0.3 are aspired. For comparison: small flies lie at k ≈ 4 (B).

LITERATUR: Junge, M. (2002): Kinematische und strömungsmechanische Untersuchungen über das Fächelverhalten der Honigbiene (*Apis mellifera L.*). Diss. Math. Nat. Fak. Univ. d. Saarlandes. – Nachtigall, W.; Mitarb. v. A. Wisser (2003): Insektenflug. Springer, Berlin. S. 266–279.

Non-stationary aerodynamics in the honeybee wing

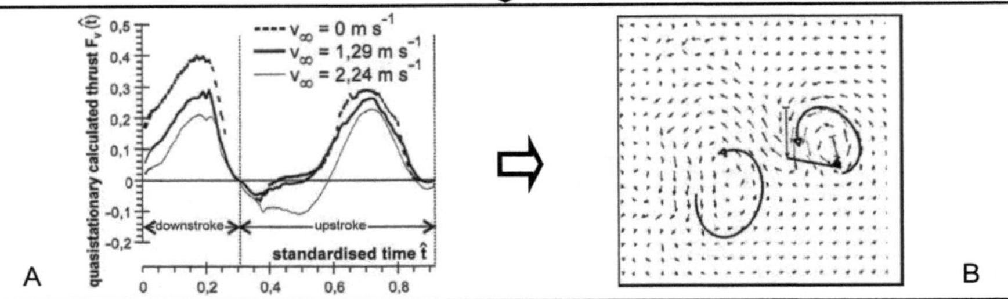

The bee-wing, oscillating at approximately 180 s^{-1}, is efficient due to specific kinematics and non-stationary dynamics.

BIOLOGY:
On hot summer-days bees stand individually or lined up to chains at the entrance of their hive and "fan". The oscillating wings act on this occasion as a stand propeller, whose airflow aerates the hive. For the single bee, the mean of velocity amounts up to 1,93 m s^{-1}, the degree of shear-load 1.24 to 2.05, the net power 0.70 milliwatt, the "propeller power" 0.88 milliwatt and, consequently, the "propeller efficiency" 0.80.

PRINCIPLE:
In part of her promotion at the working-group Technical Biology and Bionics of W.N. at the University of the Saarland, M. Junge has quasi-stationary calculated (amongst others) the thrust-distribution during a down- and upstroke for different airflow-speeds. Apart from small negative values at the beginning of the upstroke, always positive thrust values were present (A). These were compared with data from a shear-platform, on which the bee stood.

COMPARISON:
In the comparison, it was determined, that the time-courses of the thrust forces admittedly for the quasi-stationary calculation and for the output of the shear-platform were very similar. However, the absolute values for the quasi-stationary calculation were clearly smaller. Therefore, the difference must be led back to non-stationary flow-processes. With two-dimensional non-stationary calculations (A. Shekhovsov), these were attributed to special vortex shedding (B).

LITERATUR: Junge, M. (2002): Kinematische und strömungsmechanische Untersuchungen über das Fächelverhalten der Honigbiene (*Apis mellifera L.*). Diss. Math. Nat. Fak. Univ. d. Saarlandes. – Nachtigall, W.; Mitarb. v. A. Wisser (2003): Insektenflug. Springer, Berlin. S. 266–279.

Mechanism for a MAV based on fly-wing-joint

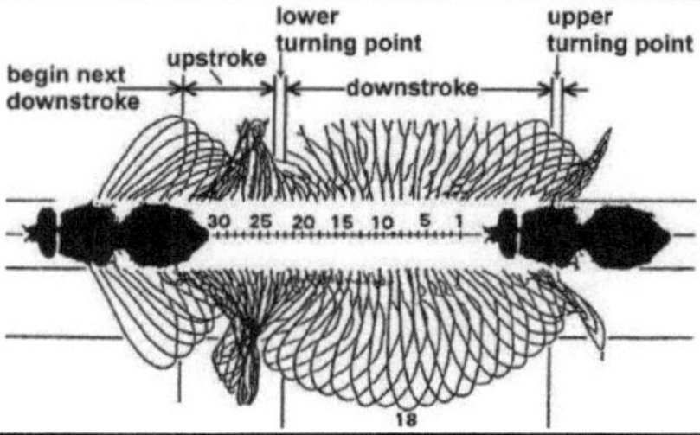

A

B

A special mechanism was developed for the technical realisation of the fly-typical combined beat rotating oscillation.

BIOLOGY:
The wing beat kinematics of the bluebottle (A), with its typical combination of a beating- and a phase-shifted rotating oscillation, are permanently steered by the wing-joint. This represents the probably most complex movement mechanisms in the animate world. It is neither possible nor meaningful, to rebuilt it technically in a similar manner. Instead, a technically autonomous mechanism was developed that works in a comparable way.

PRINCIPLE:
For the technical wing joint, it is only important, *that* it makes wing kinematics according to the type of the bluebottle wing possible, not *how* it does this. Naturally, a technologically meaningful autonomous solution that corresponds to the possibilities of technical mechanisms was the aim. This gear was developed by Žbikowsky et al. at the Royal Military College of Science of the Cranfield University, Shrivenham.

TECHNOLOGY:
The mechanism (B) overlays a beating oscillation with a rotating oscillation. So it allows adjusting the position of the beating plane to an animal defined plane (usually the xy-plane). This adjustment was achieved by a parallel-propulsion-mechanism (B, right-below), that couples the adjustment of the angle of attack between the two named planes phase-shifted to the rotating-oscillation.

LITERATUR: Žbikowsky, R. et al. (2000): Current research on flapping wing micro air vehicles at Shrivenham. Symp. unmanned vehicles. Ankara. – Nachtigall, W.; Mitarb. v. A. Wisser (2003): Insektenflug. Springer, Berlin. S. 215–220.

SENSORS

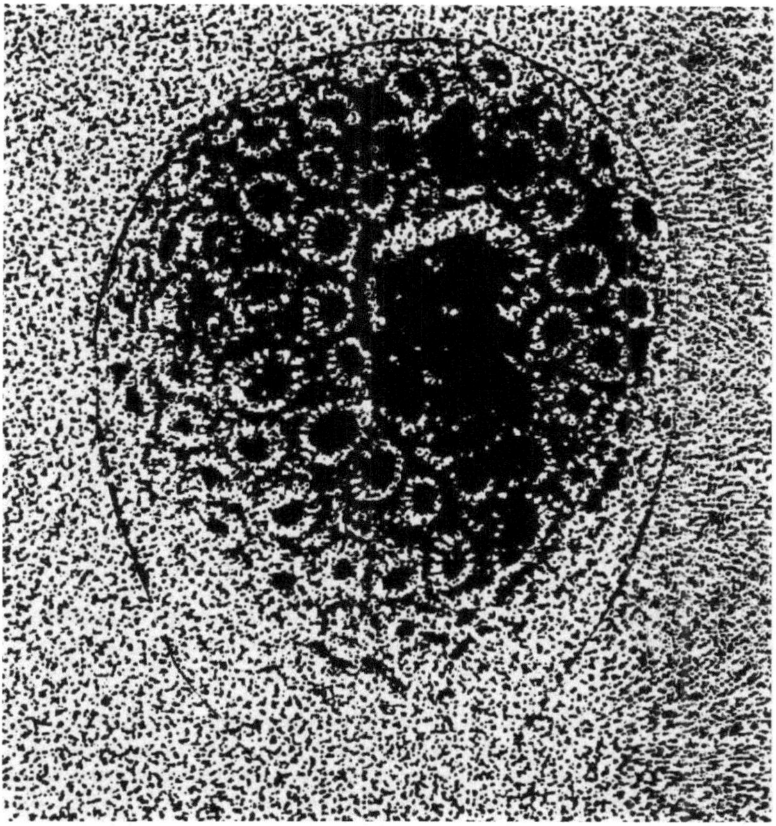

AND NEURONAL CONTROL

W. Nachtigall and A. Wisser, *Bionics by Examples*,
DOI 10.1007/978-3-319-05858-0_9

SENSORS AND NEURONAL CONTROL

Biological sensors – Technical sensors – Neural biological systems – Neural technical control and regulation – In-coupling of biological subsystems.

Matters of monitoring of physical and chemical stimuli, target finding and orientation in the environment belong to this area. For example, the problem, to monitor chemical substances in the body of the human being (keyword: diabetes) or in big-technical converters (keyword: biotechnology), becomes more and more important. Biological sensors that are configured for all possible chemical and physical inputs were already early studied and are now increasingly analyzed under the point of view of a transfer-possibility to the technology of today. Data-analysis and information-processing using intelligent circuits are in a stormy development. Especially the interconnection of parallel computers and the development of "neural circuits" have got decisive stimulation from the fields of neurobiology and bio-cybernetics.

Biosensors work like biological membranes

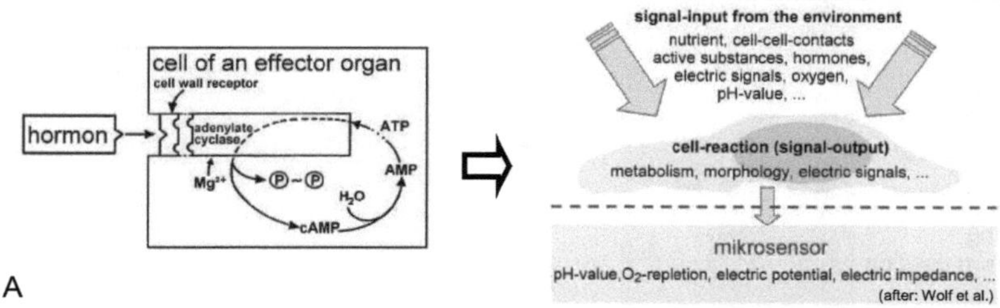

A B

The docking of specific molecules on biological membranes is imitated analogously by surfaces of biosensors.

BIOLOGICAL MEMBRANE:
The biological cell membrane carries surface-groups, on which specific substances can dock on according to the key-lock-principle. From this a membrane-process results, that implements on the other hand specific processes in the underlying lumen of the cell. So, for example, a specific peptide hormone can be attached to appropriate binding sites on the outside and thus via adenylate cyclase localized inside can induce the binding of cAMP (A).

BIOSENSOR-"MEMBRANE":
If a certain molecule shall be monitored in a bioreactor, on a suitable holder, for example a little glass-plate, a membrane-like coating must be generated, that offers possibilities of attachment for this and only this molecule. Because of the attachments, the reflection-qualities, for example, of a laser-illumination onto the underside change. Ideally, the measurable alteration of the reflection is proportional to the molecule-concentration (B).

VARIETY:
The named basic principle is undoubtedly bionic (realization of the abstraction of a biological model), however, from this, already a multiplicity of technological modifications has developed. The utilization of biosensors for the physiology, biomedical technology, biotechnology and monitoring of the environment is anchored in a stormily developing own discipline; in 2011, the 7. German Biosensor Symposium has taken place.

LITERATUR: 7. Deutsches BioSensor Symposium (DBS). BIOforum 1/2011, 11. – www.dbs2011.de – Wolf, B. et al. (2010): Zell-basierte bioelektronische Chips u. Systeme für Diagnostik, ... Biomed. Tech. 55(Suppl. 1). De Gruyter, Berlin/New York. DOI 10.1515/BMT.2010.492

Bending-based hair-sensors for flow-monitoring

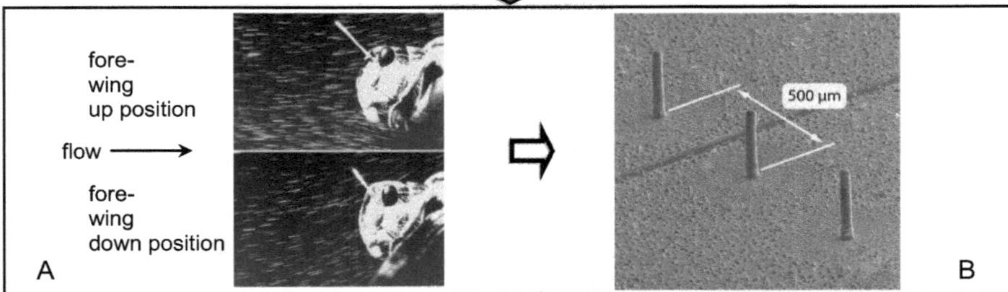

Flows around arthropods as well as flows in technology can be monitored by fields of bending based hair sensors.

BIOLOGY:
The individual hair-sensors (trichobothria) of the spider *Cupiennius salei*, which monitor finest air-flows in the surrounding, are studied well. Also very sensitive are corresponding sensors in scorpions. The animals can localize obstacles, as well as predators and objects of prey by these sensors. Several hair-fields in the head-region of locusts monitor for example wing-positions (A) and serve the flight-control.

PRINCIPLE:
Hairs work as stiff or deformable lever-arms. In the end, the membrane of an innervated sensitive neuron becomes distorted (comp. p. 163). As particle velocity receivers, they are optimally adjusted as high-pass-filters on certain frequencies according to their geometry (400-1150 µm); a field of different hairs sensitively covers a bigger frequency-area. To this group belong also whiskers, vibrissae and shear stress receptors.

TECHNOLOGY:
Today, bio-inspirited hair-rows or -fields with heights of about 1 mm and slenderness of about 20 can be produced out of high-flexible silicone-materials (B). Their bending caused by a flow-field can be measured optically. There are already numerous other lever-constructions and measuring-bases by which it will be possible to monitor quantitatively, for example, wall-near boundary-layer-sections with an approximately linear velocity-increase.

LITERATUR: Große, S.; Schröder, W. (2012): Deflection-based flow field sensors – examples and requirements. In: Barth, F. et al. (Hrsg.): Frontiers in sensing ..., 393–403. Springer, Berlin. – Barth, F. (1996): Dynamics of arthropod ... Phil. Trans. R. S. Lond. B, vol. 351, 933–946.

Monitoring by artificial side line organs

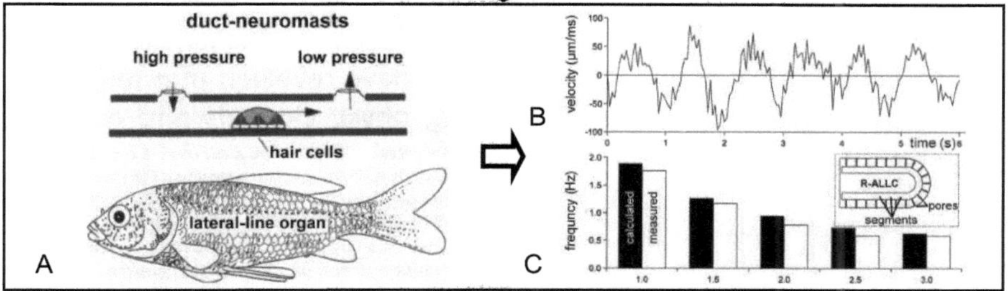

With sensor-formations analogue to side-line organs of fishes objects can be discovered and localized in flows.

BIOLOGY:
Side lines of fishes, at which the receptors sit chain-like in fluid-filled canals, are pressure difference receptors. Over openings between these so called duct-neuromasts pressures of the environment are perceived. These bend gelatinous enclosed hair-bundles (cupulae) of sensory cells in the direction of the lower pressure (A). The resulting shear represents the adequate input. The side-lines of variety of fishes are well examined.

ECOLOGY AND TECHNICAL INSPIRATION:
"Fish use side-line-information for prey-finding, intra-specific communication, swarm formation, object-discovery and -differentiation, entrainment and rheotaxis, primarily however for orientation in the nearby surrounding." Since v. Campenhausen (1981), one tried to transfer this principle that is usable in many different ways, onto technical sensors under silent-water-conditions. Now, it is also proved under flow-conditions.

FLOW MEASUREMENTS:
Bleckmann et al. (2012) report, that the vortex-street (B), generated by upstream localized cylinders in a flow, can be detected with an artificial side-line-duct. Used cylinders of different diameter differ in the frequency of their vortex shedding, and, according to their position to the detector, in their vortex-distribution. The detector can distinguish (C) cylinders by their size and situation.

LITERATUR: Bleckmann, H.; Klein, A.; Meyer, G. (2012): Nature as a model for technical sensors. In: Barth, F. et al. (Hrsg..): Frontiers in sensing. From biology to engineering. Springer, Berlin. S. 3–18. – Campenhausen, C. von et al. (1981): J. Comp. Physiol. A 143, 369–374.

Bio-inspired sonar improvements

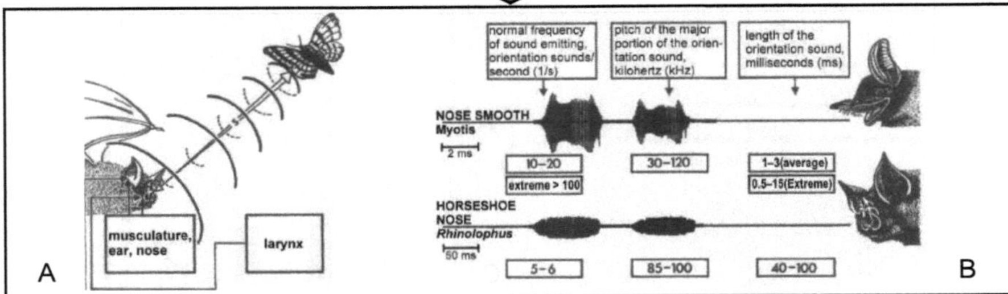

In comparison with the slow technology, the bat sonar quickly wins more information from a single impulse.

BIOLOGY:
The echo orientation of bats with their ultrasound sonar (A) is based on different bearing sounds. While the mouse-ear-bats send out approximately 10 singles-tones of 5 ms during a 100 ms series, the horseshoe-bats use typically approximately 1 single tone of 50 ms duration (B) that makes the utilization of the Doppler-effect possible. They do not contact 0.2 mm thick threads. From the distance of 30 cm, a distance-difference of 1cm is recognized.

THE BAT-SONAR AND THE TERM "BIONICS":
In the year 1960, a congress: "Bionics symposium. Living prototypes - the key to new technology" took place on an airbase in Dayton / Ohio. Predominantly the question was how one could learn something from the echo-bearings-system of the bats to the improvement of the technical radar. H. Foerster talked introductory about "BIONICS – new frontiers ...". This concept equivalently means: "Things that have to do with the animate world."

TECHNOLOGY:
H. Peremans et al. (2012) have studied how the bat *Phyllostomus discolor* with the echo-impulses of their sonar wins especially quickly and especially precise distance- and angle-information of moved objects. According to this they designed a bio-inspirited 3-D-localisation-system and implemented this into a bin-aural robot bat head. With this, multiple reflectors could be localized quickly with one single measurement.

LITERATUR: Peremans, H.; De Mey, F.; Schillebeeckx, F. (2012): Man made versus biological in-air sonar systems. In: Barth, F. et al. (Hrsg.): Frontiers in sensing. From biology to engineering. Springer, Berlin. S. 196–207 – Herder Lexikon Biologie, Spektrum 1994.

Electric sense organs of fishes & technical monitoring

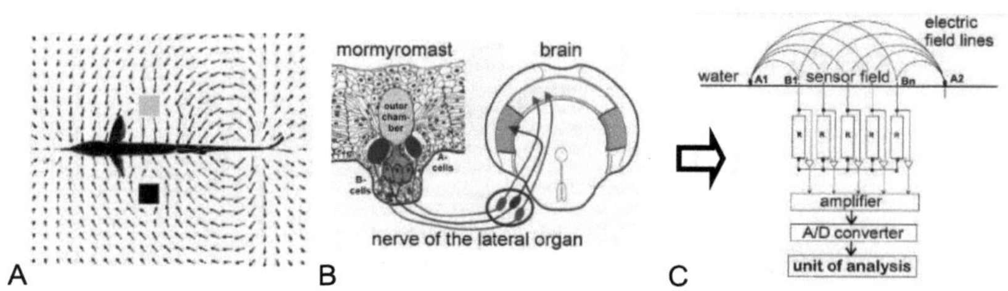

Some fishes localize objects in their environment with electric sense organs – models for technical electro location sensors.

BIOLOGY:
Especially, in cloudy-water the ability of some weak-electric fishes become important for localizing actively leading and not leading objects with their electro-organs. The classic example is the "Nile-pike", *Gnathonemus petersi*. An electro-emitter – here at the tail-stalk – generates an electric field (A) that is distorted by nearby objects. Electroreceptors (mormyromasts) on the fish-surface (B) monitor this distortion.

TECHNICAL TRANSFER OF THE PRINCIPLE:
In a formation of electrodes (C), outside lying electrodes A_1, A_2 produce an electric field that can be monitored by inside lying electrodes B_1-B_n. Each B-electrode measures the local voltage that is influenced by objects within the field-area. Over amplifiers and converters, the information is supplied to a central unit. This is analogue to the supply of the information of the mormyromasts over the side-line-nerve to the brain (B).

POSSIBILITIES OF USE:
Technical instruments could be used, for example, to detect, according to the above mentioned principle, a target in the macro area. Object-sensor-distances between approximately 0.5 and 3.5 mm can be monitored by output voltages between 0.093 and 0.400 V. Leading and not-leading objects can be distinguished, in the close-up range even Cu from Al and plastic from wood. Another field of use would be the close range distance measurement.

LITERATUR: Emde, G. von der (2012): Remote electrical sensing; detection and analysis of objects by weakly electric fishes. In: Barth, F. et al. (Hrsg.): Frontiers in sensing. From biology to engineering. Springer, Berlin. S. 314–326.

Slit sense organs ⇒ high efficient tension sensors

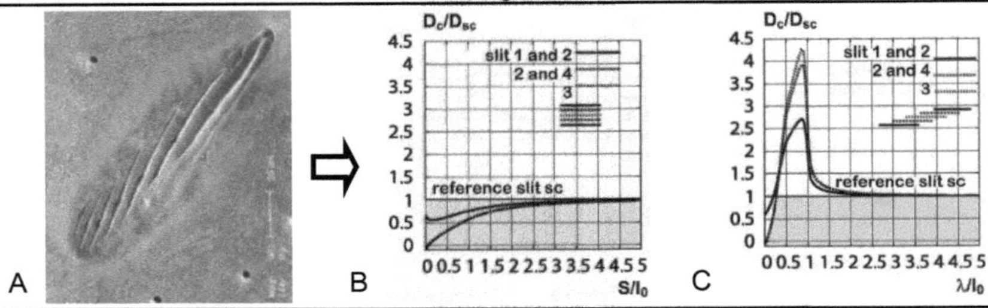

Spiders measure smallest distortions and therewith alterations of mechanical tension with their slit sense organs.

BIOLOGY:
Slit sensor organs (A) work due to F. Barth as "biological strain gauges." Their sensor cells in the exocuticula, that are associated with up to 30 approximately parallel clefs or slits of 8-200 μm length and 1–2 μm width, monitor slit-compressions of only 1,5 nm. Here, the sensitivity of lateral forces f_S = elongation / lateral force is as high as ≥10 με/mN (!). So, spiders detect, among other parameters, finest vibrations of the substratum.

MODELLING:
Because of the difficulties of direct measurements in these small dimensions, model-simulations are important. So it turns out, for example, that at 5 not staggered parallel slits the total distortion D_c related to the distortion of a singles-slit s_{sc} is diminished with the slit distance D, standardised to the slit length l_0 (B). With a (biology-typical) likewise standardised smaller slit-deposition λ it reaches a maximum at $\lambda/l_0 \approx 0.75$ (C).

TECHNOLOGY:
The possibilities of a successful manufactured technical implementation would lead to extremely sensitive strain gauges and would open up versatile new dimensions for measuring. Wicaksono et al. (2005) report about a first practical realization. Vincent et al. (2007) give a list of technical possibilities. Among other things, this includes thin-film-technology, caustic-technology, silicate micro-treatment and photo-lithography.

LITERATUR: Barth, F. (2012): Spider strain detection. In: Barth, F. et al. (Hrsg.): Frontiers in ... sensing. Springer, Berlin. S. 251–273. – Wicaksono, D.H.B. et al. (2005): J. Bionic Eng. 4, 63 –76. – Vincent, J. F. V. et al. (2007): J. Micromech. Microeng. 15, 72–81.

An infrared-detector according to fire-bugs

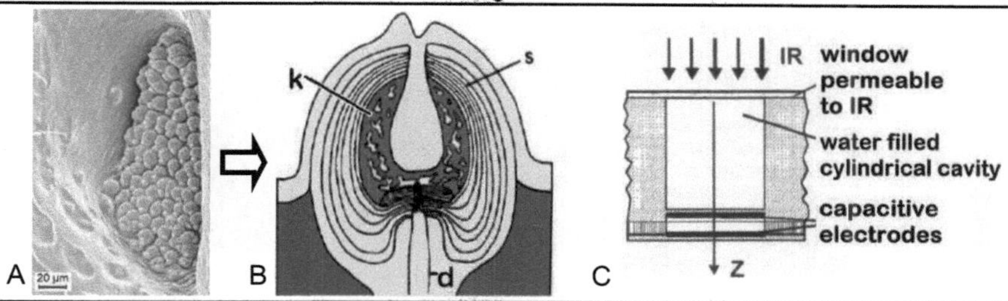

With this sensor, fire-bugs discover forest-fires to deposit their eggs – technically an uncommon principle for fire alarm.

BIOLOGY:
The wood-boring *Buprestid* beetles of the genus *Melanophila,* known as "fire-bugs", possess on each body-side a field, sunk in a pit, of about 70 infrared sensitive sensillae (A). In an onion skin-like, heat-insulating cover of cuticles, there sits a fluid containing globule k (B) that extends under heat radiation. This expansion represents the adequate stimulus for the dendrites d of one single mechano-receptive sensor cell.

TECHNICAL TRANSFER OF THE PRINCIPLE:
The uncommon principle that a thermal parameter is monitored mechanically found entrance into the development of an analogously built technical sensor. Instead of such a globule, a liquid filled cylindrical cavity is used, whose expansion is monitored by a capacitive electrode under IR-radiation (C). The membrane of the biological dendrite is still excited by deflection ≤ 1 nm (!); that of the technical electrode is much more insensitive.

PROBLEMS OF REALISATION:
The operative point for the development of a sufficiently sensitive, miniaturized, inexpensive appliance is the read-out of so less mechanical deflection. Such a capacitive system is in development. It should be supported by the application of a liquid of high thermal expansion and low heat-storage-capacity (e.g. methanol) and a wall-material with a low capacity of heat-conductivity (e.g. silicone).

LITERATUR: Schmitz, H.; Bousack, H. (2012): Designing a fluidic infrared detector based on the photomechanic infrared sensilla in pyrophilous beetles. In: Barth, F. et al. (Hrsg.): Frontiers in sensing. From biology to engineering. Springer, Berlin. S. 301–311.

Artificial compound eye for capturing pictures

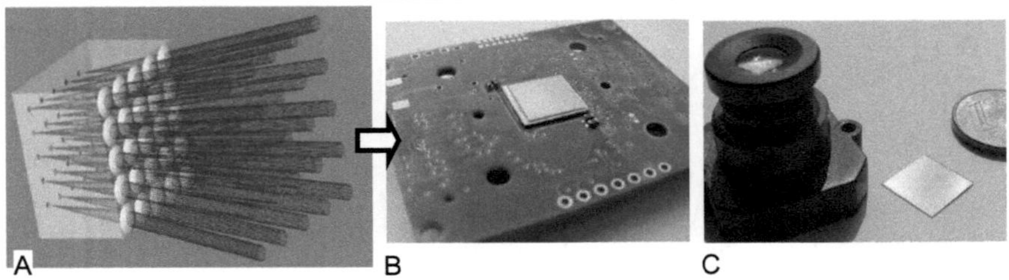

Nowadays, artificial arrays of receptors are planar. Therefore, affiliated picture-generating-systems should also be planar.

BIOLOGICAL AND TECHNICALLY-ANALOGOUS PICTURE-GENERATION:
The facet eye of arthropods consists of several dozen up to several 10.000 (dragonflies) single eyes with hexagonal (insects) or quadratic (crayfish) cornea: Their *picture-generation system* (here, it doesn't go about *receptive* aspects and *picture-processing*) consists of a cornea and optical fibres. – The planar artificial facet eye lens (A), produced in the wafer-standard, is only 0.2 mm thick and has an F-number of 2.2.

PRINCIPLE:
The thin lens-layer contains a cornea-analogous UV-polymer-layer, a glass-substratum with a thickness corresponding to the focal length and a focus-analogous layer with pin diaphragms. In 2005, an objective with a resolution of 60 x 60 pixels was built, which has a visual field of 20° x 20°. It can be mounted as a hardly noticeable thin layer directly at an integrated CMOS-sensor (B) and can replace a conventional 24-mm-objective (C).

MEANING:
Today, picture-processing systems can be integrated into very thin circuit boards. However, the total system becomes bulky by prefixed picture-giving systems like conventional lenses. A high resolution is not necessary from time to time, especially at the face recognition. Here, the planar artificial facet-eye could lead to space-saving concepts. Test shots show, that faces or patterns are satisfactorily reproduced.

LITERATUR: Duparré, J. et al. (2005): Thin compound-eye camera. Applied Optics 44(5), 2949–2956.

Insect eyes and control of aircrafts

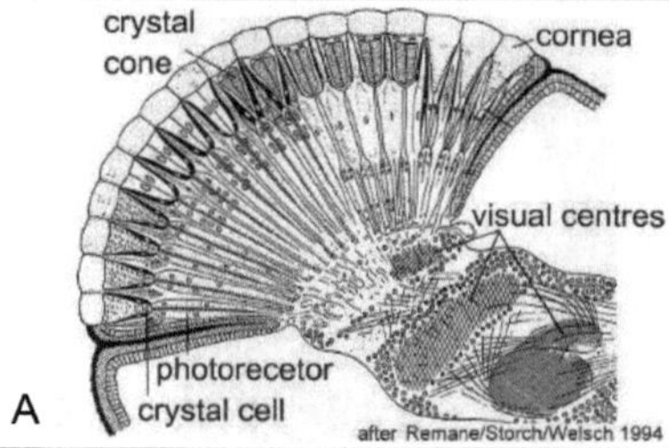

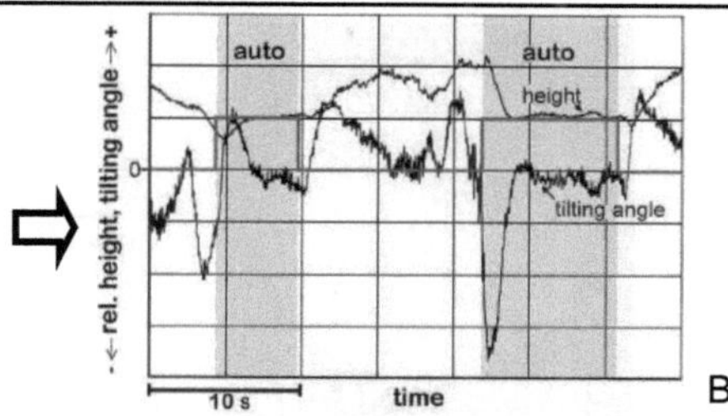

The facet-eye of insects monitors with a high temporal resolution "technically favourably" the optical flow.

BIOLOGY:
As optical sensors, insects possess compound eyes (facet-eyes) and ocelli. The first ones are complexes of cone-shaped single eyes (ommatidia) (A), the later are single elements and usually found in the forehead-region between and over the compound eyes. The optical information of the ommatidiae runs over three clearing and integrating optical ganglia (visual centres) to the comparatively small brain.

CAPABILITY:
Because of their construction, the spatial resolution of the compound eye (A) is only relatively low (≤1°), but their temporal resolution is good (≤4 ms). The low neural capacities of the tiny brains have given rise to quick, alternative algorithms. During quickest flight (bluebottles up to 3 m s^{-1}, hornets, dragonflies more than 12 m s^{-1}), insects can avoid very well collisions, pursue ways and aims and perform quick and safe landing approaches.

TECHNOLOGY:
Srinivasan et al. (2012) cite a series of transfer-possibilities, among them panorama-like picture-generation (comp. p. 164) safe crossing of bottlenecks, terrain-following overflight, heights- and landing-control, hodometry (honeybee-inspired integration of the optical flow). Also the performance of ocelli gave stimulation, in fact for the stabilization of the roll and tilting moment of small-airplanes by tracing the horizon-line (B).

LITERATUR: Srinivasan, M. et al. (2012): From biology to engineering: Insect vision and applications to robotics. In: Barth, G.; Humphrey, A.; Srinivasan, M. (Hrsg.): Frontiers in sensing. From biology to engineering, Springer, Berlin. S. 19-39,

Insect eyes and the increase of light efficiency

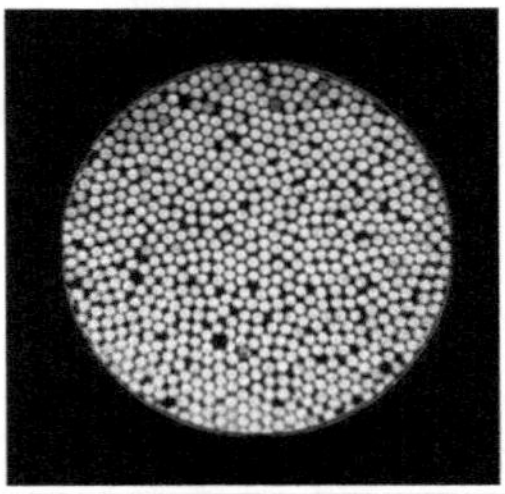

A

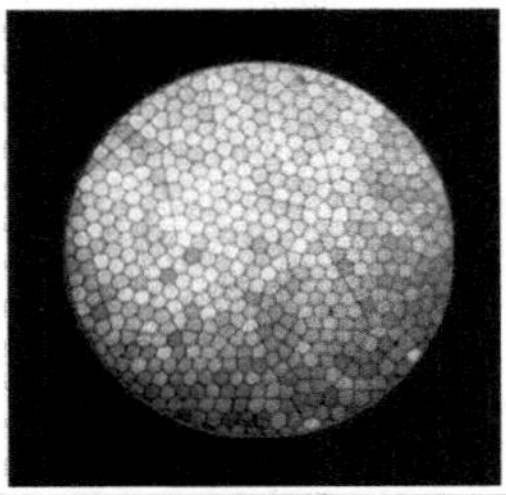

B

Insect-eyes, honeycombs and furs of polar bears inspire the construction of translucid elements for technology.

BIOLOGY:
In the insect eyes with hexagonal cornea, concentric layers can be found in the crystal-cones, that focus because of total reflection, also diagonally and obliquely incident rays of light to the retinula. In the polar bear fur, total-reflection takes place at the inside of the hairs (s. p. 51). Through competition for space, round-cylinders form themselves to stable hexagonal forms (honeycombs). The combination results in stable elements that are also good optical fibres.

PRINCIPLE:
Translucid wall-elements for the building industry (comp. p. 51) and optical fibres especially for endoscopic purposes, work according to the principle of total-reflection. Through it, once captured light or heat-radiations (the later in translucid wall-plates) remain in the fibres. In the plates, these are formed of together-lying little glass tubes, in the optical fibres, they are made of very fine glass fibres. The principle implies however relative high scatter losses.

TECHNOLOGY:
For example, bionic stimulation has led the company H. Hund to a production-technology for endoscopic optical fibres, where round fibreglasses (A) are pressed to a hexagonal form (B). This results in a bigger stability of the optical fibres according to the honeycomb-principle. Simultaneously, the refractive properties were modified by core- and cover-glass according to the principle of insect-eyes, so that a light efficiency higher than 20 % emerges.

LITERATUR: Fa. Hund, H. (2009): Bionik steigert Lichtausbeute. In BIOforum 3, 40. – Anonymus (2009): Translucide Gestaltungselemente für Innenräume. Industry News, Quelle: Bayer Material-Science AG. – Nachtigall, W. (2002): Bionik. Springer, Berlin.

Systematic approaches for auto-sensitive materials

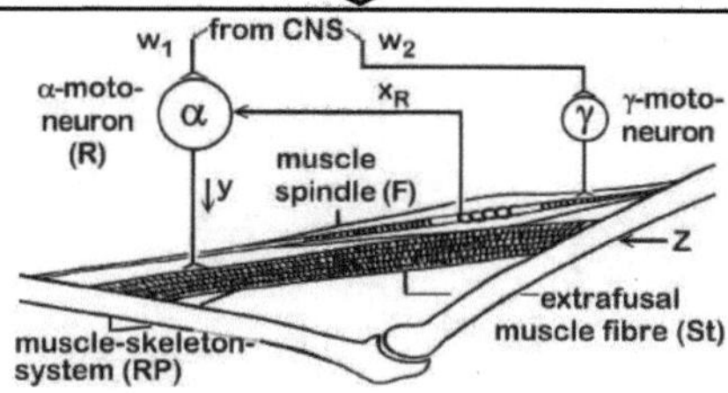

Muscle length regulation via endogenous sensors (spindles)

Biological structures are always supplied sensorial. Technical materials should be able to "feel" accordingly.

BIOLOGY:
Rarely, biological structures or materials are passive, especially then, when they are conglomerating as depositions, for example in lime-skeletons. Otherwise, these can be actively regulated, based on sensory influences. In muscles, different stiffness can be put in, skin-cuttings heal, righting reflexes provide fast unloading by "twisting one's angle", heat-sensors in the skin of the hand protect this of scalding by reflective withdrawing from the heat

PRINCIPLE:
The continuously recognizable biological principle therefore is: *Biological materials are – often via sensors integrated in the material – reactive.* A recently established "Central Scientific Establishment" at the university Bremen with the main focus ISIS (Integrated Solutions in Sensorial Structure Engineering) tries in interdisciplinary runs to transfer these excellences of enlivened structures on technical materials.

TECHNOLOGY:
Usually, average loads are taken into account. One tries to counteract high, but rarely appearing critical loads through high safety factors (→ increase of weight!) or by machine care, repair and making a machine or a system "as good as new in the wearing process" (→ increase of cost!). It would be very profitably if technical materials could absorb auto-regulative extreme loads: Structure-monitoring will be important.

LITERATUR: www.ifam.fraunhofer.de – www.materialica.de/html/fuehlen.html – S. auch Abschnitte über Sensorik und neurale Steuerung und Regelung in Lb. d. Physiologie. Vgl. auch Selbstheilung von Verbundwerkstoffen, S. 127.

Auto-dynamic running stability and control expense

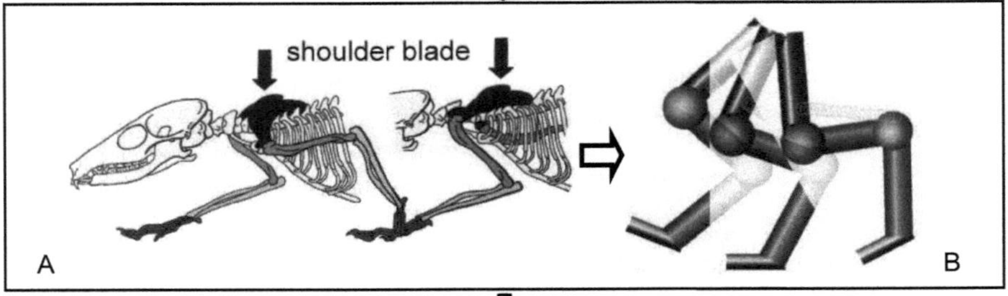

Biological principles of auto-dynamic running stability can reduce the cost of control for four-legged running machines.

BIOLOGY:
Mammal legs possess three subunits, relevant for the locomotion, beside the shoulder blade of the front-extremities: Upper leg (upper arm), lower leg (forearm) and foot (hand). In the front-extremity, the shoulder blade rotates until on ¾ of the step-length (A). The kinematics is very conservative, hardly depends from the systematic position of the animal, its anatomical specialisations or from the peculiarities of the surroundings.

PHYSICS:
Two shock systems (ligaments and passively stretched muscles) regulate auto-dynamically bending and stretching induced by gravity (B) without using excessively active muscles for "stress catching." This leads to running forms with little tiring. Especially, during running fast the typical leg swinging movements lead, in interaction with the curvature of the back, to a dynamic stability without high control expense.

TECHNICAL REALIZATION:
The inclusion of the principles of an auto dynamic running stability, as they were developed by quadrupeds, leads the designer of running machines to different procedures. Earlier, when the mechanics was laid out "quasi-statically", the development of very complex steering- and regulation-systems was the unavoidable consequence. These were reduced by the recently introduced integration of "intelligent mechanics" as optimal control strategy.

LITERATUR: Fischer, M. S. (2006): Grundprinzipien der Fortbewegung auf tierischen Beinen. Industriekongress BionikBioKoN, Berlin. S. 27–37. – Nachtigall, W. (2002): Bionik. 2. Aufl. Abschnitt 10, Robotik und Lokomotion. Springer, Berlin. S. 175–239.

Decentralized control of a snakelike robot

A

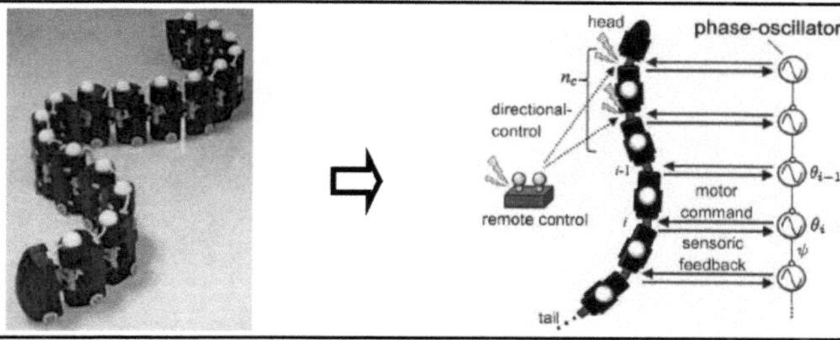

B

Autonomous, decentralized movement control – auto-adaptive and failure-insensitive – is widespread within living organisms.

BIOLOGY:
In earlier days, central time bases were discussed for the control of periodical movements in the animal-kingdom, for example to generate leg movements in locusts. Later, the research of additional decentralized regulators for the co-ordination has shown, that these were able to improve the reaction to external disturbances. So the movement of each individual locust-leg is indeed regulated for itself but it communicates with all other legs.

TONIC AND PHASIC REGULATION:
While one speaks of a tonic regulation regarding the permanent setting of the angles between two leg-limbs, the change of the angle of the leg relative to the body is phasic regulated during running. Sato et al. (2012) have transferred both principles) onto their snake robot HAUBOT (A. The stiffness of the connection between its single elements is subjected to a tonic regulation, the momentary-position of the segments to each other to a phasic regulation.

ACCOMPLISHMENT AND ADVANTAGE:
Each joint is stiffened by two servo-motors (opposite-rotate: tonic) and adjusted in lateral direction (co-rotate: phasic). These are controlled over the phase of their responsible oscillator of an oscillator chain, and this phase on the other hand is regulated over the inputs of sensors at the joint (B). Thus, the snake-robot gains an excellent momentary adaptation to roughness (small obstacles) and local inclinations of the ground.

LITERATUR: Sato, R.; Kano, T.; Ishiguro, A. (2012): A decentralized control scheme for an effective coordination of phasic and tonic control in a snake-like robot. Bioinspiration and Biomimetics 7(016995), 1–9. IOP Publishing Ltd.

ANTHROPO-

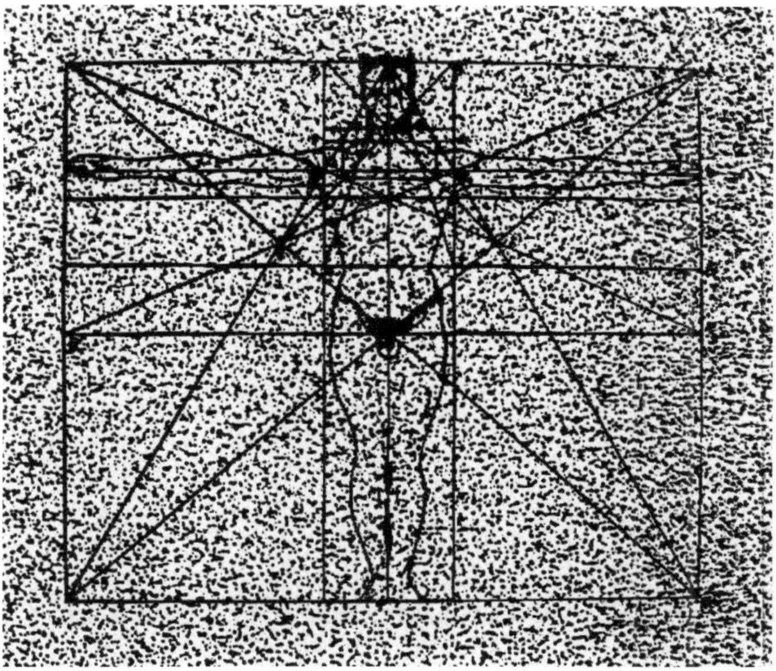

AND BIOMEDICAL TECHNOLOGY

W. Nachtigall and A. Wisser, *Bionics by Examples*,
DOI 10.1007/978-3-319-05858-0_10

ANTHROPO- AND BIOMEDICAL TECHNOLOGY

Anthropo-technology (human-machine-interaction) – Biomedical technology (without prosthetics) – Prosthetics.

The problem-circle of the human-machine-interaction, getting more and more important, and the diverse uses of robotics belong to this main focus. User friendly design of the cockpits of modern airliners which are fitted to the sensory habits of humans, would be a modern example, another would be the search of ideal-configurations of bicycles, which, for example, allow to drive the bicycles with higher muscle-efficiency, as the human creates when running(!). Problems of robotics, for example the control of gripping-arms, could be solved in an unconventional way by comparative studies of the leg-movements from invertebrates (crayfishes, insects).

Human-machine-interaction

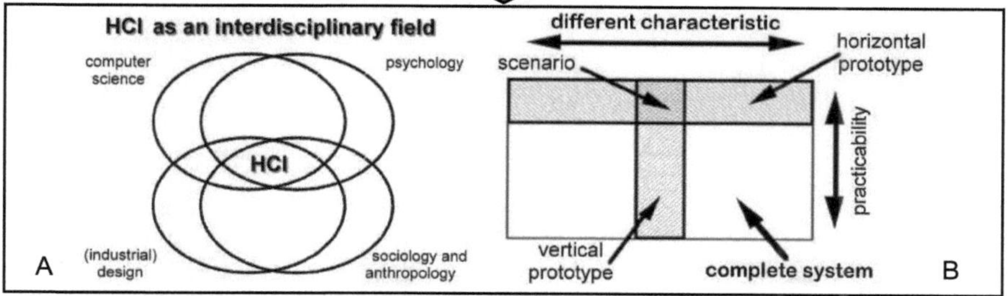

Today, no more the heavy work is the problem of operating machines, rather the stress of attention.

CLASSICS:
The energetic relief of the worker when serving machines was a main-problem of the work physiology long time. Example: Diminution of the pedal power according to pedal-optimisation. The development of a bicycle from a high-wheel to a modern appliance eventually was an adaptation to the morphological and physiological so-being of the human. Better readability of pointer devices reduces mistakes due to decreased attention.

EXAMPLE "HUMAN COMPUTER INTERACTION (HCI)":
That is one of the problems today, to which the industry dedicates most severe attention. To the improvement, at least 4 disciplines play together, namely computing, industry-design, sociology/anthropology and psychology (A). Another example is the modernisation of pointer devices in the airplane-cockpit. Since some time, here the vertically moved line-symbol became generally accepted instead of the rotating pointer.

EXAMPLE HORIZONTAL AND VERTICAL DESIGN:
Also design strategies can increase the usability. So, one can proceed "horizontally" in a characteristic(x)-functionality(y)-field (B), this means to search with given functionality in a series of characteristics, or "vertically", this means to search a given characteristic for its functionality. In a scenario-sub-field, both overlap. To search "vertically", can lead more quickly to an user friendly, function-specific optimal-design.

LITERATUR: Butz, A.; Holleis, P. (2010): Vorlesung Mensch-Maschine-Interaktion 1 für das SS 2010. LMU München, Medieninformatik. – Viele weitere Beispiele s.: http://www.medien.ifi.lmu.de/lehre/ss10/mmi1/#inhalte-contents.

Active, unconscious back support for sitting work

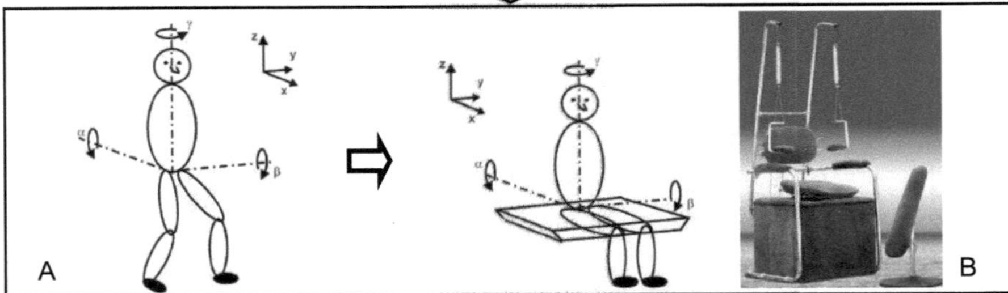

During long sitting, back-problems can be counteracted by active body musculature stimulation.

WALKING PREVENTS AND REMEDIES BACKACHE:
When walking, the pelvis is moved step-intermittently by the legs. In order to be able to hold head and shoulder-belts quietly, active, but unconscious counter movements of the entire body musculature take place with obligatorily upright bearing of the upper body. Daily half a hour of persistent walking prevents and remedies back-problems which were caused by lack of movements and crooked bearing of the upper body while sitting.

PRINCIPLE OF A MOTOR HEALTH-SEAT:
Each human being has its individual gait pattern that can be defined on the basis of his corresponding body dimensions (A). The so gathered gait pattern can be impressed to a motor moved seat – patented by G. Schon (B). So a seating person can be stimulated to perform the same unconscious activation of his body musculature as in walking, even persons unable to walk and wheelchair-drivers may be helped

IMPORTANCE FOR THE FUTURE:
At least 80 percent of the people in the industrial nations suffer from backache, the "most common illness". The here shown way could represent a very broadly applicable therapeutic base. The team-project "Spinemover" together with the inventor and several institutions under the management of H. Witte, Biomechatronik TU Ilmenau, was 2003 winner of the innovation-competition of the BMBF, supporting the medicine-technology.

LITERATUR: Schon, G. (2009): Motorisch bewegter Sitz für die Gesundheit des Rückens. Die patentierte Lösung für die Volkskrankheit Nr. 1!. In: VDI-Bericht 2067 zum Kongress Automation 2009, Baden-Baden. – Europa-Patent EP 1123025. – US-Patent 7093900.

Antidecubitus mattress according to the skin

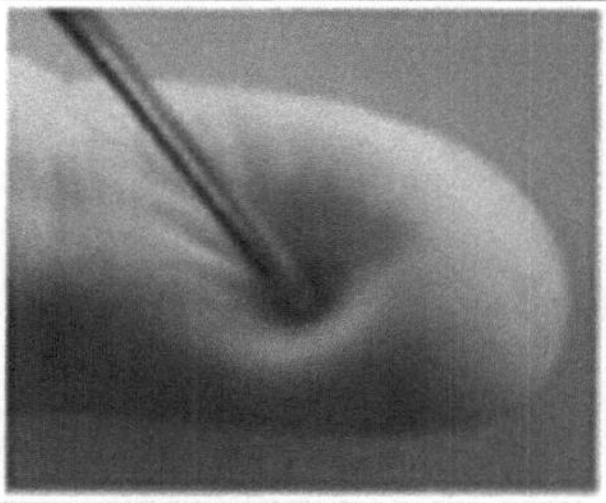

A

for the skin:
antidecubitus bedding system

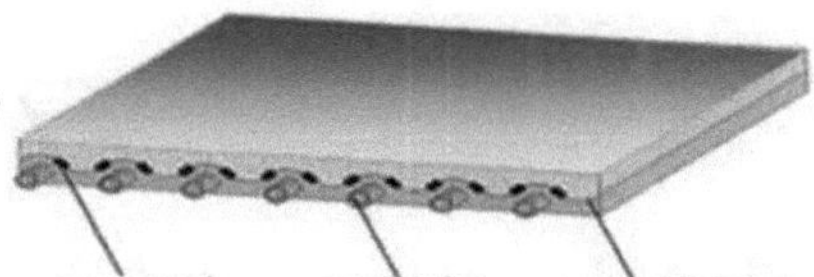

B

The skin changes auto-adaptively its mechanical qualities, basis for the development of antidecubitus mattresses.

BIOLOGY, PATHOLOGY:
The tissue "mammal-skin" can change auto-adaptive its mechanical qualities, for example the stiffness (A), according to the principle of a sensor actor interaction, because as well as sensory (tactile corpuscles) and also motor elements (muscle-fibres) are integrated in its tissue. The permanent contact pressure of patients, which are bedridden a long time, however, cannot be regulated. That can lead to bedsore ulcer (decubitus ulcerous).

COMPENSATION:
A mattress with similar auto-adaptive qualities could work against this problem. "Sensors shall detect the bruises and the temperature-increase of bedsore, and the actors will stimulate purposefully the concerned locations and will take the load off these (B). The bedding-system, that should be developed, is tested with the help of model-based examinations. Then, the prototype of an antidecubitus mattress shall be developed."

MEANING AND MARKET-CHANCES:
This project "Adaptive tissue with pressure-controlled stiffness and integrated sensoric according to the model of the skin", which is conceived in the framework of the BIONA-promotion and represented by L. Zentner, Fac. f. Mech. Engineering, TU Ilmenau, promises high social benefit in consideration of the frequency of decubitus (10–30 % of the bedridden patients in nursing homes and clinics!) and also of the demographic development.

LITERATUR: Zitat n. (2011): www.bionische-innovationen.de/#projekte/adaptives_gewebe.html – RKI (Hrsg.) (2002): Gesundheitsberichterstattung des Bundes 12, Decubitus. Berlin – Stat. Ämter des Bundes und. der Länder (Hrsg.) (2008): Demogr. Wandel 2.

"Artificial skin" produced from spider silk

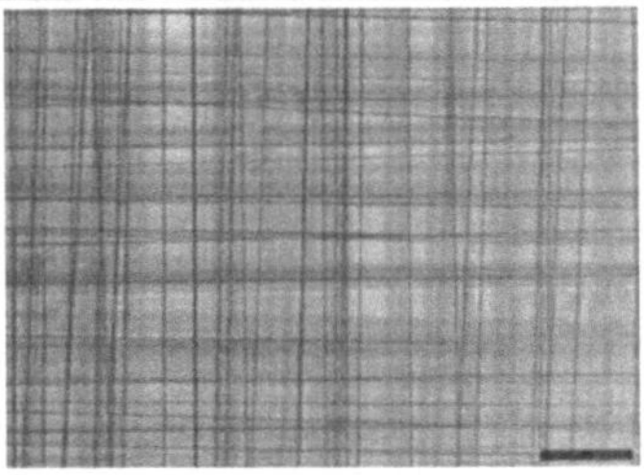

A

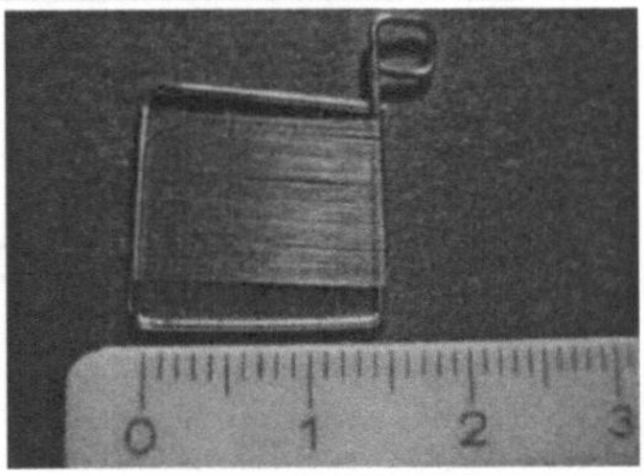

B

Own-skin-transplantation after heavy fire-injuries could be advantageously replaced by a cultured "art-skin."

BIOLOGY:
Spider silk, with its particular bio-mechanical qualities and, in addition, its good tissue-compatibility, is suitable for operation suture material scare and helps in regenerations of nerves. It also can be arranged in such a way , that flat tissues are created by vertical crossing fibre structures (A). Skin cells that can dock at the fibres can be cultured on this substratum. They proliferate there and grow-in further into the mesh network.

METHODOLOGY:
H. Wendt et al. from the Med. School and the Inst. f. Technical Chemistry, Hannover, have shown, that gripped threads of spider silk from *Nephila spec.* can be woven in a steel-framework (B) and after sterilisation can be inoculated with fibroblasts. After a two-week culture, further fibroblasts and keratinocytes are added, so that, under specific boundary-conditions, a two-layer, skin-analogous (dermis and epidermis) tissue-product can develop.

PROSPEKTIVE MEANING:
The skin cells, that proliferate into the mesh network, close up to a flat substratum within one week and form there under specific conditions a durable, separated doubles-layer. The thread-material is degradable. Therefore, the so produced biomaterial appears suitable as substratum for transplantation. It could help to regenerate skin-surfaces. Sufficient mechanical durability has been proved.

LITERATUR: Wendt, H. et al. (2011): Artificial skin – Culturing of different skin cell lines for generating an artificial skin substitute on cross-weaved spider silk fibres. PLoS ONE 6, e21833. Publ. Medical School und Institut für Technische Chemie, Hannover.

Contacts between biological tissue and technology

A 

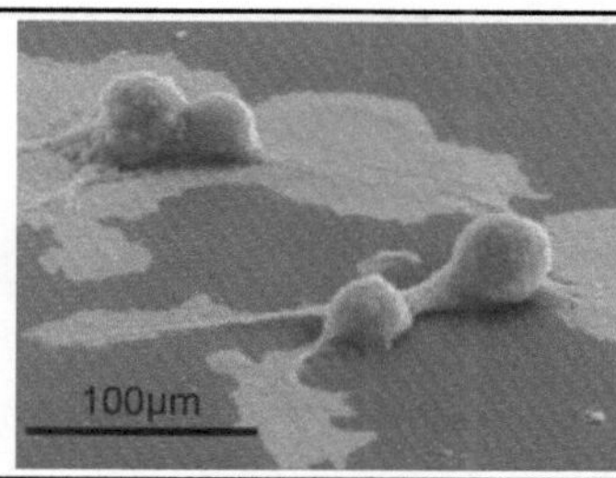

B

These two worlds can be connected for the biomedical research by micro electrode arrays and neurochips.

MICRO ELECTRODES:
Today, micro electrode- arrays (A) are produced commercially (Multi-Channel-Systems MCS Reutlingen) and put in for research-purposes at several hundred colleges world-wide. Diverse electric processes can be pursued by such arrays in biological tissue, for example in neural tissue and in cardiovascular systems. So effects and secondary effects of substances onto cell-associations can be tested by tissue breeding.

NEUROCHIPS:
For example, nerve-cells can increase on semiconductor-contacts of neurochips and they can represent partners in a combined biological-technical circuit. In 2003, for the first time, P. Fromherz/MPI Munich succeeded with neurons of the mud-snail *Lymnea stagnalis* on a SMOS-Chip (B). The NMI (scientific-medical institute) Reutlingen took over the development of such neurochips to the market-maturity.

STORMY DEVELOPMENT:
The science-journalist E. Schneider/Tubingen reports about the activities of the NMI, a spin-off-institute of the neuro-clinic Tubingen, that organised a scientific meeting about the named topic in 2010. "Wide application-fields were introduced in it for specific neurochips in the medicine-technology, the biotechnology as well as the pharmacology." The eye chip according to Zrenner / Tubingen represents a protruding new example (comp. p 179).

LITERATUR: Schneider, E. (2010): Mikroelektroden ... – Zrenner E. et al. (2010): Subretinal electronic chips ... Proc. Roy. Soc. B. Biol. Sc.. – LABO, 42 f. – Fromherz, P. (2003): Neuroelectronic interfacing ... In: Waser, R. (Hrsg.) Nanoelectronics ... Wiley-VCH, Berlin.

Cochlea and retina as prototypes

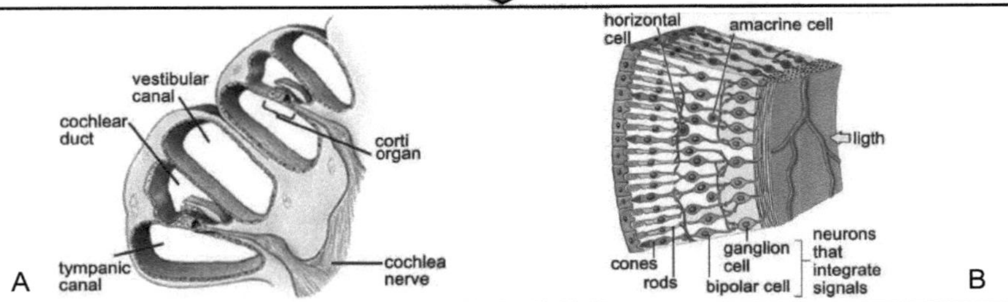

The physiology is partially well known. Its transfer into the technology can serve the humans directly or indirectly.

COCHLEA:
When the cochlea is damaged, the "cocktail-party-effect" plays a role: From a snarl of sounds, one cannot detect individual sounds. Studying the construction and mode of action of the cochlea, the Swiss neuro-computer scientist Ruedi Stoop has modelled mathematically by bifurcation theories the until now little understood active signal-production and -suppression of the cochlea (A). His artificial cochlea works like an undamaged natural one.

RETINA:
In the retina of vertebrates (B), certain neuronal connections provide the ability of motion view. The two neuro-informatics P. Lichtensteiner and T. Delbrück at the institute for neuro-informatics of the university and ETH Zurich were technically able to abstract this and to transfer it on a "silicone retina" with 16384 pixels. "The pixels of the artificial retina communicate continuously changes - like neurons in the human eye" (P. L.).

MEANING:
In the year 2010, the "artificial cochlea" admittedly was still as big as a table, however, it can be miniaturized. With fine wire-neuron-connections, it could help strongly hearing-impaired people orientating themselves well to sound-sources. The "artificial retina" is not intended as prosthesis. In the intelligent optical sensor "Smart eye", it can, for example, be applied for recording traffic-data, counting people and process-monitoring.

LITERATUR: Däniken, T. von (2010): Dossier Vorbild Natur. Schöner hören. Magazin Universität Zürich 19(3), 41. – Gull, T. (2010): Dossier Natur. Im Dunkeln sehen. Magazin Universität Zürich 19(3), 42. – S. z. B. auch ruedi@ini.phys.ethz.ch und tobi@ini.phys.ethz.ch.

Sub-retinal chip allows blind men to see letters

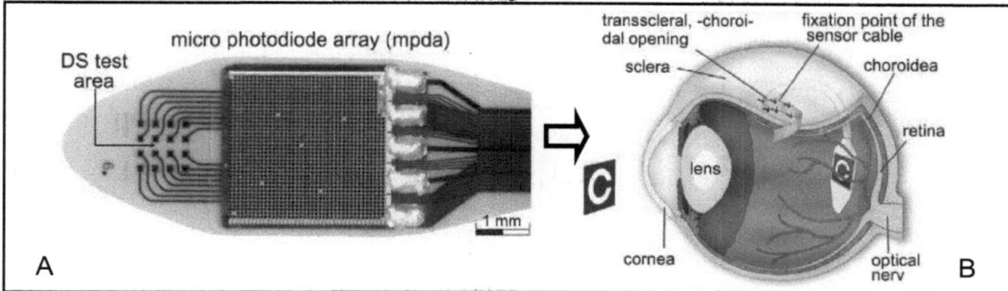

A parade-example for the possibility and the effect of biomedical research is the success of this optical sensor.

BIOLOGY:
In the retina of the mammal-eye, there are not only sensory cells – rods cells / light and shade view, retinal cones / colour view. There are also switch-cells (for example bipolar and ganglia cells), that process image definition (edge sharpening), motion view etc.. They already provide an imaging. Retina-implants must take this into account. The here named implant replaces degenerated sensory cells and stimulates the switch-cells still able to function.

PRINCIPLE:
E. Zrenner, eye-clinic Tubingen, succeeded a big step to an "artificial visual capacity" after irreversible damage of the retina (Retinitis pigmentosa or possibly strongly advanced, old-age macular degeneration). His team together with 6 further institutions developed a *sub*-retinal implantable (B) 3x3-mm-mikrochip with 1500 photo-diodes (A), everyone with its own amplifier and stimulation-electrode for stimulation of the underlying bipolars.

RESULT:
The output stimulation-voltage is proportional the light-intensity (10–100000 lux that means room-illumination until bright sunlight) and the inclination until approximately 11°. Patients could localise objects with this view implant, recognise letter rows in individual cases. This *sub*-retinal system from Retina Implant AG stands in competition with a 200-diode-system *epi*-retinal implant from Second Sight Medical Products, Sylmar, California.

LITERATUR: Zrenner, E. et al. (2011): Subretinal electronic chips allow blind patients to read letters and combine them to words. Proc. Roy. Soc. B 278, 1489–1497. – Pannu, S. S. (2010): Artificial retina: A biocompatible retinal prosthesis. LLNL Engineering, 08–10.

Intelligent sensor implants

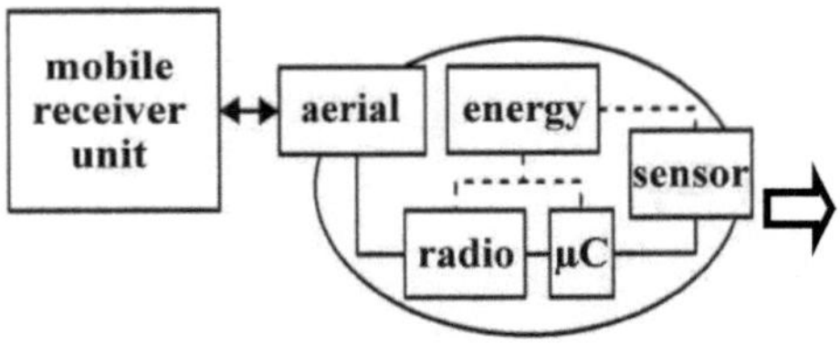

component	mAh for a month	
sensor chip	34.8	(sensor)
MSP430	0.319	(controller)
ZL70101	0.352	(radio module)
total system	35.50	

A B

Implants, which can autonomously measure the control variables, connect "sensor system" and "biomedical technology.

PRINCIPLES AND MEANING:
Such implants are designed to measure automatically vital-parameters of the body and to store, to process and to send them to the outside world. Parts of their system architecture (A) are sensor, microcontroller, radio-module and antenna; an external recipient comes to it. Today, they are restricted in the broader sense to the area of the pacemakers. In future, they shall monitor different physical and chemical parameters.

CONSTRUCTION-CHARACTERISTICS:
Preferably, the implants shall be miniaturised for the acceptance of the patient and the implant-technology. For a high lifespan, they should have low energy needs (B). Therefore, the maximum current of the sensor shall be low and the measuring-time short. The clock frequency of the controller shall be small and its architecture simple. The transmission-frequency should lie at 403 kHz (MICS-band) and the transmission-power at < 25 mW.

EXAMPLE GLUCOSE SENSOR:
A glucose sensor is developed at the IHP (Leibniz-Institute for Innovative Micro-Electronics/Frankfurt O.), that monitors by a micro-electro-mechanical system the viscosity of a sensor-liquid (< 30 mA: 100 ms) which is proportional the glucose-concentration. With as realistically assumed duration of active-inactive-phases, a residence time of 2 years in the body can be calculated. The required total-energy then amounts to about 850 mAh.

LITERATUR: Basmer, T.; Kulse, P.; Birkholz, M. (2010): Systemarchitektur intelligenter Sensorimplantate. Biomedical Engineering/Zeitschrift für Biomedizinische Technik 55, 43–46. De Gruyter, Berlin.

Sensor-actor-regulated knee joint prosthesis

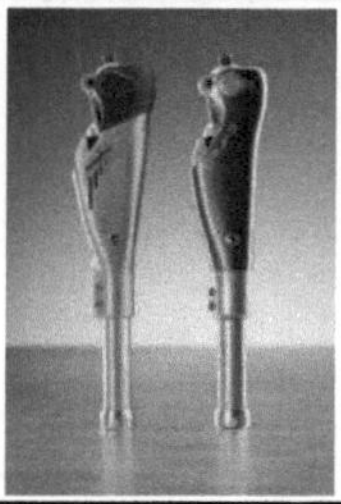

A

B

A patient can be unburdened bionically from the conscious supervision of its knee joint leg prosthesis.

BIOLOGY AND PROTHETIK:
Together with inherent damping devices, neuro-motoric feedback control system provide an unconscious steering and supervision of the knee-joint-system, so that a secure standing and fluent walking results. If these are missing, as it is the case using purely mechanical leg prostheses with hydraulic or pneumatic damping, a person must monitor attentively and consciously the prosthesis position and its lapse of time.

PRINCIPLE OF A BIO-ANALOGOUS SUPERVISION:
Sensory systems grasp the bio-mechanical relevant data in every stand or walk-phase. The typical angles for the individual walk and moments are measured within 20 ms-intervals and the hydraulic movement-resistances are calculated. By the sensor-output, servomotors adjust the hydraulic valves so, that a secure stand-phase and fluent extension and inflection of the artificial knee-joint are the result.

TECHNOLOGY:
The monocentric knee joint construction C-Leg® from Otto Bock HealthCare/Duderstadt consists of a Carbon frame with a knee-angle-sensor, hydraulics, servomotors, electronics and accumulator. It is positioned at the thigh-stump above and carries below a tube-adapter with strain gauges for the torque measurement. The control of the damping of the stand-phase (A) and the movement of the swing-phase (B) can be regulated finely.

LITERATUR: Hafner, B. J. et al. (2007): Evaluation of function, performance and preference as transfemoral amputees transition from mechanical to microprocessor control of the prosthetic knee. Aus Studien zum C-Leg® (1998–2007). 2. Aufl. –www.ottobock.de.

Biomimetic ceramics and new implants

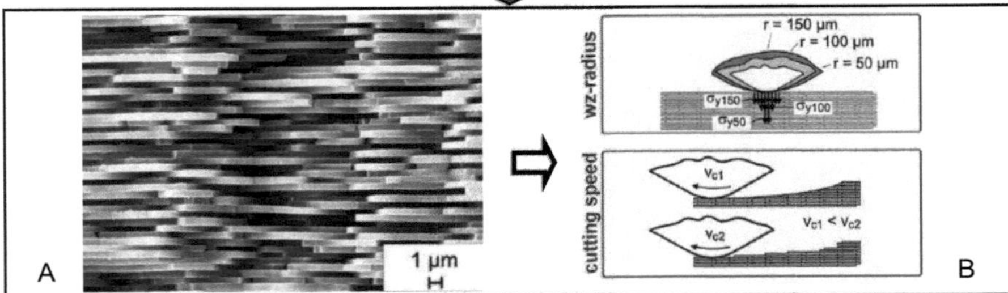

Biomimetic hybrid materials according to the prototype of *Haliotis* can improve the elastic formability of implants.

BIOLOGY:
The mother-of-pearl layer of the peel of sea-ear-snails of the species *Haliotis* consists of hard polygonal aragonite-platelets ($CaCO_3$) that are embedded into a soft organic matrix and can be moved a little during a load. Breaking edges show the construction (A). Material-hardness is combined with -tenacity. Micro rips end at the contact-border of the materials (comp. p. 59). Application-oriented tests were outstanding until now.

DETAIL-EXAMINATION:
In the subproject D9 ("biomimetic ceramics") of the SFB 599 of the DFG, such tests are executed comparatively to understand functionally the biomechanical construction on the one hand – the fracture toughness of the aragonite-layer in the shell is 3000-x higher than this from pure aragonite! – and on the other hand as basis for the development of bionic composites or hybrid-materials from inorganic nano-particles in organic polymer-beddings.

MEANING:
Materials for tooth implants must be hard. However, they break easily with high point-loads – bite on a hard cherry-kernel. Materials for bone-implants must be mechanically workable. Through different parts and orientations of the named components both parameters can be changed. Also the stress-distribution and the surface-change induced by tools can be recorded in the tests (B).

LITERATUR: DFG (2010): SFB 599, Teilprojekt D9 (Biomimetische Keramiken). – Denkena, B.; Koehler, J.; Moral, A. (2010): Ductile and brittle material removal mechanisms in natural acre – A model for novel implant materials. J. Mat. Proc. Techn. 210, 1827–1837.

Bone material interaction with an endoprosthesis

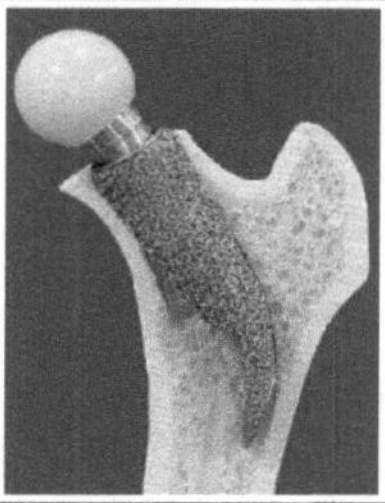

⇨

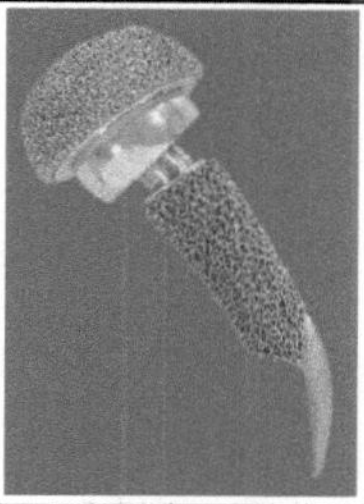

A B

Cement free hip endoprosthesis is based on safe growing of bone material at and into the prosthesis-surface.

BIOLOGY AND ENDOPROTHESIS:
Bone substance is able to grow at and into suitable surface-structured endoprosthetic-materials (A). By this contact osteogenesis, a biological-technical integral structure can be built, that also provides a safe fixing without application of cement containing binders and further a long implant-fixation with good long-time-stability. Prerequisite is a good operative primer fixation by press-fit.

SECURITY:
The fixation is more secure, if the peculiarities of the bone-growth have been taken into account during the design of the implant. So, beside favourable material-qualities (alloys), a "spongiosic" metal-surface (B) is advantageous for a connection with the bone-spongiosa since this can grow into the cavities. Manufacturers reached and reach such surface-formations with different technical approaches.

TECHNICAL DESIGN:
Classic surface structures show for instance skeletons (Mittelmeier), sintered balls (Lord) or cuttings ("Fibre-mesh" after Harris-Galante). In the modern endoprosthesis of the type CUT, A® from ESKA (A), both the prosthesis stem fielding in the femur and the hip-pan fielding into the pelvis carry a metal spongiform structured surface (technique after Grundei) (B), in which subsequently bone-spongiosa grows in.

LITERATUR: Diehl, P. et al. (2010): Zementfreie Hüftendoprothetik: Eine aktuelle Übersicht. Biomed. Techn. 55, 251–254. De Gruyter, Berlin, DOI 10.1515/BMT.2010.037. – Ender, S. A. et al. (2006): Schenkelhalsendoprothese Typ CUT. Der Orthopäde 8, 33/11, 1243–1248. Springer, Berlin.

Bone material interaction with [illegible]

[illegible] bone material [illegible] into the [illegible]

[illegible]

[illegible]

[illegible]

PROCEDURES

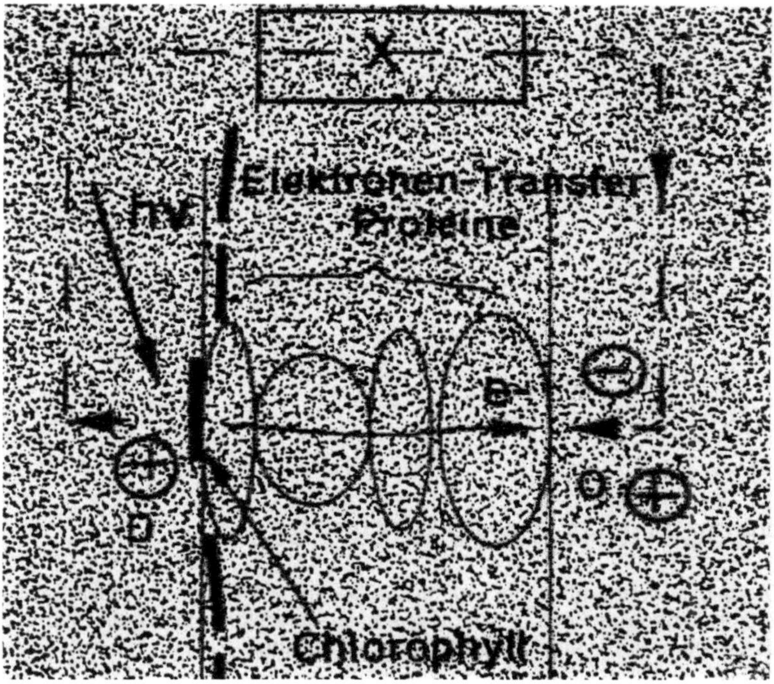

AND PROCESSES

W. Nachtigall and A. Wisser, *Bionics by Examples*,
DOI 10.1007/978-3-319-05858-0_11

PROCEDURES AND PROCESSES

Organic bio-energetics (incl. photosynthesis) – Artificial photosynthesis (incl. hydrogen-technology) – Passive solar-utilization – Wind-utilization – Use of geothermal energy and geothermal cooling.

One can not only examine carefully natural constructions for their technical usability but also procedures, with which the nature steers the processes and activities for their particular advantages. One of the most essential prototypes is the photosynthesis with reference to a future hydrogen-technology. Further, aspects of the ecological research could be examined at a big profit with respect to the control of complex industrial and economic ventures. Finally, the natural methods of total recycling and complete avoidance of dump material are worth to be carefully examined in great details for their transferability. Additionally, all procedures, that have to do with surfaces, adhesion, non-adhesion, antifouling, self-cleaning and drag-reduction, are to be carefully studied. Furthermore, also energy- and industry-plants and their procedures to manufacture high-energy products belong to this section.

Further development of vault structures

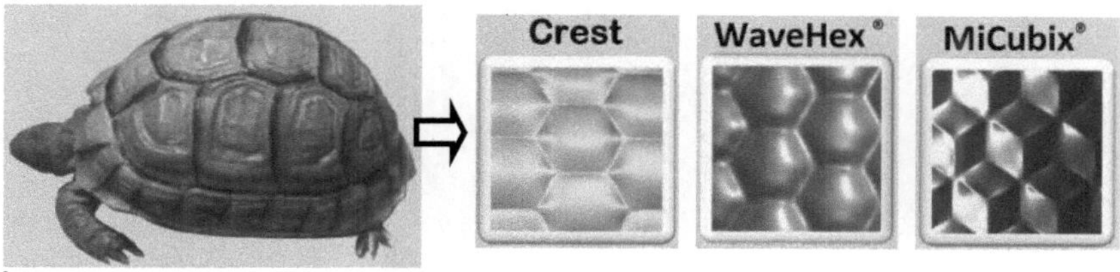

A B

Changes of the boundary conditions in the vault structuring led to important modifications for new applications.

FURTHER DEVELOPMENT:
Recently, "Hexagon" of the Dr. Mirtsch Ltd., whose basic structure is similar to the turtle shell (A), was developed further to the structures "Crest", "WaveHex$^{®}$" and "MiCubi$^{x®}$" (B) which have different applications. One can speak of a regular evolution of vault structures and 3-D-facet structures, which can now be applied to processes and tools as well as to materials (metals, plastic, cardboard / paper).

CRITERIONS FOR FURTHER DEVELOPMENT:
The development was based on criterions, that reflect the benefits of nature with respect to information, energy, time, space, structure and material:

- consistent application of the energy minimization principle,
- utilization of the controlled self organization (information),
- utilization of the synergetic qualities of structures,
- modification and integration into technical production-processes.

EXAMPLES FOR NEWER APPLICATIONS:
Successful secondary, structure-oriented transformations as edge processing and edge bending smoothed the way for practicable new applications:

- vault structured roofs (sport-palace Odessa),
- vault structured, perforated-plate-facades (skyscrapers New York),
- vault structured stair rails (citizen-house Teltow),
- vault structured lighting fittings (Siteco Hexalleuchte) etc..

LITERATUR: Mirtsch, F. (2011): Vom Schildkrötenpanzer zum Leichtbauelement – Effektivere Formenbildung versteifender dünner Wände durch kontrollierte Selbstorganisation. – Wölbstrukturen. Kongress Nanotechnik und Bionik – Hightech i. d. Bauwirtschaft, München.

Adaptive cross-stream-filtration as in sponges

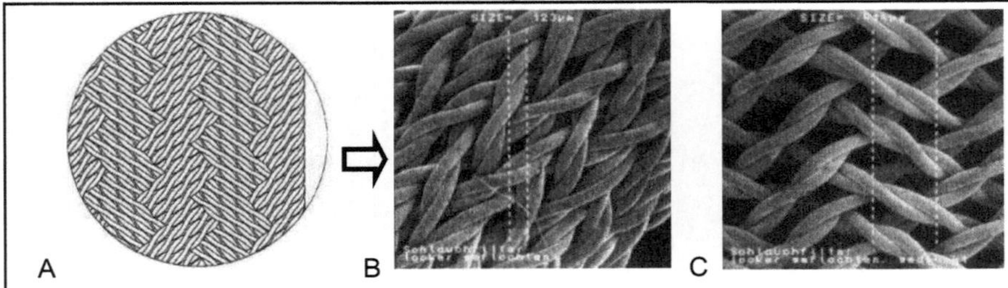

Similar to sponges filtering out particles through variable pores, the principle can be transferred to technical fibre-filters.

BIOLOGY:
Sponges, especially freshwater-sponges, are made up of a porous structure that consists of chambers, connecting passages and "chimneys". Contractile cells (myocytes) in these passages can change the pore size of this system thus regulating the water flow-rate. Sponges live on microscopic particle, which they filter out of the water-stream. The flow-rate can amount up to 50 percent of the body-volume per second (!).

PRINCIPLE:
In sponges, only a low pressure-difference from 10^{-4} to 10^{-5} bars is needed for the indicated type of perfusion, which is generated partially actively by ciliary motion, partially passively by the fluid dynamic type of the circulation. On the other hand, technical filters normally generate a large drag, whose overcoming requires a larger pressure difference. Occasionally this can be reduced by the adapted increase of the pore diameters.

TECHNOLOGY:
Filters of filamentous tissue were developed by application of twisted mono-filament-threads on the basis of a standardized braided-hose (A). This can be stretched and relaxed by tangential stress (B, C), whereby the pore-size is changed and with it the amount of the particles filtered out of a water stream. With a given filter setting, the filtrate depends on the initial distribution and the diameters of the particles.

LITERATUR: Scherrieble, A.; Stegmeier, T.; Arnim, V. von (2006): Umsetzung von Oberflächeneffekten mit faserbasierten Werkstoffen. Bionik-Industriekongress. BIOKON, Berlin. S. 95–115.

Fog collectors in animals/plants and their realization

A

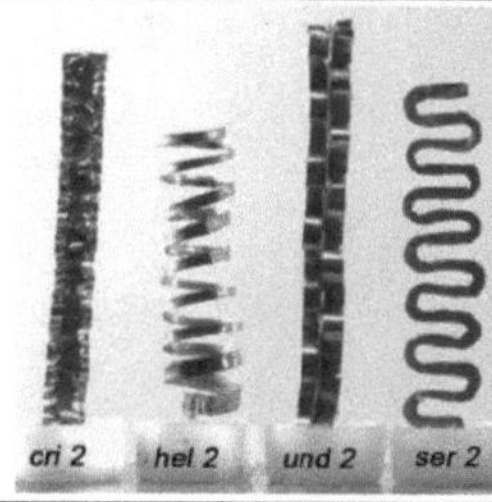

B

Novel fog collectors according to the desert bug principle and models from helical desert plants were tested.

FOG-NETS DEVELOPED:
This institution, described on p. 65, works with simple wire-nets and has a limited crop. A particular micro - and nano-structured 3-D-tissue, developed by the ITV Denkendorf in cooperation with the Inst. for Geosciences/University Tubingen, with the solar-energy Stefanakis/Stadecken-Elsheim as well as with Mattes & Ammann/Meßstetten, increases the crop onto the triple, namely approximately 3 (Namib) and 55 (South Africa) l $H_2O/(m^2$ day).

WEIGHING UP OF LOW-TECH- AND HIGH-TECH FOG COLLECTORS:
The simple low-tech wire-nets are cheap and easily to obtain. They always will be held clean on each side by the condensed droplets and will deliver little, but sufficiently good, germ-poor water. Higher profit can be obtained by longer fog fences. The high-tech tissue is expensive and must be imported; with a thickness of 2 cm, internal sterility must probably be proved first. However, the crop is much better.

CURVED GROWN DESERT GEOPHYTES:
In the semi-deserts of the cape-regions, the geophytes of different families, dependent from the dew-utilization, usually grow in a spirally bent or a serrated form (A, B). As a result, the dew-collecting surfaces and extension of the edges are maximized. Analogously sculptured metal-models collect in comparison with not sculptured models of the same size up to 66 % more moisture. A technical verification and realization didn't take place up to now.

LITERATUR: Nebelfänger nach dem Vorbild des Wüstenkäfers. Denkendorfer News, Sept. 2011, 5. – Vogel, S.; Müller-Doblies, U. (2011): Desert geophytes under dew and fog: The „curly-whirlies" of Naquamaland (South Africa). Flora 206, 3–31. www.elsevier.de/flora.

Biological glue as basis for new developments

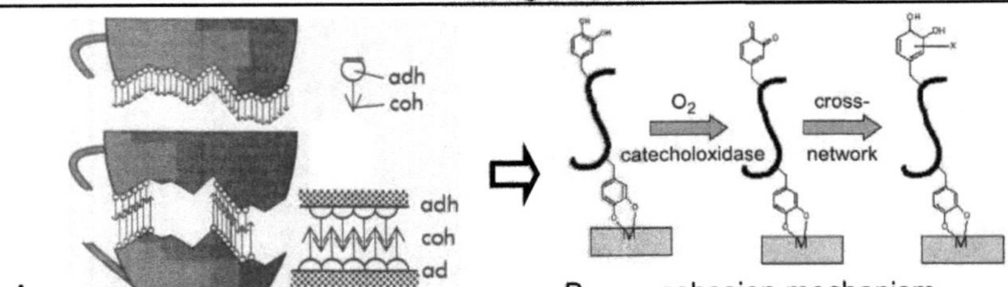

A B cohesion mechanism

The way, how mussel-larvae stick themselves to slimy underground, can lead to novel glues.

HISTORY AND VARIETY OF TECHNICAL GLUING:
Gluing is an ancient technology; already the Egyptians used animal-glues and tree-resin to stick objects together. There are synthetic glues since 1870, commercial glue in big standard since 1923 (Co. Henkel). Filling-glues overcome clefs. Reaction-glues cure over chemical reactions, for example cyanoacrylate by reaction with humidity, 2-component-glue by internal reaction. Further, there are spray-glues and adhesive foils.

PRINCIPLE:
Glue joins with the surfaces to be stuck together by adhesion and holds internally together by cohesion (A). However, glues very frequently are solvent-containing. They bind off by evaporating the solvent. This costs time and can be unhealthy. Hot-melt adhesives ("hot-glue", up to 300° C) cannot be applied universally. A bio-inspired development of novel glues therefore has good market-chances.

"MUSSEL-GLUE" AS MODEL:
- *Mytilus edulis foot-protein:* The mussel-glue works by the interaction of DOPA-containing proteins with a catecholoxidase and basic protein-groups (B). It is imitated organic-chemically.
- DNA-*immobilisation in the micro-biology:* H.O. Ham and collaborators of the Northwestern University in Evanston/USA have coated supporting structures (glass etc.) with such a foot-protein, at which DNA-molecules then dock safe.

LITERATUR: Anonymus (o. J.): Klebe-ABC. Werbeschrift der Firma Henkel, Düsseldorf. – Ham, H. O. et al. (2011): Facile DNA immobilization on surfaces through a catecholamine polymer. Ang. Chemie 50; © Wiley-VCH. Nachdruck mit frd. Genehmigung.

Hollow-fibres for self-repairable composites

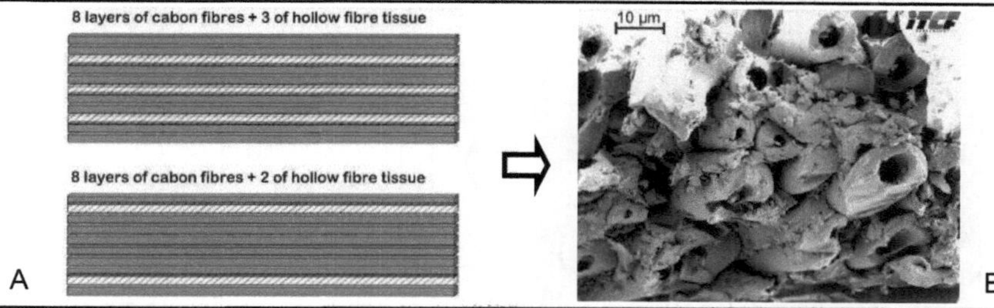

Micro rips in highly stressed composites can heal by substances that emerge from broken hollow-fibres.

COMPOSITE MATERIALS:
Usually, "fibre composite materials consist of a polymer-matrix (usually epoxy or polyester-resin), and high-module-fibres. Additionally, fibreglass (GFK), polyamide-fibres (AFK) or, in the high-price-segment, carbon fibres (CFK) are used according to the market-segment and the request-profile. Those materials are used in cases, in which either weight-savings and/or energy-savings are of importance (comp. p. 71)."

DANGER OF MICRO-CRACKING:
"Composite-materials usually suffer from micro rips in the polymer-matrix by mechanical load. When these structure-deficits dilate to a macroscopic crack, the fibres either become sheared by the break of the matrix, or it comes to a delamination between two fibre-layers, and the component breaks or tears. Until now, the failure behaviour ... is to be predicted only insufficiently, therefore the components mostly are oversized".

PROCEDUR OF SOLUTION:
"Hollow-fibres were filled with a cross linkable monomer. From these the hollow-fibres, tissue-layers are produced. A tissue-layer of hollow-fibres containing cross linkable monomer is incorporated (A) between several reinforcement-fibre-bundle-layers of a CFK-laminate. When a crack occurs in the CFK-laminate, the hollow-fibres are damaged (B). So the monomer leaks out into the micro rip and cures by polymerisation with a catalyst."

LITERATUR: Zitate nach: Frank, E.; Neumann, J. (2011): Projekt „Bionische Hohlfasern mit Monomerenkern für selbstreparierende Verbundwerkstoffe". AIF-Nr. 15777 N/1 der ITCF Denkendorf.

Analyses concerning to gecko inspired adhesion

A 

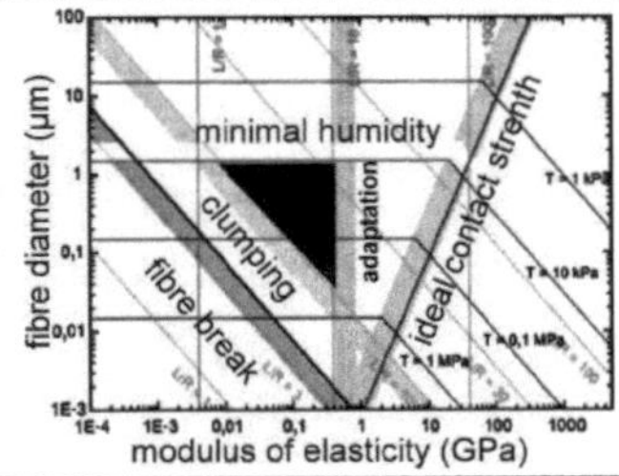

B

Only multi-disciplinary basal research can lead to industrial applications and can surpass even the nature.

BIOLOGY AND PROBLEMS OF IMPLEMENTATION:
The adhesion principle of the toes of Geckos was discussed on p. 47. The Gecko is the heaviest animal, that is sticking on smooth surfaces mainly according to the Van-der-Waals-principle; accordingly, its adhesion mechanism is well examined. However, the realisation can simply not "imitate the Gecko." It must rather take into account not only different materials and manufacture-procedures but also different technical request-profiles.

NECESSITY OF DIFFERENT APPROXIMATIONS:
E. Kroner et al. from the Saarbrucken Leibniz-Institute for New Materials have pointed out the different aspects of the named problem and the necessity of a multidisciplinary approach. Two problems of many: 1. Van-der-Waals-forces work also *between* artificial setae, they clump these (A) and so reduce the strength of adhesion. 2. According to request, different areas are to be headed for in the technical adhesion-diagram (B).

DEVELOPMENT-LINES:
The authors mention several lines of future research: Experiments with different surface-roughness under controlled temperature and moisture-conditions, an exact analysis of the adjustment between structural realities and adhesion by temperature-, moisture-, electric- or magnetic-effects, switchable adhesives as well as theories of contact. In summa, it could well be possible to surpass the nature.

LITERATUR: Kroner, E.; Kamperman, M.; Arzt, E. (2011): From science to industrial application. Adhesives & Sealants 1, 2–6. – Arzt, E.; Gorb, S.; Spolenak, R. (2003): From micro to nano contacts in biol. attachment devices. Proc. Nat. Acad. Sci. USA 100(19), 1063–1066.

An adhesion tape from the front-tarsi of water bugs

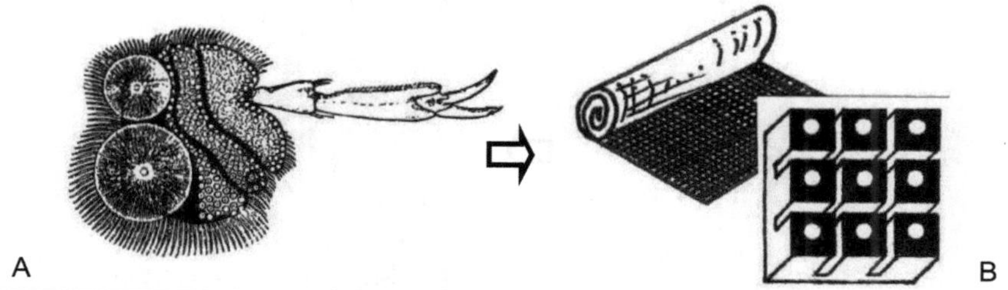

The front-tarsi of the great diving beetle work as digitized adhesion surfaces with multifunctional, adhesion elements.

BIOLOGY:
The males of water bugs of the genus *Dytiscus*, up to 4 cm long, to which the great diving beetle (*Dytiscus marginalis*) belongs, possess on the first foot-limb (Protarsus) of their front-legs widened adhesive-organs, with which they hold on to slippery loot and pin themselves at the smooth pronotum of the female during copulation. Among other things, these contain a field of about 120 stalked miniature suckers (A), which simultaneously can attach adhesively and by secretion.

PRINCIPLE:
The adhesion-surface doesn't work uniformly according to a technical adhesion-tape but it is digitised, that is dissolved into individual adhesion elements, of which not each individual needs to be attached. The suckers carry a wide adhesion ring around the suction-deepening and have free gaps between themselves, that can be filled with a greasy secretion. The adhesion elements therefore are multi-functional (sucking, wet adhesion and a type of gluing).

TECHNOLOGY:
A digitalisation of technical adhesion-ribbons or adhesion-tabs (B), that is the dissolution into single-elements, means, that not every surface-element needs to be attached. It is enough if a certain number of adhesive-elements work by chance. The multi-functional design increases the number of the adhesive-possibilities and with it the adhesive-security. Such ribbons or tabs could be adhesive on greasy or watery ground and could apply pharmacies.

LITERATUR: Nachtigall, W. (2005): Animal attachments: Minute, multifold devices. Biological variety – Basic physical mechanisms – A challenge for biomimicking technical stickers. In: Rossmann, T.; Tropea, C. (Hrsg.): Bionik. Aktuelle Forschungsergebnisse ... Springer, Berlin.

Bone-analogous metal-foams for impact-protection

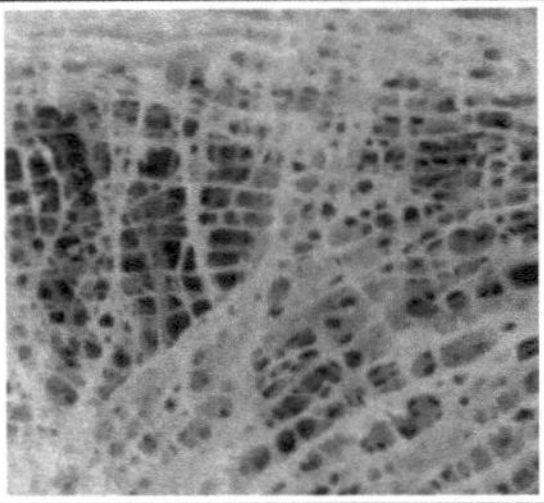

A

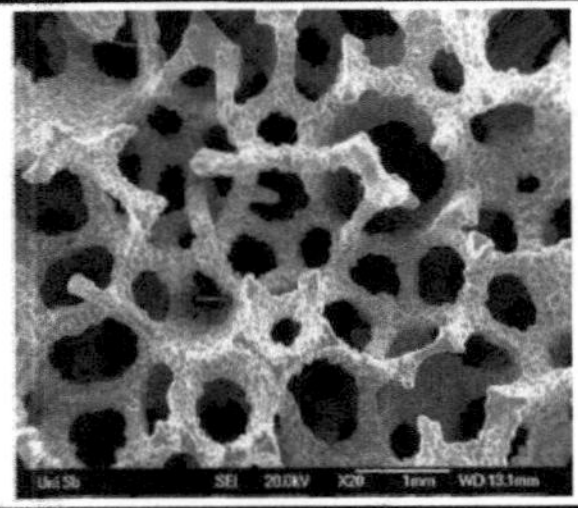

B

Like spongeous bone + periosteum, metal-foams + coatings can absorb high energies: Effective crash protection.

SPONGEOUS BONE STRUCTURE IN BIOLOGY AND TECHNOLOGY:
The spongeous bone material of long bones runs along stress trajectories, s. p. 23. On the other hand, in small volumes the bony beams appear directed by chance (A). They correspond to porous aluminium-foams (B). Both are distinctively pressure-proof, however, also somewhat tension-poof and can absorb remarkable energies. Therefore, spongeous bone and lightweight-constructions of aluminium-foams work as shock absorbers.

ENVELOPES (COATINGS) IN BIOLOGY AND TECHNOLOGY:
Bones are wrapped by a tough flat structure, the thin periosteum. This strengthens the bone-material and absorbs pushes, so that the total-structure can absorb more energy than the bone material alone. An analogous effect emerges with an only 1/10 mm measuring coating of the fine bracing elements on metal-foams. It consists of nanocrystalline nickel and is applied electro-chemically. This coating is astonishingly effective.

COATING-EFFECT:
According to R. Hempelmann/University of the Saarland, the thin nickel coating increases the energy absorption capacity of a porous aluminium-foam not less than 10 times. The author compares this effect also with the influence of the tough epidermis, that wraps a bamboo-stem. The new material could be put in as shock absorbers. Pore-filling with a polymer-foam additionally increases the compression strength.

LITERATUR: Jung, A. et al. (2011): Nanonickel coated aluminum foam for enhanced impact energy absorption. Advanced Engineering Materials 13(1–2), 23–28; DOI: 10.1002/adem.201000190.

Tissue of nature fibres for earthquake protection

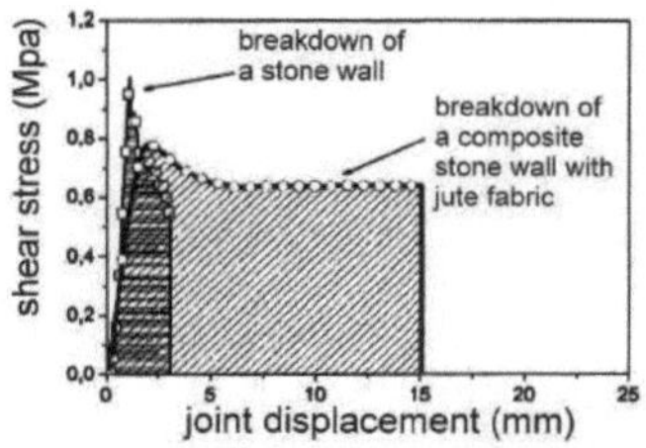

A

B

In poor regions in Iran, the fitting of old rice-sacks made of natural "jute" makes stonework more earthquake proof.

BIOLOGY:
"Jute" is a nature-fibre for the production of interlaces of all the type, from canvases up to coarse fibre-tissue for rice and potato-sacks. The material consists 1,5 to 2,5 m long bast-fibres of tropical linden-plants, semi-shrubs of the genus *Corchorus*, that usually grows in flood-regions. Its fibres can be gained in native production. They combine tensile strength with weathering insensibility.

PROBLEM:
Walls from air-dried clay-bricks, sometimes also burned bricks, are in use in poorer arid regions. Their limited tensile, bending, and shear strength make them earthquake-susceptible. By connection with a material, in which the indicated parameters are more favourable, the system can be made more earthquake-proof. It should be a material, which is as cheap as possible and producible in native production. That would be i.e. jute mats.

SOLUTION:
At the University Kassel, the mechanical composite-behaviour of such systems were examined. It turned out, that a stone wall already failed at 2 mm joint-shift, a wall-jute-system, in contrast, not until 15 mm (A). For testing, small norm-houses were produced in the Persian Arg é Bam (B). In these, the inner wall was laminated with jute-tissue, additionally covered with a mineral filled two component epoxy resin.

LITERATUR: Emami, A. D. (2011): Kleben von Naturfaserverbundwerkstoffen auf Mauerwerk zur nachträglichen Verstärkung erdbebengefährdeter Bauwerke – von der Werkstoffprüfung bis zur Anwendung. Diss., Univ. Kassel, Schriftenreihe Baustoffe und Massivbau.

On the way to an "artificial leaf"

A exhibition poster

after a slide from a talk

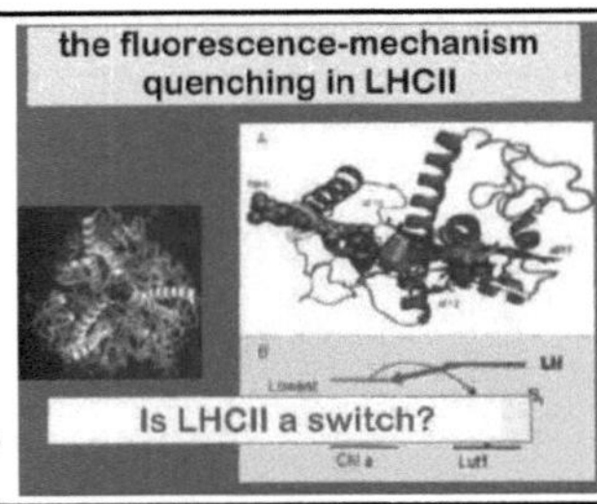

B

A solar-powered hydrogen-production can learn from the green leaf; however, the product will look different.

CONCENTRATION OF RESEARCH FOR ARTIFICIAL PHOTOSYNTHESIS: "Germany is leading in the organic photovoltaic, however, in the photo-catalysis-research only a few are active" (F. Würthner, Wurzburg). At a meeting (A) in Berlin, the state of research (B) was discussed. It is "a system that collects light in any way and produces an energy-rich compound directly, that is without passing through an intermediate stage, as it is the case for example in electricity from solar cells" (A. Holzwarth, Mulheim).

PRESENT MAIN-STRATEGIES:
How does the photo-reception work, which are the first reaction-steps for light-reception and how can one copy them? Three approaches are pursued: "Either the components are built completely artificially, or parts of biological systems are embedded into synthetic materials", or microbes were changed by genetic engineering so "that they become efficient, light-driven fuel factories" (W. Gärtner, Berlin).

ARTIFICIAL LIGHT-CAPTURE-SYSTEMS:
Imitating the antenna-pigments of the chlorophyll, artificial antenna-molecules were already produced (A. Holzwarth, F. Würthner). Hereby one went back to molecular self aggregation; the molecules settle down together to "active light antennas automatically". At present, the resisting to aging is problematic. Now, for the necessary follow-up steps – catalytic cleavage of water – cost-effective catalysts are necessary.

LITERATUR: http://www.biotechnologie.de/BIO/Navigation/DE/root,did=128918.html. – Graf, P. (2011): Photosynthese-Forschung: Das künstliche Blatt als Spritquelle. – S. auch Mttg. MPI. Mülheim, http://www.vbio.de/informationen/alle_news/e17162?news_id-12310.

Bionic light antenna for artificial photosynthesis

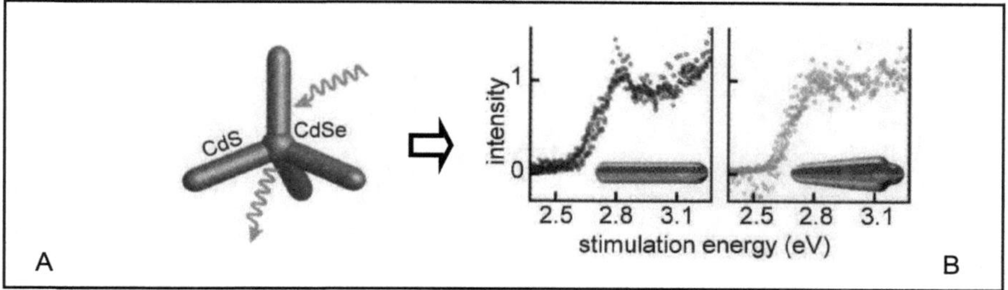

Basis for an artificial photosynthesis is the technical reconstruction of biological photon absorption systems.

BIOLOGY:
Antenna-pigments for example in the chlorophyll of the green plant are sitting on the thylacoid membrane. In the outside-area they consist of molecules of the photo-system II, that form triplets (trimers). The connection with the reaction-centre represents another trimer, together with elements of the photosystem II an and further elements. Such structure-chains are already present within the cyanobacteria.

PRINCIPLE:
Antenna pigments absorb incident light-quanta and pass on their energy through stimulation of neighbouring pigment-molecules as far as to the reaction-centre. Direct impacts of quanta on this centre are on the other hand extremely unlikely. The antenna-pigments therefore provide a drastic increase of the probability of an energy-transfer. Artificial photosynthesis needs corresponding light-collective-complexes, too.

TECHNOLOGY:
J. Lupton from the Inst. for Exp. and Applied Physics of the University Regensburg and his team developed a bio-analogous artificial light-collective-complex on semiconductor-basis. It consists of 4 "antennas" of cadmium sulphide and a core of cadmium selenide (A), that grows in a solution. Hereby efficiency depends on the geometrical aggregation of the particles; resultant club-shaped structures absorb clearly better than small rod-like (B).

LITERATUR: GIT Labor-Fachzeitschrift (2011): Künstliche Photosynthese mit formvollendeten Lichtantennen, Heft 1/. – Bolys, N. J. et al. (2010): The role of particle morphology ... Science 330 (6009),1371–1374. – Lüttge, U.; Kluge, M.; Thiel, G. (2010): Botanik. Wiley, Weinheim.

Nano nipples on butterfly eyes and thin film solar cells

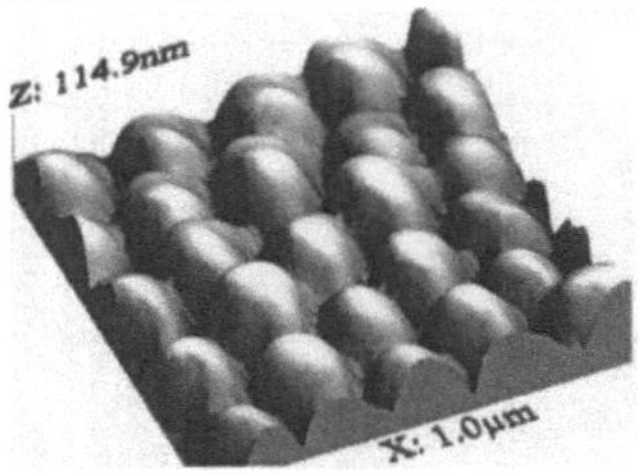

A

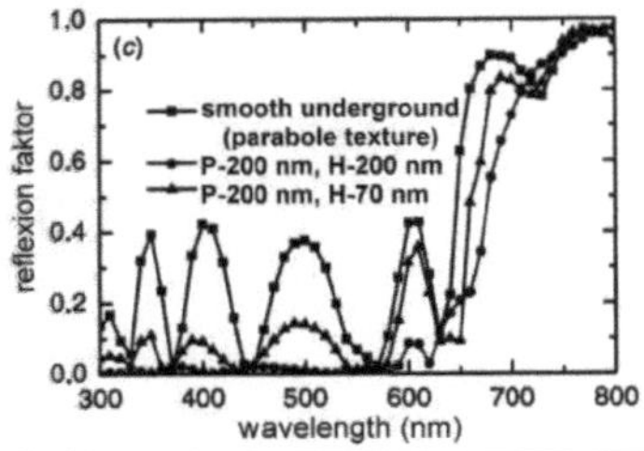

B

The transfer of biological wide-band anti reflection qualities to thin film cells increases their efficiency by 40 percent.

BIOLOGY:
The cornea of the eyes of many moths is coated (A) with directional series of nano nipples. In the leaf miner moth (*Cameraria ohridella*) these are 200 nm broad and 70 nm high. According to models, their reflection factor R depends on the nipples' form and height and on the wavelength. In comparison with a smooth surface (R = 0.043) it can sink in the UV- to the green-region to 0.005, so that the eye of the moth gets considerable more light.

DEPENDENCE ON THE ANGLE OF INCIDENCE:
The values are applicable to vertical irradiation (inclination-angles α of the nipples compared to the direction of irradiation 0°). Further, the dependence of the reflection-factor of α was tested for parabolic modelled nipples of different height. Practically no alteration resulted up to $\alpha \approx 30°$, after it a more steep increase to 0.150 at $\alpha \approx 70°$. The system therefore works approximately uniformly effectively at not too large oblique irradiation.

TECHNOLOGY:
The results were transferred (B) on a coating for thin film solar cells on the basis of amorphous Si. Whilst the degree of reflection of smooth coating had four inter-maxima between UV and IR until $R \approx 0.4$, it remained very low $0.01 < R < 0.02$ for a height of 200 nm. In this area, the quantum-efficiency then rose with smooth course (without inter-minima) up to by approximately 0.9 (!), the short-circle current from 12.00 to 17.65 mA cm^{-2}.

LITERATUR: Dewan, R. et al. (2012): Studying nanostructured nipple arrays of moth eye facets helps to design better thin film solar cells. Bioinspiration & Biomimetics 7(015003), 1–8. IOP Publishing Ltd.

Membranes for auto adaptive gas passage

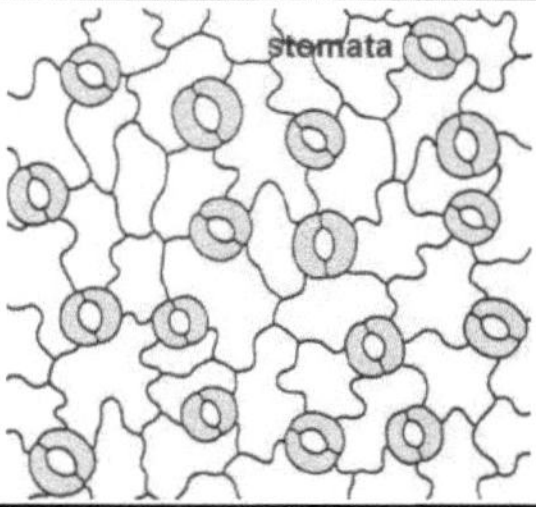

A

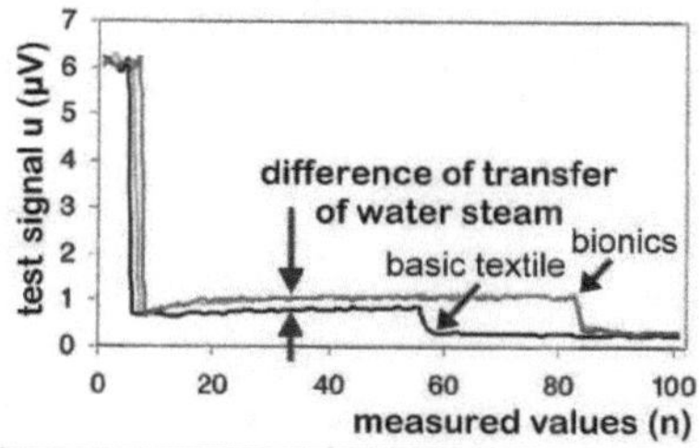

B

The water vapour passage can be regulated by self-adjusting pores in a membrane.

BIOLOGY:
Stomata (slit openings) are found in the membrane-like outer cell-layer of plant-leaves (A). These regulate the passage of CO_2 from the outside-air into the inside of the leaf by different pressure-production in their cellular mechanism for closing; the CO_2 contained in the outside-air is a factor of deficiency for the photosynthesis. Inevitably, a loss of moisture from the leaf-inside to the outer atmosphere is connected with this function of stomata.

CRITERION OF TRANSFER:
Within functional textiles, it would be desirable, for example, to change the pore-wideness with the body-temperature. If a person gets more and more hot, in general, it sweats more and more, and specifically then, the resulting water vapour should be carried off faster. Also here, it would be a type of connected regulation: Trigger would be the increase of temperature or humidity, however, a stronger passage of water vapour would be connected with

TECHNOLOGY:
In cooperation of the Institute for Textile- and Procedure-Technology Denkendorf with the Institute for Geosciences of the University Tubingen one tries to implement auto adaptive pores to textile "membranes" (fine-tissues). Already, measurements of the transport of water vapour with overlaid transport of heat showed noticeable effects, when comparing the unchanged basis-textile with such bionic changed substrata (B).

LITERATUR: Scherrieble, A.; Stegmaier, T.; Arnim, V. von (2006): Projektpräsentation des ITV Denkendorf. Bionik-Industriekongress „Innovationsmotor Natur". BIOKON, Berlin, S. 95–115.

"Antifouling" without chemistry - a bionic approach

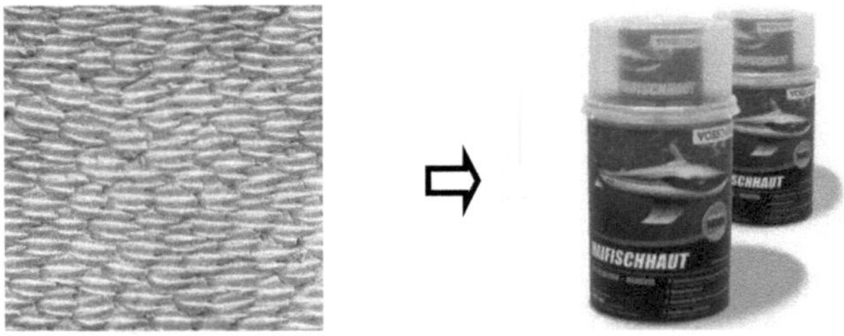

A B

It is not longer allowed to fight drag-increasing, fuel swallowing ship fouling by toxic coatings.

BIOLOGY:
Sea-smallpox (*Balanus spec).*), Barnacles (*Lepas anatifera*) and other sessile crustaceae are mobile as larvae. These take root on each suitable surface, cement themselves at and grow in groups to an all coating fouling. Not attacked are for example bone-fishes because of their mucus coating and sharks because of their micro-structured surface (A), the scales of which are embedded elastically.

PRINCIPLE:
Bone-fishes are not suitable as bionic models for an anti-fouling ship-coating, because their mucus must be subsequently delivered. The observation that washed up dead small sharks are without fouling brought A. Kesel the idea to develop a shark-analogous technical coating. Various mechanical surface structures were tested; however, at least, a permanently elastic body-coating was successful.

TECHNOLOGY:
In the project "Poison free antifouling according to a biological model" of the HS Bremen, that ran out in 2011, five scientists researched at non-toxic surface coatings to avoid fouling. A first product was marketed by the Vosschemie/Uetersen (B). This permanently elastic coating gets its bio-analogous micro structuring by suitable additional materials and self organization during drying.

LITERATUR: HS Bremen (2011): Giftfreies Antifouling nach biologischem Vorbild. www.hs-bremen.e/internet/de/forschung/projekte/detqail/index-1. – Europ. Patentanmeldung Antifouling coating EP 06018001.5. – S. auch Liedert, R. und Kesel, A. zum Thema.

Lipids of Archaea: Antifouling and self cleaning

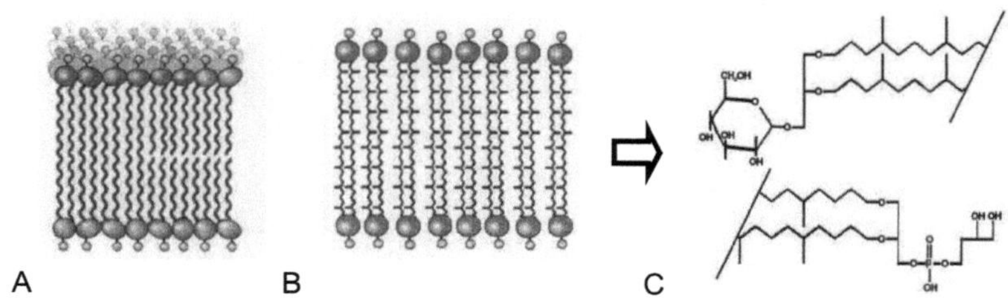

A natural membrane phospholipid (MPL), insensitive against depleting, adheres at technical surfaces and prevents fouling.

BIOLOGY:
The thermo-acidophilic species *Thermoplasma acidophilum*, (T_{max} = 90° C, pH_{min} ≈ 05), which belongs to the Archaea (Archaebactera), does not have a cell-wall. The cell-membrane consists of relatively thermo-stable compounds containing membrane-crossing, amphiphile tetraether lipids (A), that stabilize themselves under fluidic condition by pentacycles (B). As the main phosphor lipid, as shown in fig. C, a depleting insensitive molecule can be extracted.

PRINCIPLE POSSIBILITIES:
Because of its sugar as well as its glycerol phosphate end groups this molecule can be densely packed and coupled to technical surfaces. By functionalisation of the outer head-groups, it can be further adjusted to specific requirements. An anti-adhesion-effect can be reached because of the sterically caused repulsive forces between coating and bacteria. By modification of the free end groups a self cleaning effect can be attained, too.

TECHNOLOGY:
The effect was applied on the one hand already in peritoneal dialysis catheters to minimize catheter-induced infections and to extend stand-times. It is quite generally applicable in the medicine-technology. On the other hand, it was successful on surfaces of sensors for the environmental technology, specifically for monitoring of waters. It reduces the formation of strongly disturbing bio-layers by formation of a sterical water-barrier there.

LITERATUR: Liefeith, K.; Rothe, U.; Dölling, K. (2006): Archaeale Tetraetherlipide und ihre Anwendung in der Antifoulingbeschichtung. Bionik-Industriekongress Bionik.BIOKON, Berlin. S. 118–121.

New bionic antifouling research I

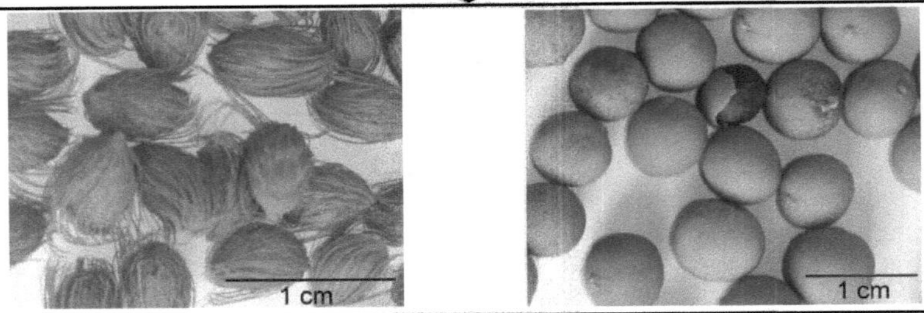

A B

The colonization of technical surfaces by organisms leads to corrosion and, at ships, to an increase of drag.

STRUCTURED BIOLOGICAL SURFACES:
The immense economic damage due to these effects forces increased further research. The Biona-project "Bio-inspired of antifouling" is located under the supervision of A. Kesel at the HS Bremen. Meanwhile (comp. p. 200) 12 species were found with reduce fouling, which are examined for their surface design and antifouling strategies. These are among others floating seeds of *Dypsis rivularis* (A) and *Acoelorrhaphe* wrightii (B).

ANTI-MICROBIAL SUBSTANCES:
Especially at ships' hulls, antifouling by paintings containing broadband toxins, e.g. TBT, was widespread. Since 2003, however, such coatings are forbidden. The development of toxic free coatings is therefore of great importance. Some bio-based substances with anti-microbial effect were found, further, their effectiveness and environment-impact are now being tested in standardized laboratory- and in field-studies.

ECONOMIC IMPORTANCE:
Both because of the aforementioned ban and due to an expected ban of copper containing paint and eventually because of the increasing shipping traffic and the pressure of competition to save fuel, new methods for drag reduction by antifouling are of eminent economic importance. There is a worldwide search for such methods, especially bio-based ones. Also corrosion reduction of "inner surfaces" of closed systems is significant (comp. p. 203).

LITERATUR: BMBF (2011): http://www.bionische-innovationen.de/#media.html. – Liedert, R.; Kesel, A. B. (2007): Giftfreies Antifouling nach biologischem Vorbild. In: Kesel, A. B.; Zehren, D. (Hrsg.): Bionik: Patente aus der Natur. B-I-C, Bremen. S. 99–106.

New bionic antifouling research II

A
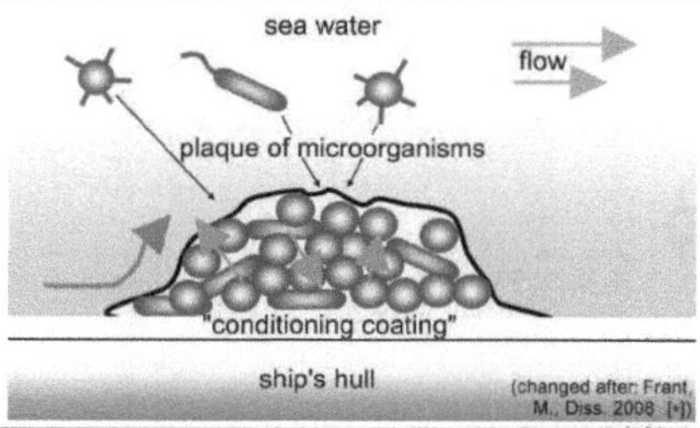

B
C
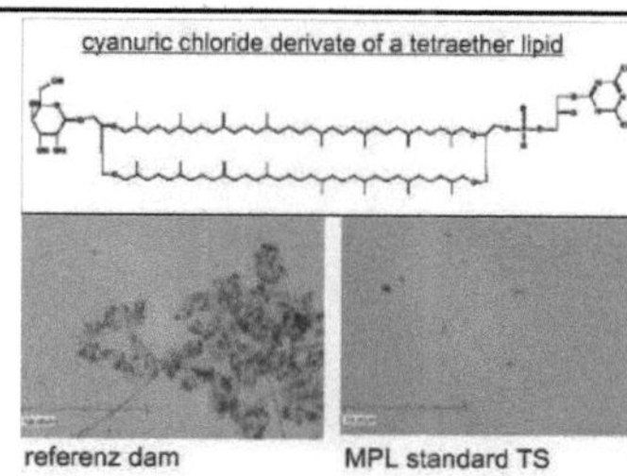

Test sets in waters, especially in sewage, frequently fail due to formation of bio-films. Here, antifouling-research helps.

BIO-FILMS:
The process of formation of bio-films can hardly be detained after microorganisms settled on technical surfaces. The strata of bacteria, fungi and algae cement to plaques (A), that stop the contact of sensor surfaces to the medium, therefore corroding them. In order to avoid this, exceptional attention was directed toward stopping the first processes of adhesion by antifouling treatment.

PREVIOUS PROBLEM-RELATED ANTIFOULING-STRATEGIES:
At first, chlorinate- and tin-containing compounds releasing toxic substances were applied. Today, "inorganic-anti-bacterial leaching techniques and organic adhesive systems" are used. Recently, antifouling interventions in the physicochemical mechanisms of primary colonization seem particularly promising. As a biological model, the archaebacterium *Thermoplasma acidophilum* has been established.

TECHNOLOGY:
A cooperative project (see lit.) was successful to isolate polar, membrane crossing, tetra ether lipids from the membrane of the bacterium mentioned (B). These can be fixed on technical substratum-surfaces using self-assembling-processes. There, they form "stable, inert, bio-compatible, arranged and impermeable monomolecular" coatings. In the laboratory test, there was a strong antifouling effect (C).

LITERATUR: Verbundprojekt Antifouling – Biomim. Konzepte. Partner: 4H-JENA engineering (Koord.) – Innovative optische Messtechnik – Sensortechnik Meinsberg – Inst. f. Bioprozess- u. Analysenmesstechnik – Inst. Physol. Chem.; Uni Halle. http://sundoc.bibliothek.uni-halle.de/diss-online/08/08H084/t1.pdf [*].

Water rolls off; new procedures for self cleaning

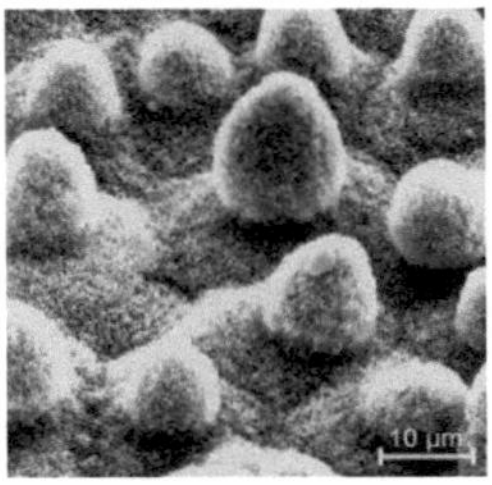

A

B

The well known effect of the Lotus-leaf works also with new materials, which provide, for instance, rapid drying.

BIOLOGY:
The basics of the Lotus effect were described on p. 65. Accordingly, it depends on the surface of the lotus leaf structure (A), which is realized with biological materials. Water spiders or spiders hunting on the water surface as the native species *Argyroneta aquatica* and *Dolomedes fimbriatus* can immerse without wetting. Their body is covered with a fur of water-repellent hairs, providing an air-bubble that wraps the body.

TECHNOLOGY:
For the bionic realization of the technology, new approaches have been made. They are based on analogously functioning inorganic structures, which have technological advantages as compared with biological materials. Practically, a water-repellent coating, for example, for swimwear could allow the wearer to come out of the water in a practically dry fashion. Using this technology, satellite dishes could be made dirt-resistant.

APPROACHES:
S. Seeger from the Institute for Physical Chemistry of the University of Zurich and co-workers developed a surface-coating that is based on a polymerisation reaction of Si-compounds. This leads to a water-repellent undergrowth of nano-filaments (B). In the Laboratory for Functional Materials of the ETH Zurich, the researchers succeeded to produce fine-crystalline cobalt-oxide from the reaction of cobalt-chloride with urea, which functions the same way.

LITERATUR: Nie mehr nass. Magazin der Universität Zürich19(3) (2010), 27–27. – Der Natur abgeguckt. ETH-Globe der ETH Zürich 3 (2010), 6 f. – Funktionelle Beschreibung des Lotuseffekts in: Nachtigall, W. (2002): Bionik. 2. Aufl. Springer, Berlin. S. 339–344.

Development: Self cleaning of metals surfaces

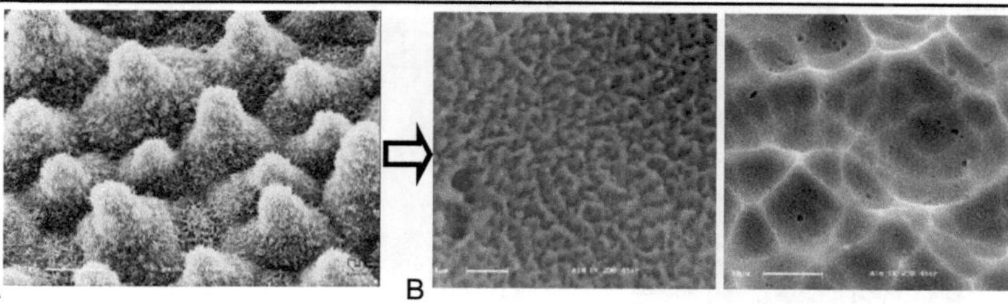

Anti adhesion develops to an important technical principle also on metal surfaces, for example for corrosion protection.

BIOLOGY:
Occurring at leaves of the Indian Lotus *Nelumbo nucifera* and named lotus-effect after this species (A; comp. p. 65, 257), this effect gains increasing importance also for the processing of metal surfaces. An example: The BIONA-joint project of the Botanical Institute of the Technical University Dresden (C. Neinhuis) with the Nehlsen BWB aircraft electroplating Dresden runs under "Anti-adhesive micro- and nano-structured metal surfaces."

PRINCIPLE:
One succeeded in applying the Lotus-principle, that is based on a combination of micro- and hydrophobic nano-structuring, to metal surfaces: "A micro structure of aluminium and stainless-steel by sandblasting (B) or flame spraying with subsequent electro-chemical treatment (C) results in a hierarchically built bio-mimetic surface, to which even highly viscous and sticky liquids do not adhere ..."

AIMS:
Novel metal-based materials and coatings are wanted, that have low surface energy and usable mechanical stability, whose surface structures can be chemically modified and thus functionalized to sustain different purposes. Physical parameters are, for example, wetting and adhesion, corrosion, friction and radiation reflection; technical parameters are lifespan, modes of processing and fields of final application.

LITERATUR: Zitat (2011) nach: http://www.bionische-innovationen.de/#projekte/antiadhaesiv.html – Vgl. dazu Literatur auf S. 65 und Website des Instituts für Botanik, TU Dresden: tu-dresden.de/die_tu_dresden/fakultaeten/fakultaet...und.../botanik

Self cleaning plastic surfaces

A

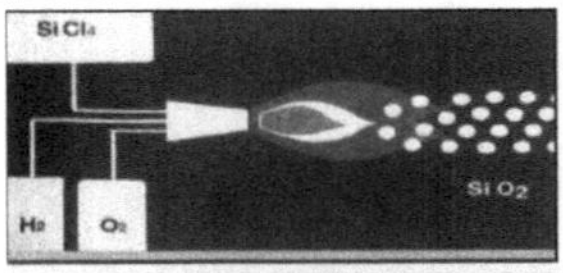

B

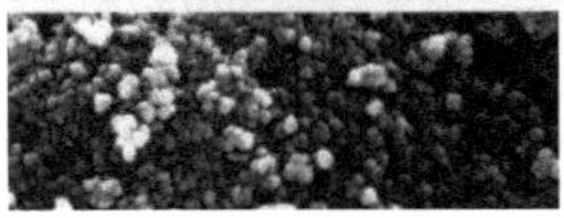

C

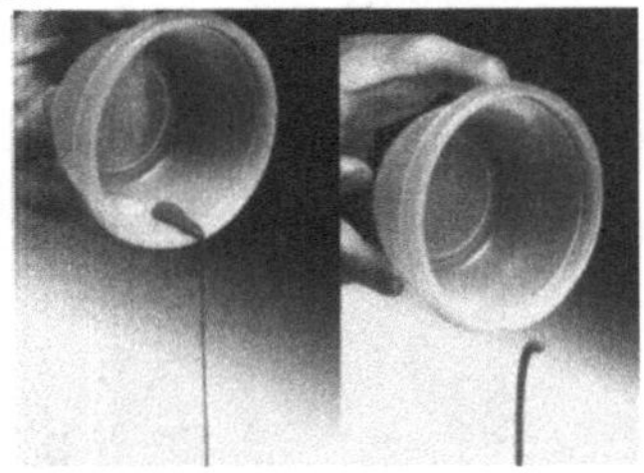

The lotus effect is well examined. However, its technically adequate transfer to plastic surfaces is rather difficult.

BIOLOGY AND TECHNICAL ANALOGY:
The Lotus effect® (comp. p. 65) presents the biological basis. The surface of the lotus leaf reaches the self cleaning effect by nano/micro-scaled hydrophobic nipples. Since 2004, the Degussa INC. actuates the manufacture and commercialisation of the Lotus effect on plastic surfaces. A promising procedure is the application of nano-scaled, amorphous SiO_2 that is superficially made hydrophobic by dimethyl u. trimethylsiloxy groups, on these surfaces.

PROCEDURES:
Primarily particles are produced (A) in a flame-reaction of an oxyhydrogen blowpipe from $SiCl_4$. They are deposited automatically to a layer of together-closing spherical nipples, which is somewhat similar to the surface of the lotus leaf. The problem of adhesion to plastic surfaces was dissolved as follows: The inner surface of the tool is sprayed with a layer of these particles. This layer is fixed by the then injected plastic melt.

LABELING OF THE NEWLY FORMED SURFACE:
After the curing of the plastic melt, the particle-surface is firmly tied up (B) and shows the Lotus effect. From such an interiorly coated cup a viscous liquid runs off free of residues (C). The contact angle of a water drop on this surface amounts to 140–160°. That means the drop is practically rounded. Typical unrolling inclined surfaces lie between 1°–10°. This surface is relatively resistant against wiping and approximately 1 year resistant in outdoor environment.

LITERATUR: Michel, W. (2006): Die Herstellung superhydrophober Oberflächen mittels nanostrukturierter Partikelschichten. Bionik-Industriekongress. BIOKON, Berlin 2006. S. 19–26.

Self cleaning of fibre based materials

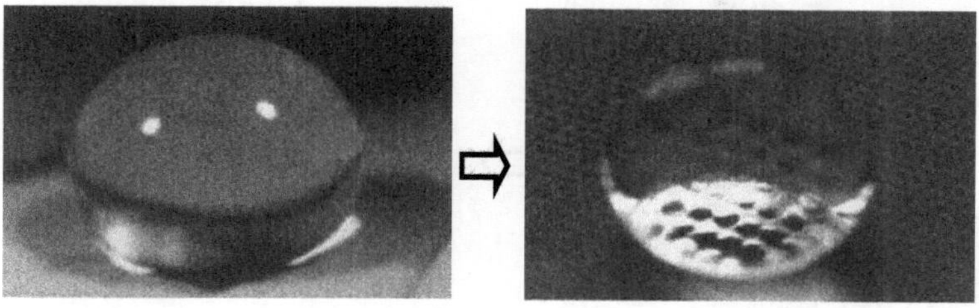

A B

By hydrophobic fibre coatings, self cleaning according to the principle of the lotus effect can be obtained.

BIOLOGY:
The Lotus effect® (A) has been described in previous examples. Summed up, it is based on the combination of three principles: 1. a micro-structured surface with an overlaid nano structure; 2. a hydrophobic surface design (1. + 2.: "super-hydrophobic"); 3. an occasional sprinkling or dew. Due to the combination of these three parameters a surface is self cleaning: good for facade coatings, awnings etc..

LOTUS-EFFECT AND TEXTILE-SURFACES:
At the ITV Denkendorf, the Lotus-effect has been transferred to textiles (B). These surfaces are structured by the arrangement of the stitches, by the weaving technology, by defined preferred directions, by the structure and the type of nonwoven and by the fibre and yarn structure. This together with a nano structured hydrophobic fibre coating results in a large number of parameters, whose ideal combinations are looked for. This requires extensive experimental variations.

TRANSFERS:
The possibility of self cleaning was tested by application of particles followed by sprinkling and counting of the remaining particles. As expected, a larger hydrophobic and self cleaning effect was shown for nano-structured hydrophobic fibres (up to 80 %) contrary to merely hydrophobic fibres (60 %). Effective structure parameters are, for example, fibre-curl, -rotation and -hairiness and the details of the spinning process, for which special machines are used.

LITERATUR: Scherrieble, A.; Stegmaier, T.; Arnim, V. von (2006): Projektpräsentation des ITV Denkendorf. Bionik-Industriekongress „Innovationsmotor Natur“ .BIOKON, Berlin. S. 95–115.

Air-retaining, super hydrophobic boundary layers

A 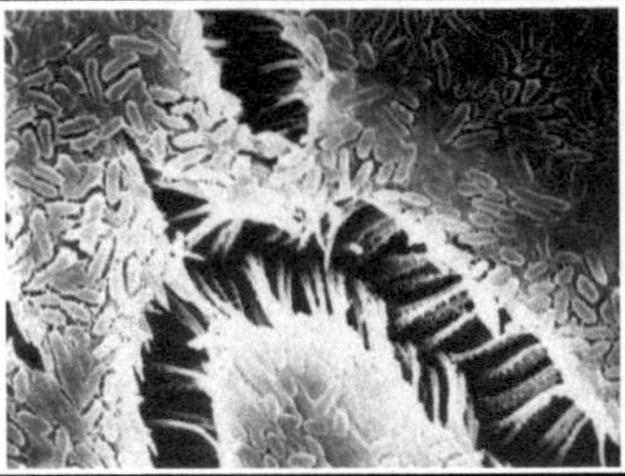

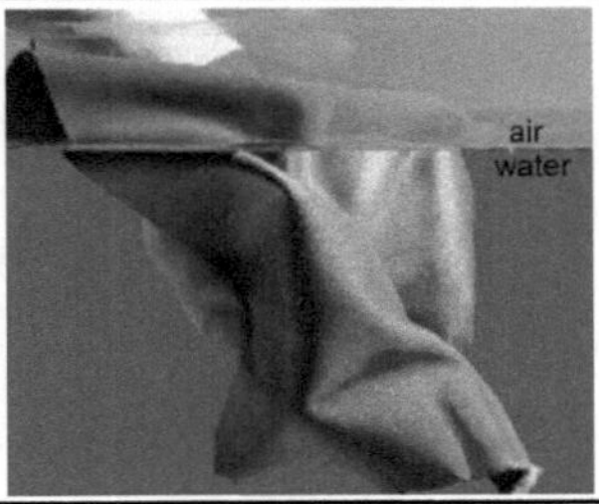

B

Micro structuring according to the surface structures of water-insects leads to the retaining of air layers that prevent wetting.

BIOLOGY:
Water-insects can retain air-layers due to the fine hairiness of their body surfaces. The stream-dwelling bug (*Aphelocheirus aestivalis*) with approximately 10^4 tiny hairs per mm^2, that are bent at the top, is able to retain an air layer, through which it breathes ("Plastron"; A). The "water bee" (*Notonecta glauca*) keeps "air-hoses." The water strider (*Gerris lacustris*) and the raft spider (*Dolomedes fimbriatus*) prevent themselves from sinking by their air-cover.

PRINCIPLE:
If a thin air layer separates a solid surface from water, the wetting of the solid surface is reliably prevented. Whereas this is rather a secondary effect (respiration-function) for bugs living in water, in the "water bee", that moves quickly by strokes, it should decrease also the friction and with it the energy consumption. For water striders and spiders this ability is obligatory to move on the water surface.

TECHNOLOGY:
The super hydrophobic effect would be significant for swimwear (B), for example, that doesn't become wet very much and dries fast. Because of the lower friction to be expected, this would result in a reduction of the drag for professional swimmers or ship hulls equipped with such surface properties. In cooperation with the Nees Institute of the University Bonn, the ITV Denkendorf developed basic principles and actually cooperates with industrial partners for their practical application.

LITERATUR: Kallenborn, H.-G.; Wisser, A.; Nachtigall, W. (1990): 3-D SEM-atlas of insect morphology. Vol. 1: Heteroptera. BIONA-report 7. Fischer, Stuttgart. – Scherrieble, A.; Stegmeier, T.; Arnim, V. von (2006): Bionik-Industriekongress. BIOKON, Berlin.S. 112 f.

"Sandfish" scales ⇨ surfaces insensitive to corrosion

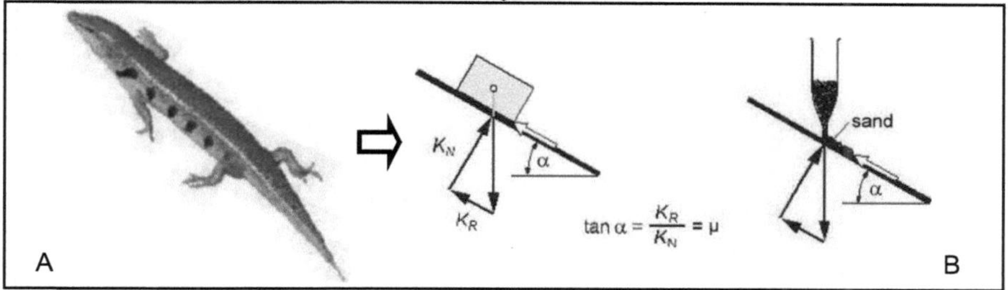

Grains of sand slip easier from textured surfaces. So they can less corrode them.

BIOLOGY:
The "sandfish" *(sand-skink, a lizard) Scincus scincus* (A) can stand in the "sandblast" of desert winds or can "swim" in the "sand-sea" with winding movements downright, without corroding its scale-surface markedly by the friction of the fine sand grains spangled with fine, glassy splinters. In an experiment, 10-hour sand-irrigating from 30 cm of height didn't show any abrasion. Contrary, on glass, steel and plastics there was distinct abrasion.

PRINCIPLE:
From the scale, sand brought up glides down already at an inclination-angle of α = 18° (B), from polished steel only at an angle higher than 40°. The coefficient of sliding friction $\mu_g = \tan \alpha$ is accordingly lower. This is explained by a nano-composite-like construction of the surface, a "micro-ornamentation", that offers a lower contact area to the sand-grain and furthermore free-combs it from its corroding splinter-edging.

TECHNOLOGY:
In analogy to the scale a technical structure was built, in which hard mineral-spikes are embedded in a polymer-matrix. The spike protrude a little over the surface of the matrix, so generating a "micro-ornamentation". Current studies try a bionic improvement of such surfaces. During electrical charges (sand storms) grains of sand adhere more strongly. A fine comb at the scales reduces this by electron leakage.

LITERATUR: Rechenberg, I.; El Khyari, A. R. (2003): Reibung und Verschleiß am Sandfisch der Sahara. Bericht zu Festo-Stipendium, unpubl. S. neben zahlreichen weiteren Darstellungen im Internet unter „Rechenberg Sandfisch".

What about the "shark skin effect"?

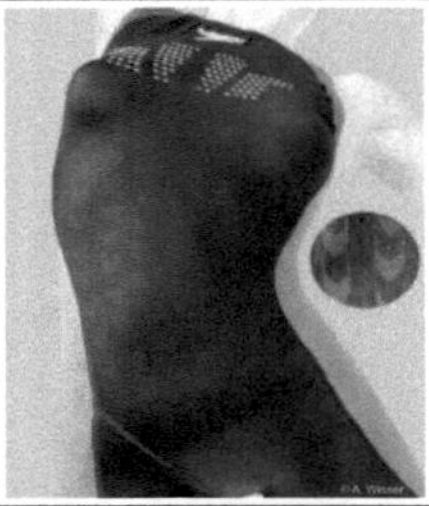

A

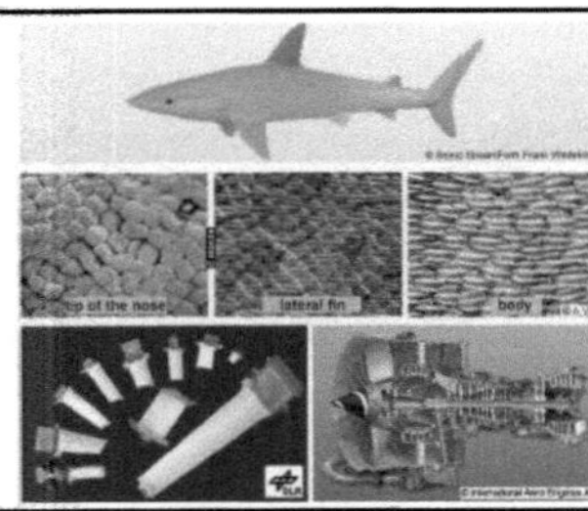

B

This effect , which is the best known classic bionic approach beside the Lotus effect®, found further applications.

BIOLOGY AND CLASSIC BIONIC APPROACHE:
Shark scales, that cover the bodies of fast open sea-swimmers according to streak lines, reduce the frictional resistance by reduction of turbulent shear stresses. The coating of big-airplanes with analogously working riblet foils reduces the fuel consumption about 2 percent (comp. p. 60). The disadvantages of such foils (their brittleness, necessary to occasional exchange, long time for gluing) prevented the general deployment in the aeronautics.

APPLICATION IN SWIMSUITS (SPEEDO):
Specifically woven products generally generate turbulent boundary-layers on the one hand, which peel off less easily. On the other hand, especially at risk of separation regions (for example in the armpit-area (A)) the structures are lined up parallel to the flow, so that drag-increasing cross-flows are hindered. After spectacular swimming-records, these suits were not allowed any more in competitions.

APPLICATION IN HIGH PERFORMANCE TURBINES:
In the DFG priority programme "Adaptive surfaces for high-temperature-applications: The skin-concept", that runs since 2007 with 13 projects, there is also a cooperation-project managed by Chr. Leyens/TU Dresden that is based on the shark skin effect (B). For turbine-shovels, that get hot as far as 1200° C, an analogous treatment of the surface is to be developed, that reduces fuel consumption and exhaust emissions.

LITERATUR: Bechert, D. et al. (2000): Fluid mechanics of biological surfaces and their technical application. Naturwiss. 87, 157–171. – Speedo-Schwimmanzüge: www.swimfaster.de. – DFG-Schwerpunktprogramm „Adaptive Oberflächen ...“ s. christoph.leyens@tu-dresden.de.

Super hydrophobic air coating on underwater surfaces

A 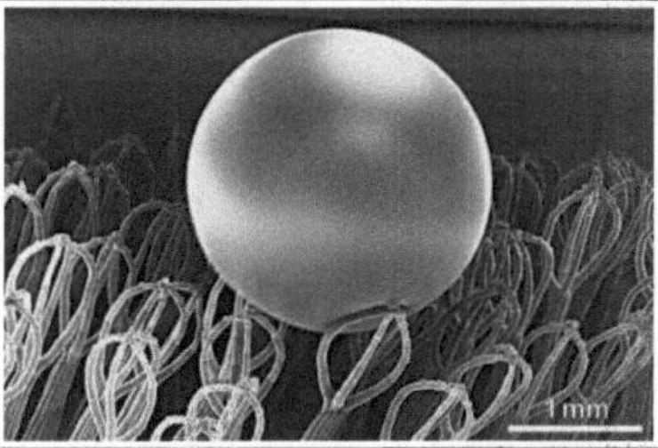

B

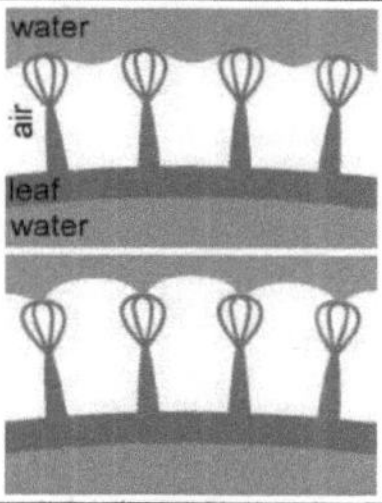

C

Ship hulls with adherent air coatings according to principle of the floating water moss could possible reduce the drag.

BIOLOGY:
The stream-dwelling bug (*Aphelocheirus aestivalis*) living on river-grounds holds tight a thin air coating consisting of approximately 10^5 per cm^2 tiny hairs, bent at their end, used for respiration. The corresponding retention mechanisms of the floating water moss (*Salvinia molesta*) are structured more roughly. That fern wraps itself into a thin air-layer when submerged. These consist of whisk-shaped hydrophobic hairs standing together (A), whose fine tops are hydrophilic.

PRINCIPLE:
Between the hydrophobic hairs – on the upper side of the swimming-leaves – an air-layer (B) remains when submerged. This could, of course, escape easily, if shear stresses work on the fine hair-forest when attacked by water flow. However, the hydrophilic tops "dive into the surrounding liquid and tack the water so to speak in regular distances to the plant (C)". This is also to the fact in *S. biloba, auriculata and oblongifolia*.

TECHNOLOGY:
Ship-construction-institutes highly classify the potential of such a bio-analogous surface, with a stable held air-layer, on ship hulls. The drag resistance could be reduced by approximately 10 percent what would correspond to a world-wide fuel-saving of approximately 1 percent. If the effect is true, that would be one of the most important bionics ideas. Moreover, an antifouling effect is to be expected (comp. p. 200). The realisation just begins.

LITERATUR: Barthlott, W. et al. (2010): The *Salvinia* paradox: Superhydrophobic surfaces with hydrophobic pins for air retention under water. Advanced materials 22, 2325–2328. – Bionische Beschichtungen helfen Sprit sparen. www.vbio.de/informationen.

Energy and industry plants in Germany

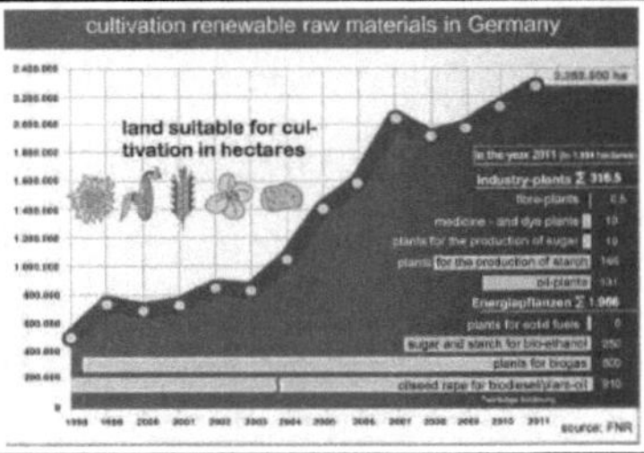

A

B

For delivery of agricultural commodities such plants are grown increasingly in Germany. That is not without controversy.

RENEWABLE RAW MATERIALS IN GERMANY:
As substitutes for fossil energy sources energy and industry plants will be grown, namely above all raps, corn and China grass (Miscanthus) increasingly also in Germany. The dependence from import is reduced on the one hand and the trade and the life cycle assessment is increased. On the other hand industry-, food- and animal-feed-use compete against them; water-consumption, necessary nitrogen fertilisation and energy-intensive processing are negatives.

DIMENSION OF THIS CULTIVATION IN GERMANY:
In the year 2011, such plants were grown on roughly 2,3 million hectares. This already corresponds to 19 percent of the entire farmland (FNR = Agency for Renewable Resources; A) and an increase of 7 percent (150000 ah) compared with 2010. The same area of 150000 ah is used for the production of biogas, and 10000 ah is used for the production of bio-ethanol. Cultivation of raps for bio-diesel and plant-oil are the most important one.

TENDENCY:
The market does not accept the addition of bio-fuels to technical fuels well. Nevertheless, the Federal government further favours to substitute E 10. Furthermore, also culture-plants are increasingly grown for the biogas production, namely corn, wheat, millet and Silphie (*Silphium perfoliatum*) from N-America: "Plates or tank?" (R.E. Schneider). An increase is found also in the cultivation of woods and recently Amflora-potatoes (B) for industry starch.

LITERATUR: Schneider, R. E. (2011): Nachwachsende Rohstoffe ergänzen natürliche Ressourcen: Große Hoffnungen in Energie- und Industriepflanzen. Labo 8 (Okt.) 2011. S. 32–37.

Energy plants and sustainability

W. Reimer:
„Our society is dependent on a reliable energy supply.

⇨

At the same time we owe it to future generations to conserve and protect limited resources in a sustainable manner."

"More energy from biomass – but please persistently produced", Agricultural College-day 2011, University Hohenheim.

BIOLOGY:
With us, the biomass-share of renewable energies amounts to 70 % today. However, not sustainable biomass-utilization leads to exhaustive cultivation. Instead, "more ecology by new variety of energy plants and their cultivation" is demanded. On this occasion, not only high energy efficiency should be the aim. Bio-diversity and climate-protection are to be taken into account. After I. Lewandowski / Uni. Hohenheim at least the following 4 aspects are important.

PLANT UTILIZATION:

1. *Increase of perennial plants*: "Perennial plants use less pesticides and especially less fertilizers."

2. Increased use of new plants with new variety: "Just in the production of plants we should use a wide spectrum."

CROP SEQUENCING AND LAND UTILIZATION:

3. *Increased use of reasoned crop sequencing*: "Selective planning of the temporal sequence of different culture-plants in the same field."

4. *Increased utilization of marginal land*: "Marginal land, also contaminated areas or grassland, that fall out from the feed-production."

LITERATUR: VBio news 26/2 (2011): Energiepflanzen & Nachhaltigkeit. Forscher formulieren zentrale Anforderungen. http://www.vbio.de/informationen/alle_news/e17162?news_id=12100.

Bio-fuels of the 2. generation as energy-source

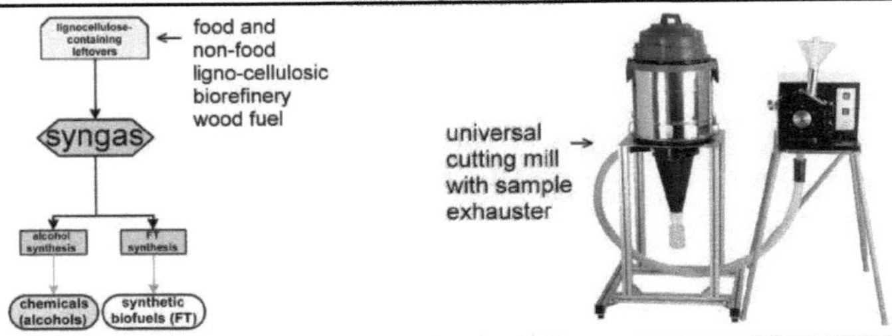

A B

The production from oil- and sugar-containing plant fractions is replaced by ligno-cellulosic fermentation.

DISADVANTAGES OF BIO-FUELS-MONOCULTURES:
The clearing of big areas in developing countries to cultivate sugarcane and others for the production of bio-fuels or synthetic btl-(biomass-to-liquid)-fuels is neither ecological nor socially acceptable. Cultivation and land-treatment still are based on the use of fossil fuels and they need a lot of pesticides and fertilisation. The lands get lost for the food-production which marks up the food-prices.

ALTERNATIVE:
Bio-fuel of the just named 1. generation carried 2011 not less than 6.1 percent fuel share in Germany; this share may rise, however, by using other methods. Bio-fuels of the 2. generation are more and more fermented from other, hardly usable, plant-components like cellulose and lignin (wood, fruit-peels etc). A classification of bio-refineries, also from the point of view of simultaneous production of food and non-food, was presented by Küppers (A).

MILLING TECHNOLOGY:
Prerequisite for an effective fermentation is the creation of a large surface of the fermentation material by suitable milling technology. Aim is the reduction of the "retention- and fermentation-period in the biofermenter. On this occasion, the end particle size is quite decisively." Therefore, grain-sizes must be produced in a favourable distribution down to < 250 µm. For this, combinations of cutting-mills and sample exhausters are tested presently (B).

LITERATUR: Brecht, H. (Fritsch GmbH) (2011): Biokraftstoff – Energieträger der Zukunft? GIT Labor-Fachzeitschrift 09/2011, 618. – Küppers, S. (Forschungszentrum Jülich GmbH) (2011): Bioraffinerien. Nachwachs. Rohstoffe ... GIT Labor-Fachz. 06/2011, 392–395.

An E. coli strain for the syntheses of bio-fuel

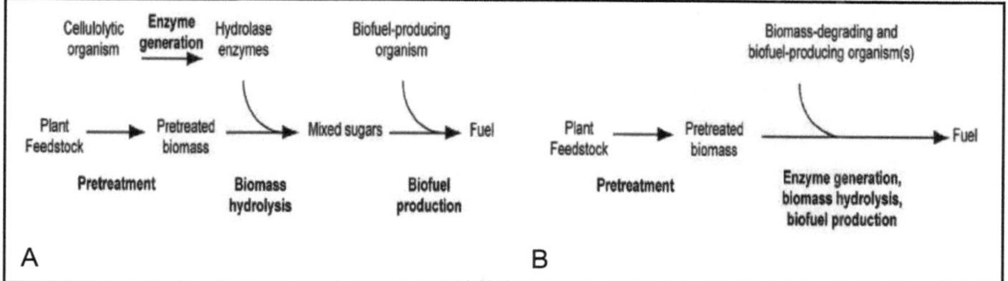

A genetically transformed strain can degrade biomass and thus can synthesize hydrocarbons for bio-fuels.

COMPLEX OF PROBLEMS:
There are already known bio-fuels; these are based on the general ability of organisms to ferment the sugar in vegetable substances to ethanol. But first, sugar must be produced from existing cellulose by addition of the enzyme glycoside hydrolase, which is to be produced very expensively and at high price. Indeed, there are Ethanol burning motors in Brazil, but in Europe they are not very widespread.

DESIRABLE:
Therefore one searches for a microorganism that produces this enzyme on the one hand for itself and can grow on cellulose-containing plant-material consequently; on the other hand it should be able to convert the inter-product sugar further to hydrocarbons that are usable by common European motors – namely from gas-, diesel- and kerosene-motors as well as for jet engines. Now, G. Bokinsky and other American researchers succeeded in doing this.

PROCEDURES:
Garbage of plants is pre-treated with ionic liquids. In the laboratory-standard, a genetically changed strain of Escherichia coli can grow on it, brake down hemicelluloses and cellulose with self generated enzymes and convert them further to hydrocarbons, which directly or as pre-stages lead to the three named alternative fuels (A). More efficient and industrially applicable micro-organisms are now being the aim.

LITERATUR: Bokinsky, G. et al. (2011): Synthesis of three advanced bio-fuels from ionic liquid-pretreated switchgrass using engineered Escherichia coli. Proc. Nat. Acad. Sci. (PNAS) Early Edition. www.pnas.org/cgi/doi/10.1073/pnas.1106958108.

Green algae as hydrogen- and voltage-sources

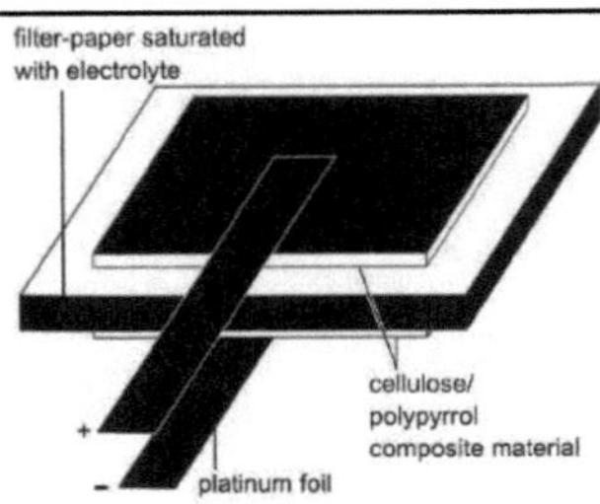

A

B

Micro- or macroscopic green algae can be breed in such a way that they produce H_2 or generate electric potentials.

H_2-PRODUKTION:
The basic research about alga-bacterium-symbiosis was discussed on p. 62. S. Peter writes about the potential of green algae: "With the help of the enzyme hydrogenase they crack water into oxygen and hydrogen. They get the energy necessary for it by photo synthesis. If one puts them on a sulphur-diet, then the algae increase their metabolism. An energy surplus occurs, which is disposed in form of H_2": Synthesis of hydrogen.

ALGA BATTERIES:
"In 2009, Swedish researchers succeeded to develop a battery on basis of a nano structured alga-cellulose without high energy-expenditure ... The scientists covered the nano structure with a 50 nm thick layer of the polymer polypyrrol. So they generated a completely new electrode-material with low weight and high loading-speed for accumulators" (A). These may substitute lithium-ion batteries for developing countries.

PROBLEMS OF REALISATION:
In practice, the breeding of green algae, for example the breeding of the microscopically small genus *Chlamydomonas* (→ H_2) or of the filamentous alga *Cladophora* (→ battery), and the interconnection with technical elements are not simple. Algae are cultivated in plate reactors (B), for example, that need continuous supervision and care. Until now, living biological elements are still hardly embedded into maintenance-free products (e.g. in facade films).

LITERATUR: Peters, S. (2011): Materialrevolution. Nachhaltige und multifunktionelle Materialien für Design und Architektur. Birkhäuser, Basel. Zum Thema: Abschnitt „Grünalgen", S. 167 f. – S. auch Institutionen wie z. B. Karlsruher Institut für Technologie KIT.

Algae breeding in deserts for production of biomass

A   B

In our latitudes, the use of the large economic potential is problematic because of high breeding and harvesting costs.

ALGAE CULTURES IN THE DESERT:
Already in 1994, Rechenberg has produced on small scale hydrogen in combined crops of algae (comp. p. 62) in the Sahara. In principle, desert regions are well suited because of the intensive solar radiation. However, the problem is the algae-breeding on a big industrial scale. In 2011, G. Klöck (HS Bremen) reported about the mass culture of micro algae in the South-Arabic desert with sewage from the petroleum production – an interesting approach.

TESTS IN 2010:
After successful breeding of strains of single-celled green-algae in the oil field Nimr (sultanate Oman), which tolerate temperatures above 50° C, an installation was tested that works with "accessory water" (5–20 l waters per 1 l of produced oil) from oil drillings. This accessory water is inappropriate for drinking and for soil irrigation. Cultures were tested in open tanks of 240 (A) and 1000 (B) litres at soil temperatures up to about 65° C and at 230000 Lux.

PRELIMINARY RESULTS:
It has been shown that under these conditions the production of micro algae as renewable raw material for the production of biomass is quite possible – which can then be used for fuel- or material-production. Also, the separated algae masses then can be quickly and at low energy, so to say cost-efficient, sun dried – a large potential, because in 2007, the world-wide production of algae masses amounted to first $\leq$ 10 000 t.

LITERATUR: Klöck, G. (2011): Nachwachsende Rohstoffe – Mikroalgenkulturen in der Wüste. GIT-Laborzeitschrift 4, 254 f. – Zur Algentechnologie s. auch (2012): http://www.renewableenergyworld.com/rea/blog/post/2010/08/a-microalgae-industry-internet-directory.

Fuels from algae

A

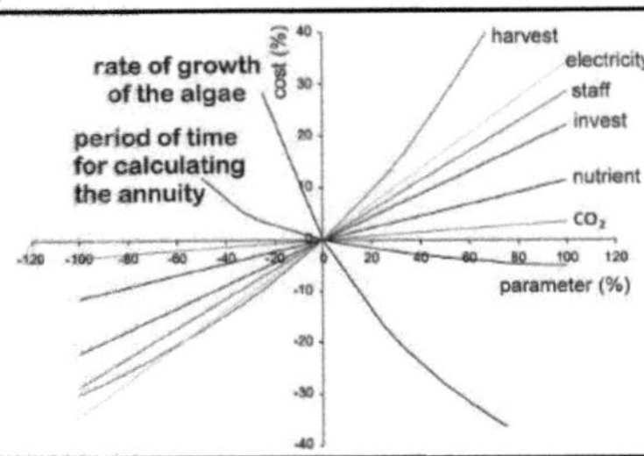

B

Fuels made of biogenic material are food-politically problematic. Algae as source materials could be a solution.

BIOLOGY AND BREEDING

Algae, as CO_2-catchers, are quantitatively as significant as plants of the mainlands. However, they grow faster and have higher oil content per reference area compared with mainland-plants - even with raps, oil-palm, and soybean - and could be cultivated in hydro-farms without competition to agricultural crop lands. Especially, this is applicable to micro-algae (size $\geq$ 4 µm). These would particularly come in consideration for the production of bio-diesel.

PRACTICE:

Micro algae populate the oceans in myriads. Therefore as to say, the important point of view that the fuel-production does not compete with the food-production does not play any role. However, one cannot win these algae from the sea but must breed them (A) technologically extensively. In the comparison to the corn land, the costs of production and especially the costs of harvest – sedimentation, filtration or centrifugation – are very high.

TECHNOLOGY:

U. Ehrenstein and B. Sayder carried out a sensitivity-model-analysis for the costs of the alga-production. Today, bio-oil from algae can not compete with such from agricultural crops. Where are the most promising saving potentials? The analysis (B) carried out at the Fraunhofer UMSICHT shows that one must begin with the maximization of the algae-growth, the minimization of the electricity costs and the improvement of the harvest-technologies.

LITERATUR: Kraftstoffe aus Algen. Fraunhofer UMSICHT Jahresbericht 2009/2010, S. 38 f. Geschäftsfeld „Nachwachsende Rohstoffe“.

First flight with algae flight fuel

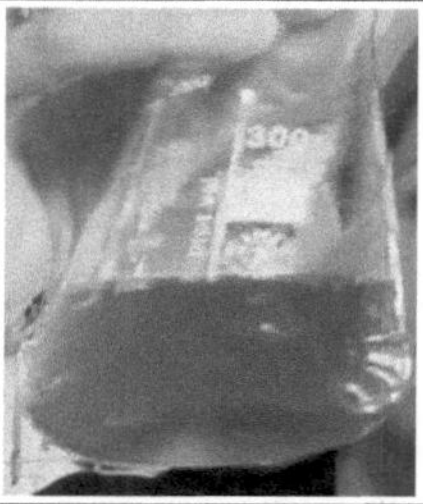

A

B

For the first time, an airplane flew only with "algae fuel": a DA42 from Diamond Aircraft, 6.–13.6.2010 in Berlin.

ADVANTAGES:
Beside the ecological advantages shown in the preceding example, there are also energetic and technical advantages. Due to the fact that algae-fuel (A) shows a higher energy density, the named airplane (B), type new generation, had used 1.5 litres per hour less compared with the usual JET-A1-fuel, as J. Botti/EADS reported. After insignificant adjustments and technical adaptations, the engines tolerated this flight-fuel.

LABELING:
"Bio-fuels of the third generation are more than just a substitution of conventional energy sources. They bring quite new possibilities for the drives of the future. Exhaust measurements indicate that the algae bio-fuel contains 8 times less carbon than kerosene from fossil raw materials. On the basis of the very low nitrogen- and sulphur-content ... bio-fuels emit up to 40 percent less of sulphur oxides and very low quantities of nitrogen oxides." (VBio).

DEVELOPMENT:
At present, the European aeronautics-concern EADS pursues an industrial pilot-project with the aim to produce more test-quantities. Partners are the named Austrian company Diamond Aircraft and the IGV-GmbH Potsdam. The basic material, algae oil, was delivered from the Argentinean company Biocombustibles del Chubut and was further-processed by the German company Verfahrenstechnik Schwendt (VTS) – a truly international cooperation.

LITERATUR: www.vbio.de/informationen/alle_news/e17162?news_id=9778 vom 16.6.2010. – www.biotechnologie.de vom (16.10.2010).

[illegible]

For the first time, [illegible] only with [illegible] UAAZ from Diamond Aircraft, [illegible] 2016 in Berlin.

[illegible]

[illegible]

[illegible]

EVOLUTION

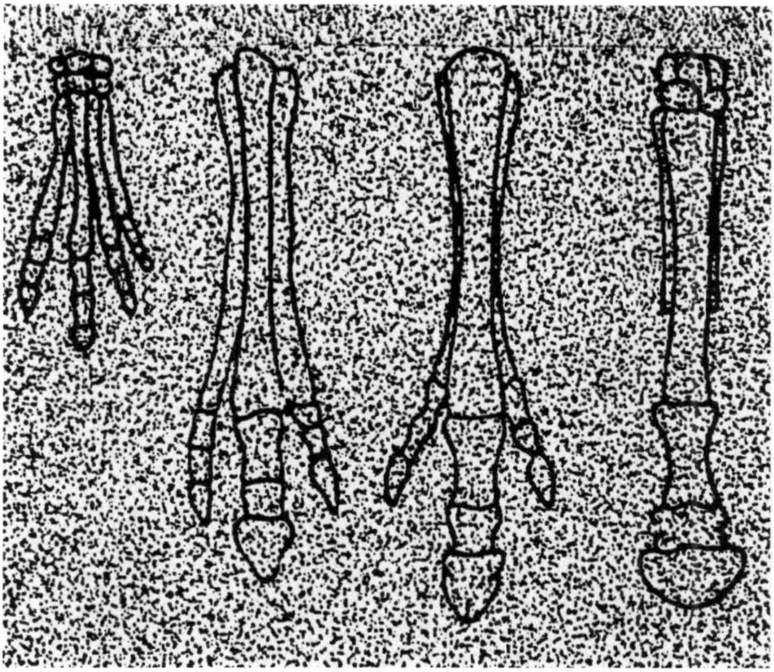

AND OPTIMIZATION

W. Nachtigall and A. Wisser, *Bionics by Examples*,
DOI 10.1007/978-3-319-05858-0_12

EVOLUTION AND OPTIMIZATION

Biological evolution – Optimization of a form and of a structural component – Evolution-strategies.

Evolution-technology and -strategy try to make the procedures of the natural evolution usable for the technology. Particularly then, if the mathematical formulation of complex systems and procedures has not yet been developed so far that arithmetical simulation would be possible, the experimental or virtual development due to trail-and-error remains an interesting alternative. Technical optimization-problems can be related to all areas, for example, to make structures as light as possible, cutting processes as sharp as possible and flat structures as stable as possible.

The biological evolution as a model

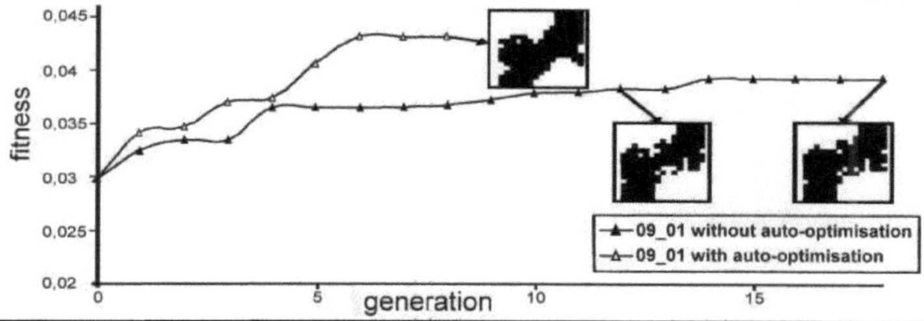

A

Today, the classical technical realisation of evolutionary ways has split into many kinds of different facets.

BIOLOGY:
This principle of the biological evolution underlying, the Berlin development of the evolution-strategy by Rechenberg, Schwefel and Bienert (comp. p. 29–35) still forms the basis: During the propagation (→ *reproduction*) accidental alterations (→ *mutations*) are made, which lead over realization-strategies (→ *recombination*) to slightly different products. From these, the "environment" selects the "fittest one(s)" (→ *selection*).

EXAMPLE:
J. Adamy (2002) as well as S. Schäfer, B. Briegert and S. Menzel (2005) understand evolution strategies and genetic algorithms as subsets of the evolutionary algorithms that belong to the bionics on the other hand. They applied a particular evolution strategy to calculate the flow at facade-models, and they defined fitness-values from the results. These are reached earlier and in a higher developed form with the auto-optimisation applied than without this (A).

HIDDEN APROACHES:
The aforementioned basic steps frequently are concealed in computer-programs, so, for the user, it does not become conscious that these go back to the strategy of the "biological evolution." But this is quite common with bionics. Frequently, it is no more recognizably in the ready product that is based on nature. But this does not matter, because bionics is indeed a "stimulation-discipline" for idea-finding and product-development, not a self purpose.

LITERATUR: Adamy, J. (2002): Fuzzy Logik. In: TU Darmstadt (Hrsg.): Bionik. Thema Forschung 2. Darmstadt – Schäfer, S.; Briegert, B.; Menzel, S. (2005): Evolutionären Algorithmen ... In: Rossmann, T.; Tropea, C. (Hrsg.) Bionik. Springer, Berlin.

Body spindle of the lowest drag

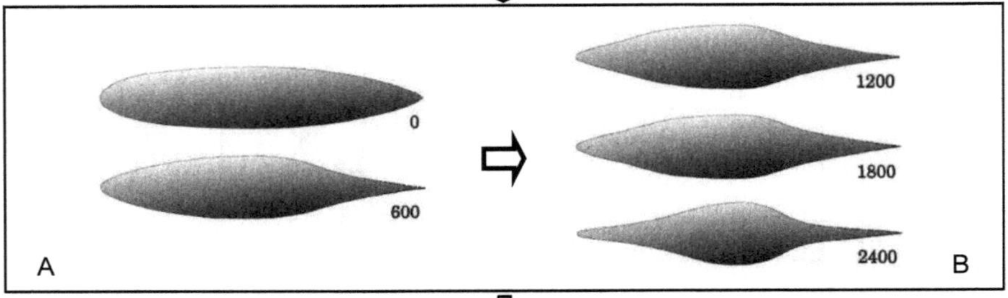

Evolution strategy leads to a flow optimal spindle form that one cannot calculate without giant expenditure.

BIOLOGY:
Penguins are known for their very low drag coefficient. The donkey-penguin (*Pygoscelis papua*) for example possesses at a Reynolds number of $Re \approx 10^6$ a frontal area drag coefficient of merely $c_{W\,fa} = 0.07$. If one replaces the slight asymmetrical penguin body by a symmetrical rotation body, even smaller c_W-values lower than 0.05 result. Can evolution-strategy reproduce these forms as well as improve them?

QUESTION AND SHARES OF DRAG:
Which is the best streamlining contour of a body spindle for a Reynolds number Re_l related to the spindle-length? At first, under application of an (1+1)-unit evolution strategy with a 1/5-rule for success the pressure-distribution was calculated for every mutative variation of the body contour. Using this, the course of the boundary layer was determined and finally via the impulse loss of thickness the drag resistance F_{WR}. To this, a pressure drag F_{WP} is to be added.

QUALITY-FUNCTION AND SOLUTION:
The pressure drag F_{WP}, mentioned above, is calculated from the displacing effect of the boundary layer. The total drag $F_{tot} = F_{WF} + F_{WP}$ should reach a minimum. Beginning with a conventional airship-form (A, contour 0), the evolution strategy, after 2400 mutation-selection-steps, leads to a very unconventional "dolphin-form" (B, contour 2400), that fulfils the criterion $F_{tot} \rightarrow$ min. with the boundary-condition $Re_{l\,penguin}$ = const..

LITERATUR: Pinebrook, W. E. (1982): Drag minimization on a body of revolution: Dissertation, University of Houston. – Rechenberg, I. (2006): Optimierung mit Evolutionsstrategie. Bionik-Industriekongress. BIOKON, Berlin. S. 218–236.

Body spindle optimum as a function of the Re-number

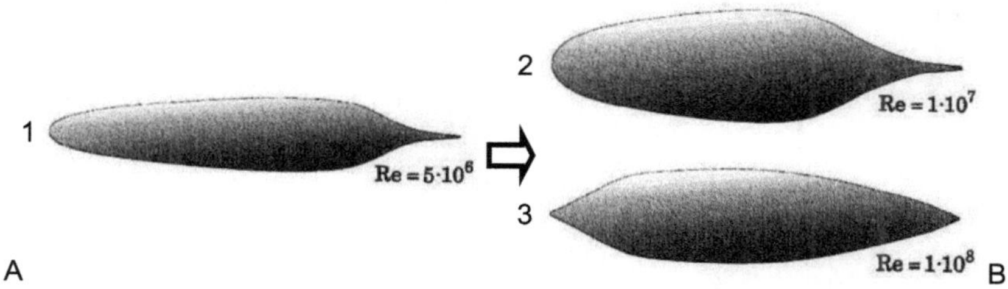

Streamlining forms of low drag, spindle shaped rotation bodies depend strongly on the Re-number.

BIOLOGY:
Penguins and other rapid swimmers move with very different speeds, which correspond to the different Reynolds numbers. Usually, the quickest movement ($Re_{max\ penguin}$) is used as basis for the finding of an ideal form. But technical Reynolds numbers go significantly beyond. The question is, how must the spindle-contour ideal for $Re_{max\ penguin}$ be changed for the adaptation to higher Re-numbers?

QUESTION AND APPROACH:
Using evolution-strategic methods, the ideal-form for $Re_{max\ penguin}$ should farther be changed, so that it becomes ideal for Reynolds numbers $5\ 10^6 < Re < 10^8$. That means, that it reaches in each case the lowest drag coefficient possible in this region. For rather "low" Reynolds numbers of $5\ 10^6$, the evolution-strategically attained optimal form corresponds, as expected, to a laminar spindle (A, contour 1).

SOLUTION:
With the contour 1, the boundary layer remains laminar over more than 70 percent of the running-length by flow-acceleration. At $Re = 10^7$, the body must be more thickened (through acceleration) to stabilise the boundary layer (B, contour 2). At $Re = 10^8$, as it is given at big airships, the running-length is reduced further, because of the necessity of an even stronger acceleration, so that a strange sharp nosed form results (B, contour 3).

LITERATUR: Lutz, T. et al. (1996): Shape optimization of axisymmetric bodies in incompressible flow. Ber. Inst. Aerodynamik und Gasdynamik, Stuttgart. – Rechenberg, I. (2006): Optimierung mit Evolutionsstrategie. Bionik-Industriekongress. BIOKON, Berlin. S. 218–236.

Energy saving flow by the "Mäander®"

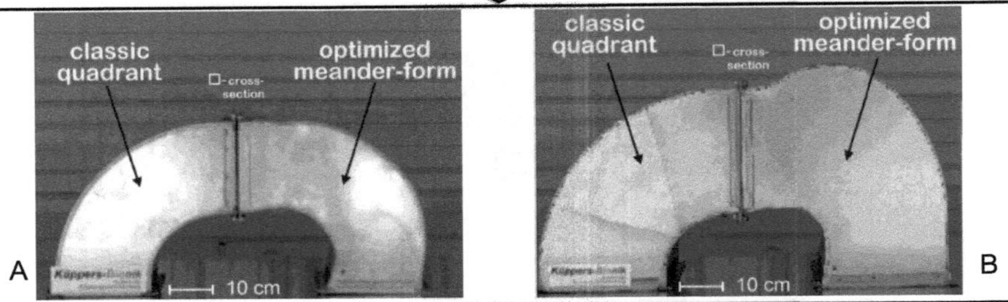

An only slight looking, evolution strategic change of contour at elbows reduces the pressure loss.

INANIMATE AND ANIMATE WORLD AS "PROTOTYPE":
Bionics, as the name says, takes the animate world as stimulation sources for technical realizations. However, sometimes it can also be the inanimate world. For example, meandering river-systems reduce their rate of energy dissipation and their transportation losses through their self-organized "adaptive windings". Similar self-optimization is found also in organisms, which can reduce thereby the total pressure losses of their blood- and gas-cycles.

PRINCIPLE OF THE OPTIMIZATION:
Beside the evolution strategic optimisation – successive variation of bending radii (comp. p. 32) over the elbow, here 90°, until the pressure loss is minimal in the form part – flow-optimal tube-bows can be achieved also by the way of direct measurement of the geometry of river meanders and bio-analogous curve-transfer. Both ways lead to comparable flow-technical advantages.

EFFECT:
The steady energy efficiency optimization leads – as well as the modelled curve-geometries of the river-meanders – to typical bow-forms with a "swerve". This "Mäander"® (A, B) leads to a reduction of the flow-loss for each 90°-bow of about 20 percent. In a complete flow-system (heater, ventilation, climate, environment, foundry technology) with many elbows the energetic-, the tribological- as well as the stand-time-advantages would be greater.

LITERATUR: Küppers, U. (2009): Mäander-Effekt® in der Landtechnik. Bauen und Planen 5, 343–345. – Zur allgemeinen Vorgehensweise s. auch Küppers, U. (2007): Natureffiziente Lösungen erobern die Technik. HLH 11. S. 58; HLH 12, S. 34. VDI, Düsseldorf.

Tool optimization according to the anteater claw

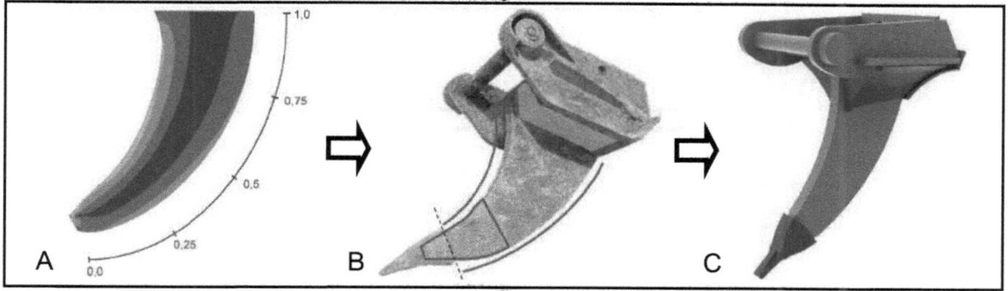

For structure optimization of high stressed work tools for raw material production, this claw served as approach.

BIOLOGY:
The middle finger of the big anteater (*Myrmecophaga tridactyla*) carries a pronounced, contoured claw (A). The animal uses the claw as a rupture tool in breaking the stone-hard constructions of termites, his preferred food. On that occasion, he folds in the claw that means, he reduces the angle between axis of the claw and the forearm by approximately 90°, and then adducts the foreleg, while he is supported by the other legs.

ANALOGY ZUR TECHNOLOGY:
The claw (→ ripper tooth of a hydraulic shovel; B) as well as the leg movement (→ kinematics of a hydraulic shovel) and finally also the supporting (→ caterpillar basis) find their correspondence at hydraulic excavators, that rupture grounds mechanically with their "ripper tooth" for raw material production. In practice, specific and recurring and therefore expensive breaks of the ripper tooth occurred. The comparison with nature helped to avoid them.

OPTIMIZATION:
In the original ripper tooth, 2.8 t heavy, (B) were high stress peaks at the outside-contour close to the dashed line. In the anteater claw, the stress was only relatively half as large as there, and its progress over the parameterised distance (A) was much more homogeneous. The optimisation process finally led to a claw (C) in which the maximum stress at the outside-contour was reduced by 58.3 percent, at the break line by 54.1 percent.

LITERATUR: Sontheim, F.; Ulrich, A. (2012): Natürliche Vorbilder für die Strukturoptimierung hoch beanspruchter Arbeitswerkzeuge in der Rohstoffgewinnung. 41. VDBUM Seminar (29. Feb.), Braunlage.

Optimization strategies at Sachs Engineering

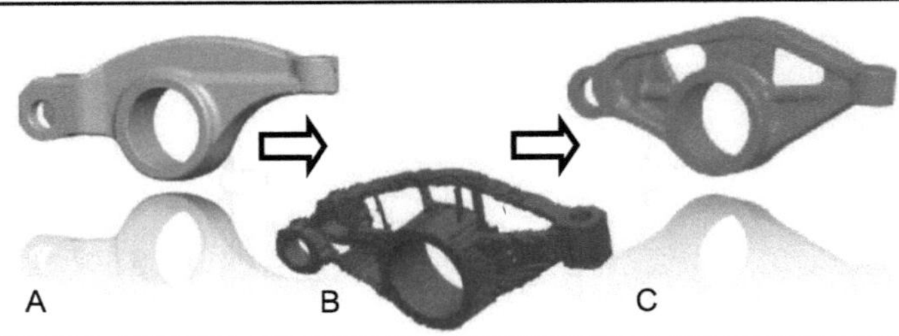

C. Mattheck's optimisation methods are applied versatility: Here some examples for topology and shape optimisation.

METHODS (comp. p. 54; in detail also referred in Nachtigall 2002):

- The SKO-(Soft-Kill-Option)-method according to Mattheck removes first carefully ("soft") not-bearing areas of components and prepares an already quite well pre-optimized approach for a lightweight design.
- Based on this, the CAO-(Computer-Aided-Optimization)-method applies biomechanical self-optimisation of nature to the further form-optimisation of the technical component that than approaches the ideal-form.

APPLICATION AT SACHS ENGINEERING:

- SKO (topology-optimisation): It comes to lightweight and material saving. "Especially there, where each gram counts, is SKO the method. For transportation and logistics, the potential of the saving is enormous."
- CAO (shape-optimisation): This is for a reduction of notch stress (comp. p. 54), homogenisation of stress and more stable components. "Extremely stressed areas become reinforced, less stressed recede."

EXAMPLE ROCKER ARM OF A VEHICLE:
To describe the way, W. Sachs, going out from the component (A) which should be optimized, makes up the following sequence. "1. Topology-optimisation (B). Aim: lightweight design. 2. Realisation with CAD. Aim: Design due to constructive and structure-mechanical requirements. 3. FEM calculation. Aim: Calculation of all operational- and failing-loads. 4. Shape optimization (C). Aim: Notch stress optimization, stress homogenisation. 5. Component design.

LITERATUR: Sachs, W. (o. J.): Optimierung der Konstruktion nach dem Vorbild der Natur. Leichtbau und Formverbesserung durch den Einsatz von Bionik. www.sachs-engineering.de/bionik. – Mattheck, C. (1992): Design in der Natur. Der Baum als Lehrmeister. Rombach, Freiburg.

Hexagonal structuring as self organization

⇨

A B

Hexagonal structure in the animate as in the inanimate world is created most by – technical transferable – self organization.

BIOLOGY AND INANIMATE WORLD:
Hexagonal structuring with its favourable qualities of stiffness always originates when roundish structures positioned in layers, flatten mutually. One can observe this at honeycombs or at radiolarian skeletons. In the inanimate world, for instance, hexagonal cells are formed self organized as thermal convection cells in hot fluids. Such formations lead to structures on low energy level.

TECHNICAL PROCEDURE:
F. Mirtsch has patented a procedure, with which metal sheets unwound from a role change under a given stress distribution self organized and continuously into hexagonal structures. The advantage compared with forcible impression lies in the "*natural*" production of *utmost stiffness with a given sheet thickness* as well as in a uniform mass distribution over the surface under avoidance of local stress peaks.

FIELDS OF APPLICATION:
Because tubes of a given wall-strength exhibit under vault structured condition a bigger stiffness, one can use thinner and therefore cheaper metal sheets, for example for cans (A). The firm Miele has used vault structuring for the drum of its washing machines ("spare drum"; B). Further applications will follow, for example, in the ventilation technology (booming reduction), the lamp-industry (anti glare reflection) as well as in car manufacturing.

LITERATUR: Mirtsch, F. et al. (2004): Vault-corrugated sheet metal on the basis of self-organization. First International Conference Bionik2004, VDI-Berichte, Vol. 249, S. 299-313 – und: www.mirtsch.com.

SYSTEMICS

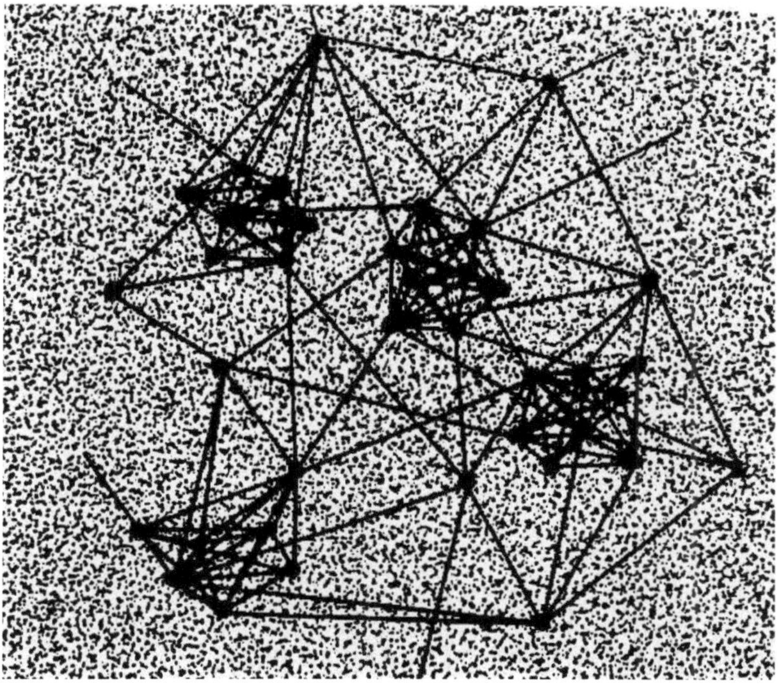

AND ORGANIZATION

W. Nachtigall and A. Wisser, *Bionics by Examples*,
DOI 10.1007/978-3-319-05858-0_13

SYSTEMICS AND ORGANIZATION

Specifics of bio-systems – Self-organisation, synergetic, system-theory – Environmental ecology – Bio-informatics, bio-cybernetics – Economic systems.

When single solutions of bionics are integrated into whole strategies, one can speak of systemics. This includes, for example, environmentally suitable packages that – if they should become effective - must be bound into a overall strategy which leads from concept to recycling. Questions, how social living organisms or ecosystem manage complex organisation problems, belong just to this as well as an effective network organisation, swarm intelligence, management close to bionics and joint-up thinking.

Contradiction oriented innovation strategy and bionics

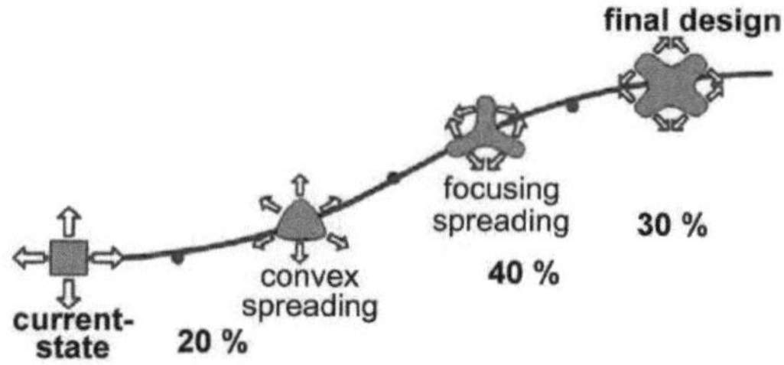

Innovation chances result from a contradiction. The strategy "WOIS" includes to a high degree bionics.

BIOLOGICAL EVOLUTION AS A BASIS:
"The biological evolution-theory offers hidden success patterns for superior developments. But not in the sense of fast, further ... What is the superior development of systems? ... Maybe more benefits, but less effort" (Linde 2008). Small mutative alterations appear very early in an evolutionary chain. According to this, WOIS structures exactly the earliest phase in the innovation process, which is otherwise often misunderstood in its meaning.

PRINCIPLE:
The innovation strategy WOIS working partly on a biological, partly on a philosophical-epistemological basis, "produces innovation jumps due to seemingly unsolvable contradictions innovation jumps, that could give the business the essential competitive-advantage. The biological analogy is the mutation that can let a system "up-jump" from a developmental valley onto an evolutionary height that may have its way in the following selection.

EXAMPLE:
One tried to improve the chisel of drill hammer (Hilti) that was easily getting blunt due to its square-like cross-section. Woodpeckers don't use any parallel surfaces (contradiction). The solution, a chisel with convex spreading, brought a 20 % higher demolition performance; focusing spreading brought additional 40 % (!). The end design with a crosswise cross-section avoided getting stuck, provided self sharpening and brought further 30 % more demolition performance (A).

LITERATUR: Seeger-Wiechers, E. (2008): Rasterfahndung nach Innovationschancen. Der F&E-Manager 03, 34–37. – Websites: www.wois-institut.de and www.wois-innovation-school.de. – Linde, H.-J.; Herr, G. (2009): Wettbewerb der Innovationsstrategien. Firmenschrift.

Systemic thinking with integrated bionic approaches

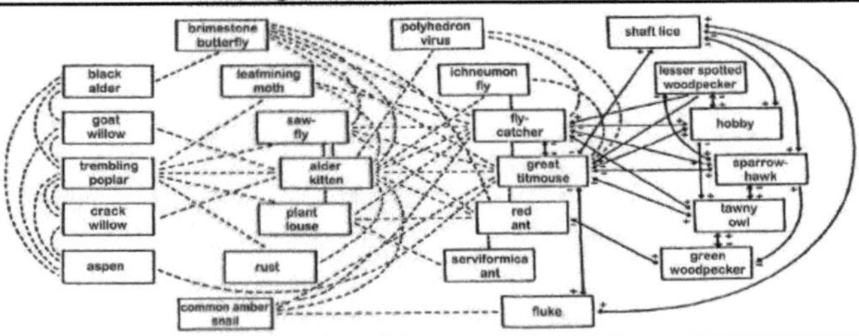

relationships in the cross-linked system "hedge" (after: Dylla, Krätzner)

Today, multidisciplinary thinking and acting "in systems" develops to a strategy of survival importance.

PROBLEM OF THE OWN SPHERE OF ACTIVITY:
"The search for new technical organisational solutions for energy saving (and consequently for cost saving) by energetic efficiency reaches increasingly the heads of trouble shooters in crisis-times as well as in times of big revolutions from the atomic- to the regenerative energy-systems. Indeed, with the numerous strategies developed solution, the view after new products and procedures often gets stuck helplessly in the own thought- and working-environment.

HARDLY BIONIC-INTEGRATION IN HOLISTIC THINGKING PROCESSES:
This lack of holistic thinking and acting hinders the valuable inspiring view over the horizon of the own science. By this, big companies are exactly as affected as smaller firms. Although an interdisciplinary view into the unlimited reservoir of efficient natural solutions would be sufficient, the 50 year old science discipline BIONIK did create only quite few internationally recognized solutions at the market.

PREVENTION OF CRISIS INSTEAD REPAIR OF CRISIS!
"The momentary finance-, economic-, and social-crisis shows the necessity to recognize early (by system thinking and acting) conflicts instead of escalating them and to have to repair them expensively. System thinking is a motivator for recognizing of connections, also for the development of lasting bionic solutions. It's about to take up suggestions for sustainable products, procedures and organisational processes and to use them practically in a bionic way."

LITERATUR: Zitate aus einer pers. Mittg. bezüglich des praxisorientierten Systemlabors „Küppers-System denken" von U. Küppers. kueppers@uni-bremen.de. – S. auch: Küppers, U. (2010): Systemisches Bionik-Management. Wissenschaftsmanagement 1, 37–42.

Systemic acting in the network organization

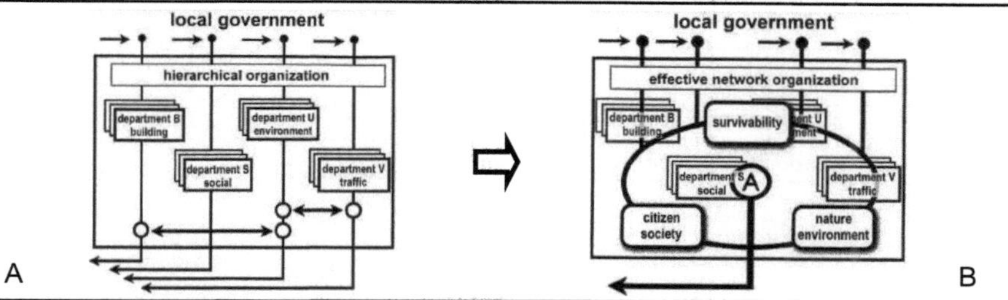

Nature-strategies for complexity handling cannot be copied, but can be reproduced analogously as functional networks.

SYSTEMIC ACTIVITY:
"It is called systemically, when solutions are developed that do sustainably justify to the dynamic processes in complex communities and strengthen their ability for development" (U. Küppers). It supposed departments to solve a problem via an early active action networking of organisational structures (divisions, departments of a local government etc.), therefore a conscious cybernetic planning. The nature offers analogous models for this.

NATURAL MODEL AND ANALOGOUS REPRODUCTION:
An analogous model admittedly doesn't have any actively dealing partners, however, it leads by the peculiarities of complex networks necessarily to self-sustaining "stationary" systems, as the example "forest border" shows (Fig. p. 234). Therefore, one must proceed the other way round: Due to the possibilities of our brains, we do not have to copy the system-maintaining-strategies of nature step by step; we can reproduce them *analogously*, instead.

FUNCTIONAL NETWORK, EXAMPLE:
What nature manages with her strategies and what man can reproduce with his thinking-strategies, can be marked as "functional networking". This could be a model for local administrations. These are usually structured hierarchically in fields of responsibilities (vertical) and functions (horizontal) and are laid out on conflict solution by compromises (A). A functional network organization of high attentiveness Ⓐ (B) would embrace the predominant municipal complexness.

LITERATUR: Küppers, U. (2011): Die Wirkungsnetz-Organisation – ein Modell für die öffentliche Verwaltung? apf (Ausbildung-Prüfung-Fachpraxis), Z. f. staatliche und kommunale Verwaltung) 5, 129–136. – Küppers, U. (2011): Die systemische Kommune. AKP 1, 52–54.

Strategies of bionic oriented package technology

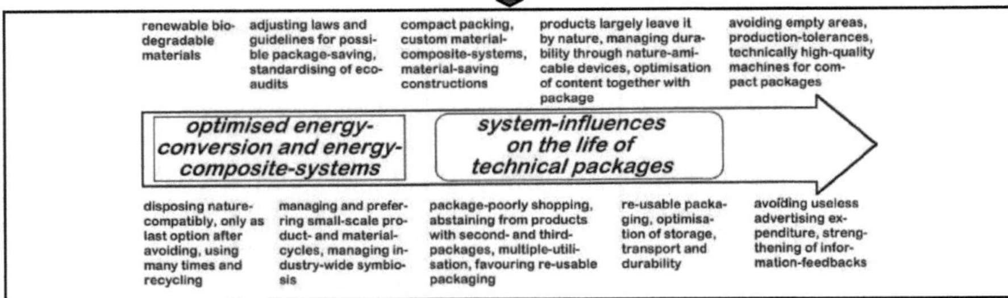

The strategies of bionic package technology come together in three linked paths: Material ⇔ organisation ⇔ community.

DESTINATION PATH 1 (TECHNICALLY-MATERIALLY):
Future packages have to be much more subjected to dynamic functions than present packages: Containment of control-oriented sensor-actor-systems – active volume-reduction after cleanout (the problem is not in the weight but the volume of disused packages!) – signalling of the self-life-durability of foods – self repair of cracks etc. – thermal adaptation when temperature alternates.

DESTINATION PATH 2 (ORGANIZATIONALLY):
Future packages have to be much more parts of cross-sector change of energy-converting and material-processes chains, than until now: Solutions of principles and organisation, accompanied by systemic functional network-analyses, reinforced integration of self-organisation processes in pathways from production to recycling – reinforced bonding of small-scale integrated networks into cross-sector companies.

DESTINATION PATH 3 (SOCIALLY):
Future packages have to require much more cross-linked activities to natural, humanistic and sociological disciplines, than until now, because they pass through almost all social fields of our life. One must get away from pure cost-efficiency and sustainability. And one must accept fault tolerance and environment-orientation as at least equivalent development targets. For this, there are many stimulation from bionics.

LITERATUR: Küppers, U. (2010): Verpackungsbionik der Zukunft. Verpackungsrundschau 9, 52–55. – Küppers, U.; Tributsch, H. (2001): Bionik der Verpackung. .. – Verpackte Technik. Wiley-VCH, Weinheim. – Nachtigall, W. (2002): Bionik. Springer, Berlin. S. 401–407.

Bionic package contra package flood

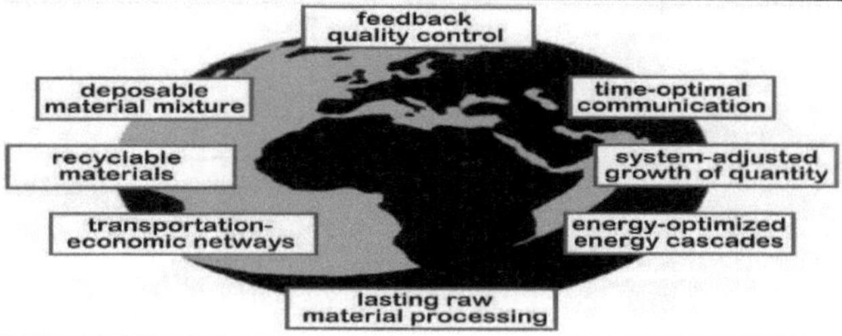

Not only packages themselves but also their circular flows "raw material ⇔ recycling" can be optimized bionically.

"PACKAGE" IN NATURE:
The covers of fruits that one can understand as „packaging“ for the protection of the embryo with its nutrient tissues are always laid out polyfunctionally. This is in complete contrast to the classic, technological packaging. Either a certain layer perceives different packaging functions, or these functions are distributed to different layers of a multilayer package (Coconut, comp. p. 64).

"PACKAGE CIRCULAR FLOWS" IN NATURE:
As in technology, in nature package specific materials, energies and information are managed in cross-linked closed loops. However, the nature stabilizes a dynamic balance in her packaging-system, in which she organises – temporarily and spatially highly linked – cross-specific packaging management. This happens by self-regulative adaptations. In the end, a stable total turnover results.

CHALLENGES FOR TECHNOLOGY:
The stabilizing, dampening effect of a cross-linked, coordinated control mechanism is still missing in the present technical package networks. At present, no balance can still be recognized in their cross-relationships; the package-flood and the involved problems are raising more and more (see e.g. the unabated growing plastic waste whirl in the north pacific). One must urgently work against this tendency by strategies of a biological evolutionary packaging optimization.

LITERATUR: Küppers, U.; Tributsch, H. (1993): Verpackungsstrategien der Natur – Vorbild für eine ganzheitlich vernetzte Materialwirtschaft. VDI-Berichte 1060, 333–343. – Küppers, U. (2010): Verpackungsbionik der Zukunft. Verpackungsrundschau 9, 52–55.

Bio-plastic as a package material on the rise

A

B

Until now, bio-plastic materials as package materials had a type of niche existence. However, this is changing now.

MOMENTARY STATUS:
While, until now, especially manufacturers of expensive products preferred petroleum-based package materials, e.g. polystyrene or air-cushion-foils, a process of rethinking took place in the course of a broadly increasing environmental consciousness. Now, more and more bio-based materials (corn-straw, sugarcane etc.) are preferred. In Nov. 2012, participants of the European Bioplastics Conference in Berlin have discussed these questions (A, B).

PROBLEMS:
To the extent, in which the bio-based materials leave their previous niche existence, problems occur of large scale favourable manufacturing, of application and finally of the question, what happens to the disused materials. Unfortunately, a still inadequate communication (green washing, proof for lower environmental noxiousness) exists. Momentarily higher prices and therefore a delayed admission into the mass market are negative aspects.

CHALLENGE:
Bio-plastics cannot use/show all of their immanent potentials yet.

- Example: For certain types of bio-plastics still end-of-life-solutions as well as utilisation streams must be developed.
- Example: Cultivation of useable biomass still can be optimized by innovations in the agricultural technology etc...

LITERATUR: Vbio 48/5: www.vbio.de/informationen7alle_news/e_17162 ?news id=13029.

Bees wax as building material; "honey bee state"

A

⇨

ventilator-chains; photos: Heilmann

B

"The bee state resembles a magic fountain; the more one draws from it, the richer it flows" (Karl von Frisch 1965).

BIOLOGY 1, HONEYCOMB-LIGHTWIGHT:
The summarizing representation of Tautz (2007 shows several interesting bionic aspects, which are only named here – true to the sentence of Frisch – but which are not scanned for already occurred technical realization. - Honeycombs of the honey bee (comp. p. 53) represent an extreme, high-loadable lightweight construction. For a nest containing 100000 cells, only 1200 g of wax are used, for what about 1.5 million wax-platelets are "sweat out."

BIOLOGY 2, HONEYCOMB-BORDER AS VIBRATION-CARRIER:
Approximately 50 percent of the wax is used for the bulged thickened upper honeycomb border. This represents the "communication-network and the memory storage for the super organism". For example, vibrations, as they are caused by a dancing bee, are distributed over the whole honeycomb by this hexagonal bulge network, which is supported by the tender honeycomb borders. They can be picked off by bees (A), sitting elsewhere on this network.

BIOLOGY 3, AUTO-ADAPTATION OF THE BEE-STATE:
The bee state is a high-complex system with adaptive abilities. Among other things, these lead to self-organisation (system-inherent ability for the attitude to environment-parameters) and to emergence formation (formation of qualities of the super-organism, which cannot be realised by a single organism, B). Honey bees are the single creatures, apart from men, who form their environment actively; evolution probably assesses at the system "state + environment".

LITERATUR: Tautz, J. (2007): Phänomen Honigbiene. Fotos: H. Heilmann. Spektrum, Heidelberg. – Frisch, K. von (1956): Tanzsprache und Orientierung der Bienen. Springer, Berlin. – Frisch, K. von; Lindauer, M. (1993): Aus dem Leben der Bienen. Springer, Berlin.

Avoidance strategy instead of treatment strategy

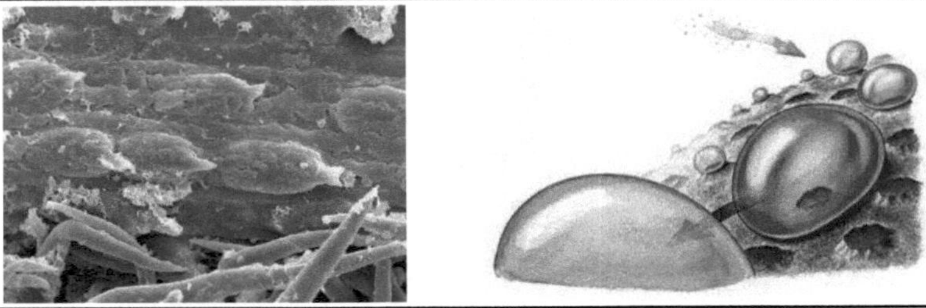

A B

Instead of treating the effects of noxa, the indirect strategy to avoid the creation of noxa is more effective.

EXAMPLE OIL-AEROSOLS:
If oil-lubricated machine tools run, oil aerosols with particle-diameters of 100-500 nm are created. These must be eliminated from the air. Until now, this was not satisfactorily successful. If such particles get into the respiratory tracts, there, they can cause conditions of irritation and illness like cough, toxic-allergic reactions and bronchitis. The optimization of elimination appears more meaningfully than the treatment of the effect of such noxa.

INCREASE OF THE EFFICIENCY OF ELIMINATION:
Until now, mechanical and electric elimination procedures were applied, that didn't lead to a near-quantitative elimination, however. In the BIONA-joint-project "3-D-bio-filters", the Inst. for Textile- and Process-Engineering Denkendorf, the Inst. of Geosciences (Tubingen) and 4 industrial project-partners try to increase the efficiency by "developing (energetic) high-efficient coalescence separators on basis of innovative three-dimensional, nano-structured filter media."

BIO-BASED TECHNOLOGY:
Several bio-based principles are converted technically-adequately:

- "Liquid-separation by the processes of combing out the fog and absorbing the dew (comp. p. 63) from desert plants (*Stipagrostis,* A), *Trianthema* etc, cistern plants (*Bromeliaceae*) and desert-bugs (*Stenocara*) (B).
- Self cleaning of the filter-media by nano scaled surface-structures (*Nelumbonaceae*"*;* comp. p. 65).

LITERATUR: Zitate nach: http://www.bionische-innovationen.de/#projekte/3dbiofilter.html. – Zum Lotuseffekt s. Lit. S. 65. – Zum Wüstenkäfer-Effekt s. Lit. S. 63. – Industriepartner des Projekts sind: Keller Lufttechnik, Junker Filter, Jetter & Herter, Hofmann Maschenstoffe.

Systemic management also for bionics

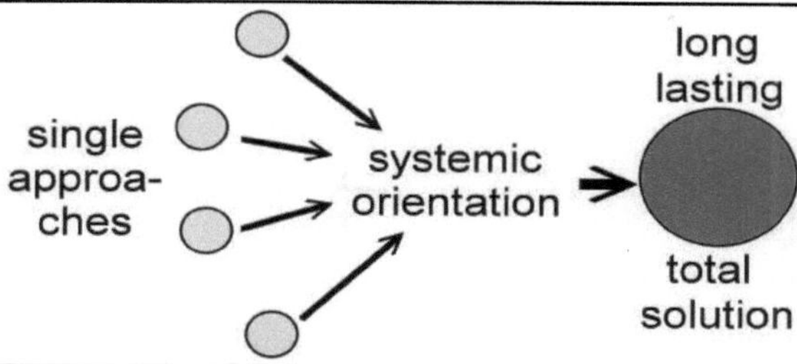

Even the "composite-science" bionics itself can be optimized to an increasing influence by systemic management.

NATURE STRATEGIES AND COMPLEXITY:
After his dedicated criticism of the momentary status of bionics (comp. p. 261), U. Küppers points out possibilities to promote the aspects of bionics as composite-science. Basis must be the willingness to learn to understand our natural environment in its complexity. "The methodical transferability of the systemic development and solution-strategy (of the nature) is applicable to every ... problem in a cross-linked complex environment."

CORRESPONDENCE OF LEARNING:
Until now, bionics was indeed fixed more or less to single products. In future, it admittedly should not turn away from it; however, open itself to a general strategic orientation. So one should study especially such aspects of the nature, which continue on this way. "In this sense the correspondence of learning is to be understood as a new system-oriented guide for lasting products, processes and organisation forms."

THE AGGREGATING NETWORK:
This concept should mark the "registration of real *cross-linked* connections in complex surroundings." In contrast to the registration-tools for *linear* relationships (tables, calculation-programs etc.) networks include only 2 basic items, namely influencing variables (in graphics: data) and directional relationships (in graphics: relationship-arrows). Therewith it is already possible to generate a picture of cross-linked effect relationships.

LITERATUR: Zitate nach Küppers, U. (2010): Systemisches Bionik-Management. Eine ganzheitliche Sicht auf nachh., umweltökon. Produktentwicklung. Wissenschaftsmanagement 1, 37–42. – Küppers, U. (2004): Bionik und Wirtschaftlichkeit. BIUZ 5, 316–323.

Animal swarms and collision avoidance

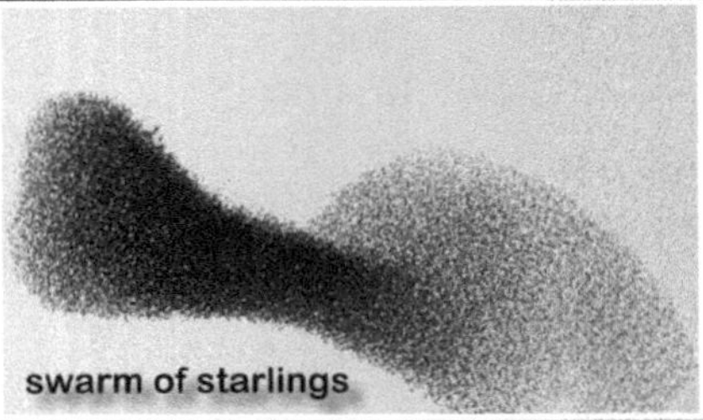

A

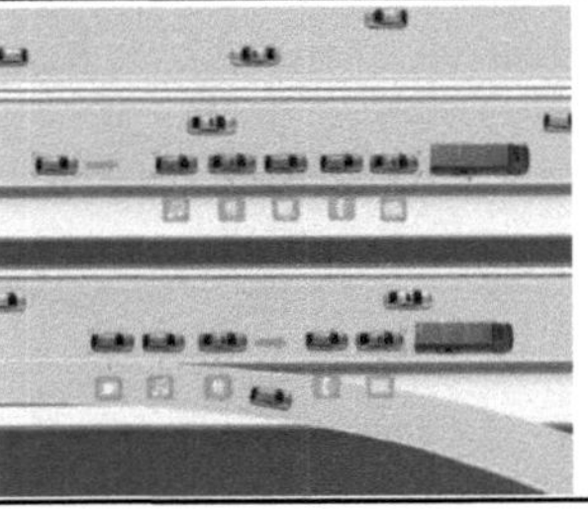

B

Also in dense animal swarms, single animals hardly ever touch themselves – possibilities for traffic-control?

BIOLOGY:
Swarms of birds (A), bats, fishes and insects can move very quickly in the area, at which they stay together also during sudden changes of direction, without colliding between individual animals. From time to time, fishes swim just to touch one another, while in the other swarms the degree of occupancy of one animal/m^3 already is regarded as "dense". But also here quick changes of direction of the swarm require very fast reactions of the "single element".

PRINCIPLE:
Ants build streets and apparently proceed coordinately to raids. On this occasion, they touch themselves only with common activities; otherwise, they hold a certain minimal distance, whereby the touch-sensitive antennas play a role. Aforementioned swarms beside optical and acoustic mechanisms possess especially "distance touch sensors" in form of pressure- and flow-sensitive elements. These act very effect to keep at distance.

TECHNOLOGY:
So-called "road trains" are driven by a professional driver. The vehicles, "docked" at the road train, for example private cars now do not drive independently from each other but drive in a convoy that is led by this one vehicle. The road train passengers can now eat comfortably and can hear easy music or look at a video, because computer sensors monitor completely independently their trip (B).

LITERATUR: Bonabeau, E. et al. (1999). Swarm intelligence: From natural to artificial system. Oxford University Press, New York. – Randelhoff, M. (2011): Das SARTRE-Projekt: Schwarmintelligenz für eine erhöhte Verkehrssicherheit. Übersichtsartikel s. Internet.

"Swarm intelligence" and management bionics

A 

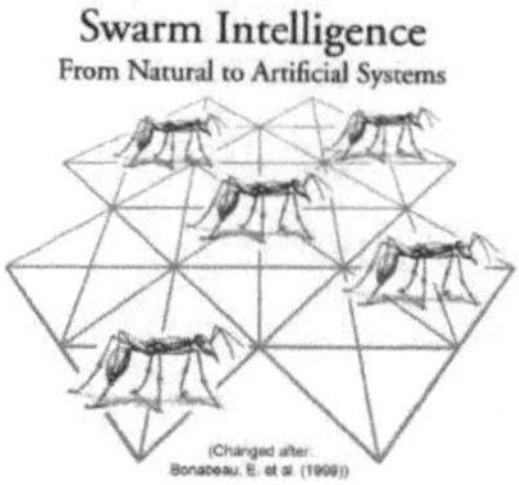

B

Recently, the concept of "swarm intelligence" was introduced, first transferred to system coherence, then to system control.

BIOLOGY:
If ants determine a way to a good food source, this happens by the track- and following-animals over unspecific communication and way marking (A). If a bee-swarm searches a new home, this happens also according to a "vote" over the most different offers of track bees by general behavioural mechanisms of the entire bee-swarm. So, the swarm normally finds the optimal one of many offers.

PRINCIPLE AND APPLICATION EXAMPLE:
For such mechanisms of animal swarms, that are usually unspecific, but lead to solutions of complex problems in the end, recently the term "swarm intelligence" was introduced. The American firm "Liquid Gas Company" optimized its distribution logistic with the help of the "ant route planning principle." So, the optimal route possibilities were sought and found (B), which were the fastest with the least possible energy consumption.

DIFFICULTY:
For example, it is difficult to discuss the control aspect of ant populations within a circle of managers. Exactly here, the term "bionics" often is misinterpreted; the transfer is taken too literally, seen in the sense of a nature copy. Principles only are transferable, however, abstracted, as the example of the "shortest-way-strategy" shows. These principles are inter-disciplinary. Testing them therefore is worthwhile. Bonabeau et al. (1999) have shown ways.

LITERATUR: Bonabeau, E. et al. (1999): Swarm intelligence: From natural to artificial system. Oxford University Press, New York. – Lloyd, T. (2002): Inside track: Organisational lessons from the ant colony. Financial Times, Apr. 30.

Bionics in the management: What's being able?

„Complexity lets it-self manage only with complexity."
(Asby's Law)
R. Asby (2005)

Therefore „biological systems supply best-practice-principles."

A series of bio-based terminologies meaningfully lead to problem solution strategies in the field of management.

BIOLOGICAL MODELS:
M. Piffner (2006) reminds on qualities of biological systems, "of which our organisations can only dream. They are fail safe, adaptive, flexible, reaction-fast and even self repairable. They know how one integrates and grows solidly, how one renews, how one disseminates, how one self organizes, how one minimizes a cost and how one simultaneously generates sturdiness." These can be bound into management methods.

INTEGRATION INTO MANAGEMENT METHODS:
From such bionic based management methods the author names a selection of five aspects as "approved models and methods from the kitchen of the management bionics, which are available to managers": 1. the viable systems, 2. the syntegration, 3. the operations room, 4. the bio-cybernetic sensitivity model, 5. the evolutionary innovation and the portfolio optimisation." These include at times specific bio-abstractions.

CLARIFICATION:
Ad 1: On the basis of the human ZNS, S. Beer (1972, 1979) has developed the model of "viable systems" for the structuring of organisation. Ad 2: To the syntegration comp. p. 251. Ad 3: Central topoi are the brain-analogues parameters: information, simulation, memory and attention. Ad 4: For F. Vester's bio-cybernetic sensitivity model (comp. p. 250). Ad 5: Rechenberg's evolution strategy (comp. p. 29–35) optimizes the expenditure of development.

LITERATUR: Pfiffner, M. (2006): Von biologischen Systemen lernen. New Managem. 12, 25–29. – Beer, S. (1972): Brain of the firm Wiley, Lond. – Vester, F. (2002): Die Kunst vernetzt zu denken. Dtv, München. – Rechenberg, I. (1994): Evolutionsstrategie 94. Frommann-Holzboog, Stuttgart.

Bionics in the management: What's not being able?

„The nature cannot be a general model for us. There was also a social and cultural evolution."

„From which solutions one wants to learn, the human beeing must evaluate this ultimatively by himself."

Analogues from biological systems will influence much the management of the future: But, there are clear borders.

BIOLOGICAL NOT MODELS:
M. Pfiffner (2006) has completely right if he points out the following three aspects: "1. The nature is no model. – 2. The nature doesn't deliver any blueprints. – 3. Superficial analogies and metaphors can be dangerous." Instead he clarifies the right way: "The goal of management bionics is not to return to nature." Rather future management simply wants to profit "from solutions, which men didn't develop themselves."

CLARIFICATION:
The named three aspects are clarified as follows. Ad 1: "The management bionics doesn't want to abet romantic ideas like biologism, ecologism, sozial darwinism or neo liberalism. Ad 2: It is not possible to transfer solutions in the direct way. Ad 3: Superficial analogies and metaphor (for example money flow of the national economy, blood circulation of men) are not only useless but also dangerous."

CONSEQUENCE:
"One must avoid the traps of the application of the bionics onto the management, but, nevertheless one must learn, what one can learn. In future, the best-practice is not being found in comparison with other companies. Whoever compares himself with others, can be at most as good as these. The competitors, who compare their existing solutions with the optimized solutions of biological systems, become better." – This already happens.

LITERATUR: Pfiffner, M. (2006): Von biologischen Systemen lernen. New Managem. 12, 25–29. – Beer, S. (1972): Brain of the firm. Wiley, London. – Beer, S (1979): The heart of enterprise. Wiley, London – Ashby, W. R. (1956): An introduction to cybernetics. Chapman & Hall, London.

Are ant strategies transferable into the management?

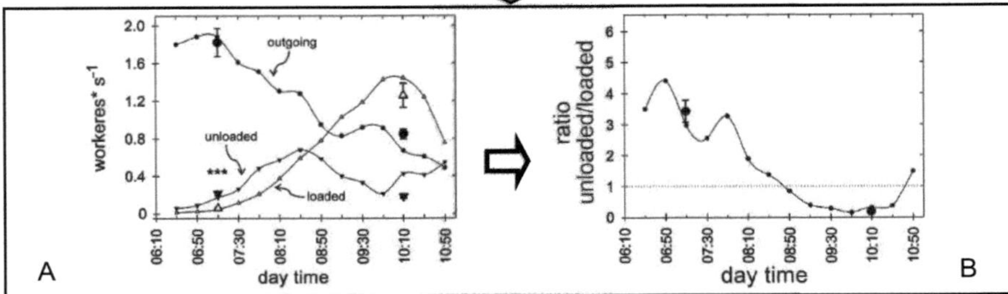

If ants find a new food source, they first inform the hive mates instead of reaping everything now.

BIOLOGY:
The leaf-cutting-ant (*Acromyrmex heyeri*) from Uruguay is specialized on reaping grass-stalks. If swarmed ants have found a suitable harvest place, only 30 percent immediately start with the harvest, but quickly run back with small and light shreds of grass. 70 percent run back unloaded, locate an odour trail on that occasion and inform other ants on the way back and in the nest, who then swarm out along the trail.

PRINCIPLE:
Apparently, after discovery of a harvest place it is more advantageous for the discoverers to direct as fast as possible different individuals there, than immediately to begin with the harvest. If a competing stock discovers the same food source, this stock is at an advantage, that could recruit the larger number of harvest helpers. It therefore is strategically more favourable to recruit as many and as quickly possible helpers by information transfer (A, B).

TRANSFER POSSIBILITY:
It is said, that strategies, with which insect states secure their survival, are of high interest for managers, who, in the end, have also the duty to secure the survival of their firms. However, an 1:1-transfer is problematic. Where can one apply the "strategy of the quick recruiting"? Maybe a competition problem with time-pressure appears, that should better be processed quickly by many than slowly by lesser?

LITERATUR: Bollazzi, M.; Roces, F. (2011): Information needs at the beginning of foraging. PloS ONE 6(3): e17667.doi:10. 1371/ journal.pone.0017667. – Oeller, K.-H. (2011): Integrated Implementation Management. Malik SuperSyntegration®

Evolution management: Evolution in the management

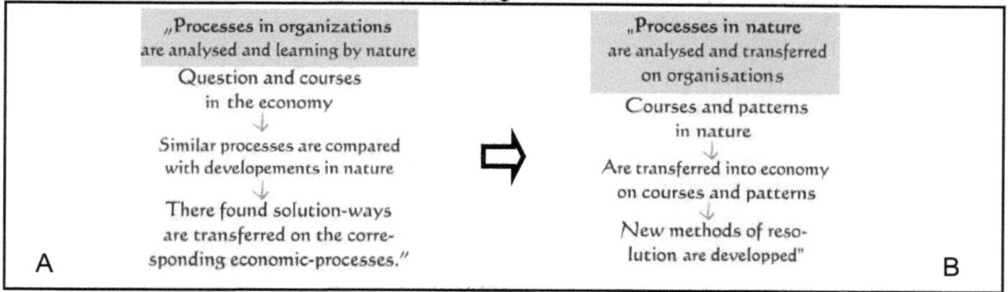

"Developments in and between organisations are executed ... after similar patterns as evolutionary processes in nature."

DEFINITION:
"Under evolution management we understand an approach to the management of organisations with which the processes in and between organizations are regarded as processes of life, which proceed according to the same ... principles and regularities of other processes, executed in nature. From these comparable natural processes, one can learn for the individual operating level of the manager and the level of the total processes."

WAYS OF THINKING:
F. Malik names 5 bases for this: 1. comparability of the patterns for developments, 2. analogy to the learning from bionics for technical solutions, 3. the human being given wilful possibility to intervene in evolutionary processes, 4. transfer of knowledge from the biology to the level of organisations as well as leading behaviour of the single manager, 5. reintegration of the economy system into biological events (e.g. sustainability).

PROCEDURE:
There are different possibilities in order to learn from evolutionary processes in the nature for the management. We can start from nature (A) and transfer the realizations found there to organisation processes, or we look at questions in organisations (B), check, which solutions the nature has selected in comparable situations, and let us inspire for the organisational work." (All quotations by F. Malik).

LITERATUR: Malik, F. (2008): Strategie des Managements komplexer Systeme. Ein Beitrag zur Management-Kybernetik evolutionärer Systeme. 10. Aufl. Haupt, Bern. – Malik, F. (2011): Strategie: Navigation in der Komplexität der Neuen Welt. Haupt, Bern.

VSM for viable organisation structures

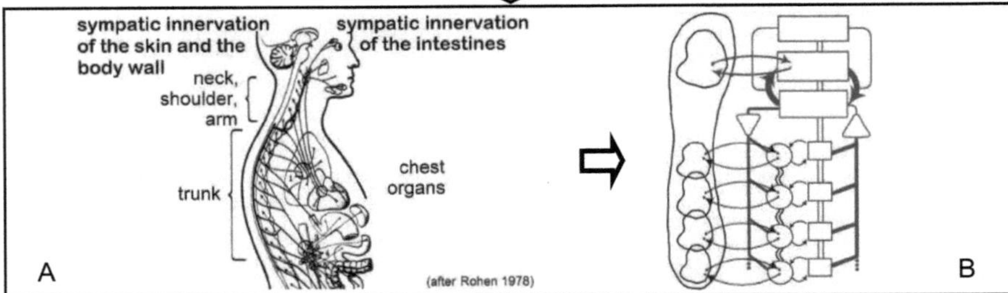

In 1972, the "Viable System Model" first was developed by Beer as an analogue and then as a strict model of the CNS.

BIOLOGY:
This central nervous system (CNS) of the human being consists of parts of the brain, which stand together in cybernetic connection, and the spinal cord. Each higher-level area gives information to diverse subordinate areas and receives information from these that influences its further information brokering. The total system therefore works hierarchically steered, as well as regulated in extremely manifold meshing.

PRINCIPLE:
The VSM of the management cybernetic Beer transfers the "law of the viability" on all types of economy and administrative organisation (B). Simplified by the management specialist F. Malik and adapted to practice, "today, it serves as "blueprint" for the diagnosis and the design of organisation structures of all types and sizes of institutions". It is essentially that the latter then are developing in a functional, sustainable and adaptable way.

EFFECTIVENESS:
The application of this system in the management doesn't provide profit-maximization, but, over customer orientation, leads to the maximisation of the viability of a firm – quite analogous to the evolutionary fitness. "One recovers all strategic and operative functions in the correct composition and evaluation; in particular, however many questions of the traditional organisation teachings and their inconsistencies are dissolved." (Oeller)

LITERATUR: Beer, S. (1972): Brain of the firm: The managerial cybernetics of organization. Wiley, London. – Beer, S. (1990): Diagnosing the system for organization. 2. Aufl. Wiley, Chichester. – Malik, F. (2000): Strategie des Managements komplexer Systeme. 6. Aufl. Haupt, Bern.

EKS for viable organisation structures

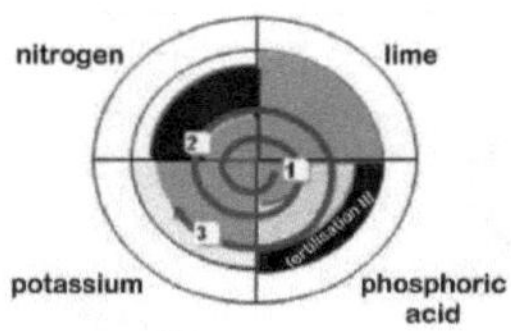

A J. v. Liebigs „minimum law"

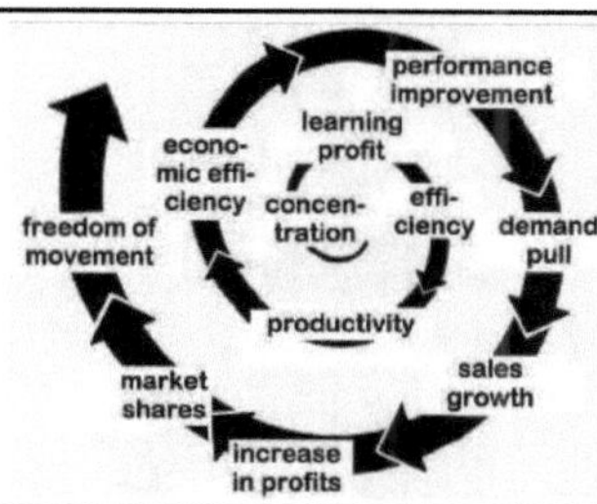

B

In 1970, Wolfgang Mewes developed the "bottleneck oriented (behaviour- and) leadership strategy" (EKS) close to nature.

BIOLOGY AND ECONOMY:
In biology, on the one hand inner dispositions (factors) decide about what develops from an element. The type of this development on the other hand, whether rather positively or negatively, is decided by outer environmental influences that this element cannot influence. In principle, this is similar in the economy; however the outside influences can be partly influenced here. The knowledge about cybernetic correlations decides about the success.

PRINCIPLES OF THE EKS-STRATEGY:
The 4 EKS-principles receding on J. v. Liebig (A) are:
- to concentrate the power to the potentials of strength; reduction of frittering,
- orientation of the strengths to an compact target group,
- to go in the gap (niche),
- to go into the depth of the problem; to target the market leadership.

Therefore: To find out the system connections and use them "from indoors" (B).

IMPLEMENTATION:
In the year 2008, Malik management took over the rights at EKS®. The contents, to which large firms like Würth and Kärcher attribute their successes, are mediated by 11 web modules: "Unique selling point and market-leadership", for self-studying, accompanied by regional presence events. They run alongside the profession of a user over 1 year, require about 100 man-hours and come to an end with a diploma of the promoter.

LITERATUR: Mewes, W. (2000): Die acht Ursachen für die überlegenen Wirkung der EKS-Strategie. (s: www.wolfgangmewes.de/) Vortrag Schönherr Bindesysteme GmbH, Maschen/Hamburg. – Malik, F. (2010): Seminarprogramm – Malik Open Management Education 2010.

The "art of network thinking" in management

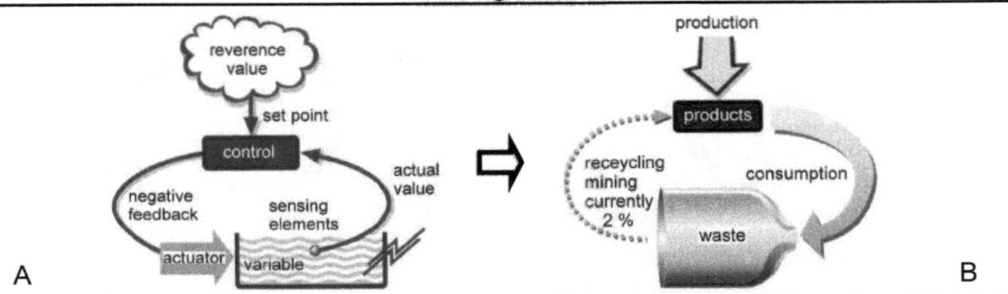

Today, Frederic Vesters approaches of the 90-er years win a special importance for economy and management.

"BIOLOGICAL RULES" FOR A MANAGEMENT OF COMPLEXITY:
F. Vester (1999) positioned 8 rules from the research of nature (partly quoted):

1. Negative feedback (-fb) must dominate (A) over positive (+fb).
2. +fb brings the things running, -fb then guarantees stability against disturbance.
3. The system must work function and not production oriented.
4. Using of existing power (jujitsu principle instead of the boxer method).
5. Multiple utilization of products, functions and structures of organisations.
6. Recycling: Use of circle-processes for the waste- and sewage-utilization (B).
7. Symbiosis: Mutual benefit of heterogeneity by coupling.
8. Biological design of products, procedures and organisation forms.

The basis of these rules is the consideration of complex economical technical associations as nature analogous systems with many times cross-linked elements by feedbacks and substitution. For practical use, Vester has developed his *sensitivity model* from this basis.

MEANING FOR THE MANAGEMENT:
F. Malik puts out the meaning of this model for the management practice: "Decisive is not the surface and the knowledge of the elements of a system but the knowledge of the relationships, which exist between them. Sensitivity models are used in the *management bionics* especially for modelling the inner structure of an institution in relation to its environment and for finding the cybernetic most effective adjusting levers."

LITERATUR: Vester, F. (1999): Die Kunst, vernetzt zu denken. Ideen und Werkzeuge für einen neuen Umgang mit der Komplexität. DVA, München. – Malik, F. (2011): Bionik im Management – Strategie. Navigieren i.d. Komplexheit der Neuen Welt. Campus, Frankfurt.

Sensitivity model and Syntegration

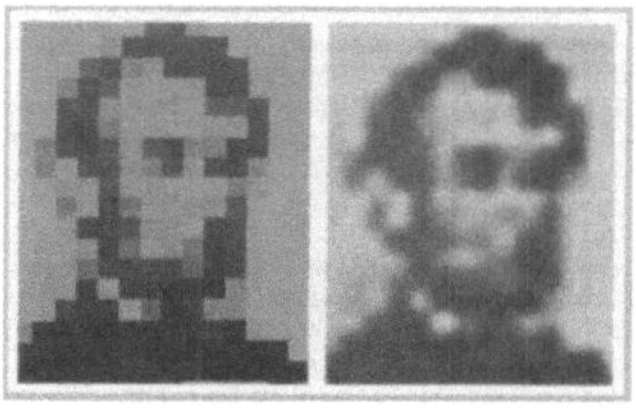

A

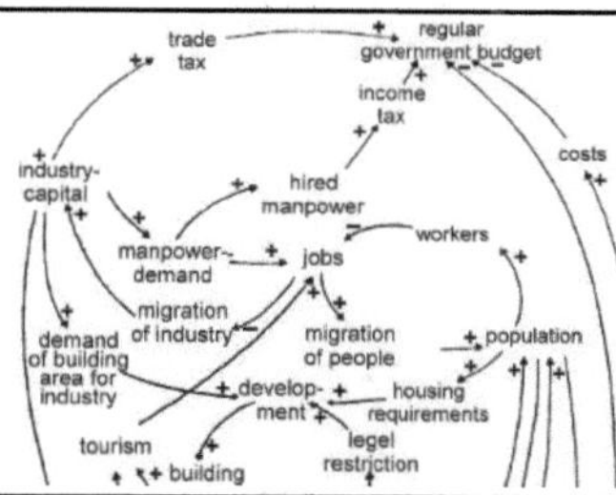

B

Due to Vester's sensitivity model (since 1985) the treatment of complex problems became practicable, basis for syntegration.

CONNECTIONS IN THE BIOLOGY:
Each biological system is topmost complex in comparison to the technology: In the micro-biological field the life-period of a bacterium exposed to many influences, in the physiological field the frequently regulated blood circulation, or in the ecological field the stability of the interaction of the extraordinarily many cross-relationships in an edge of a forest. – All single relationships cannot be "understood" in the sense of a logical working off.

PRINCIPLE:
The sensitivity model is based on Vester's approach of "cross linked thinking." This approach tries to find out the internal connections in complex systems. If this is not already possible linear-logically, it tries to approximate their tendencies and to describe them – in fact like the "fuzzy logic" – by a type of "blurred view" on the total (A) instead of a sharp view on single elements that maybe not at all so very important.

FITTING THE MODEL INTO MANAGEMENT STRATEGIES:
The model has been proved very successful, if it goes to recognize essential influences on complex operational connections (B) and to make the system crisis-proof against not predictable alterations of such influences. That means to guarantee the survival of the system. That is also the main task of any management. Today, this model as well as the super syntegration (idle-free working together and others) is integrated in the Malik management.

LITERATUR: Vester, F. (1999): Die Kunst, vernetzt zu denken. Ideen und Werkzeuge ... DVA, München. – Malik, F. (2011): Strategie: Navigieren in der Komplexität der Neuen Welt. Bd. 3: Management: Komplexität meistern. Campus, Frankfurt, New York.

CONCEPTIONS

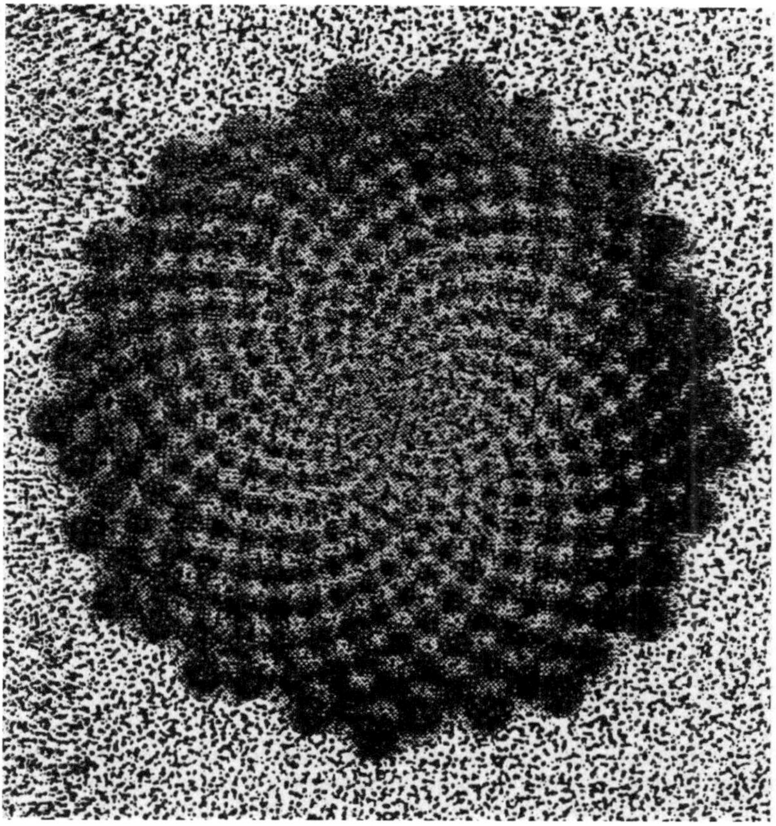

AND DOCUMENTATION

W. Nachtigall and A. Wisser, *Bionics by Examples*,
DOI 10.1007/978-3-319-05858-0_14

CONCEPTIONS AND DOCUMENTATION

Bionic approach – Bionic as socio-political challenge and multidisciplinary aspect of viewing – Bionic as creativity training – Bionic and theory of cognition – Publication – Documentation – The "bionic promise" and bionic value chains – Warning of trivial bionics.

Here on the one hand, bionic strategies are summarized and solution areas defined, which include bionic methods. These methods can also lead to superior complexes by synopsis of single disciplines. On the other hand publication-, exhibition-, and documentation-possibilities are presented.

The Lu-method: Specified approach

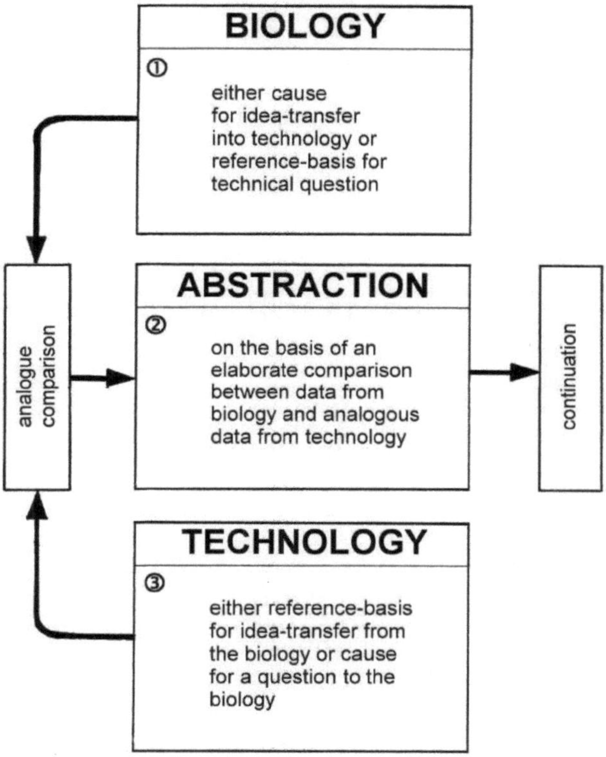

(comp. p. 256, 259)

THE PROCEEDING OF THE LU-METHOD:
In the Lu-method (Lu = "Luscinius"; latinised surname of the first author) there are only few steps, however, these can include large areas. Therefore, the selection of an analogue, that is suitable in every case, must result from a large data pool. This either requires a very pronounced detailed knowledge or inter-disciplinary cooperation. Or one installs inter-steps, in which the steps 1–3 of the above flow chart are iterated.

LITERATUR: Nachtigall, W. (2010): Bionik als Wissenschaft. Erkennen → Abstrahieren → Umsetzen. Springer, Berlin.

Lu-method: Biology at the beginning – general

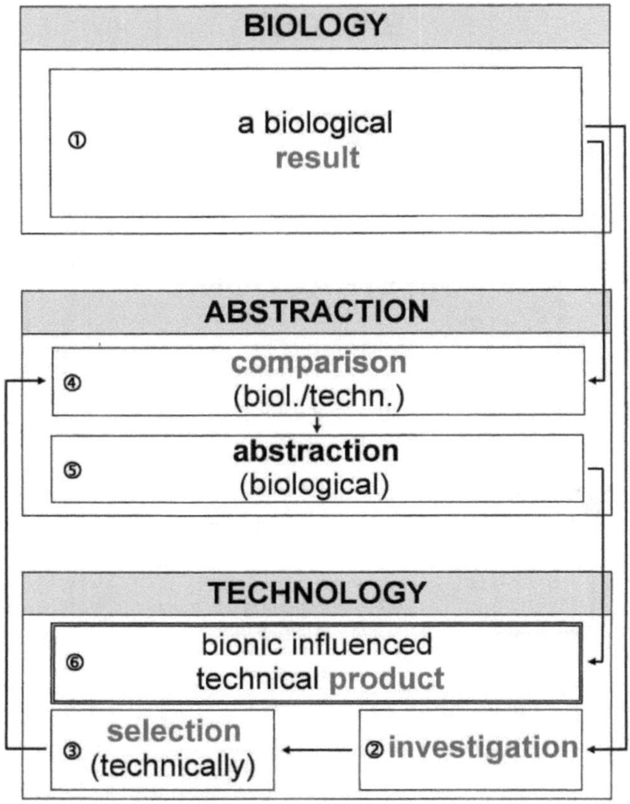

SIX-STEP-WAY:
If a discovery from the biology stands at the beginning, the question is: "What could one start with certain biological results in the technology?" Therefore, one researches in the technology and looks for a problem there that represents the lock for an existing biological key. The accessory six-step-way is sketched in the above flow chart. A (classic) example of use is on the next page.

LITERATUR: Nachtigall, W. (2010): Bionik als Wissenschaft. Erkennen → Abstrahieren → Umsetzen. Springer, Berlin.

Lu-method: Biology at the beginning – example

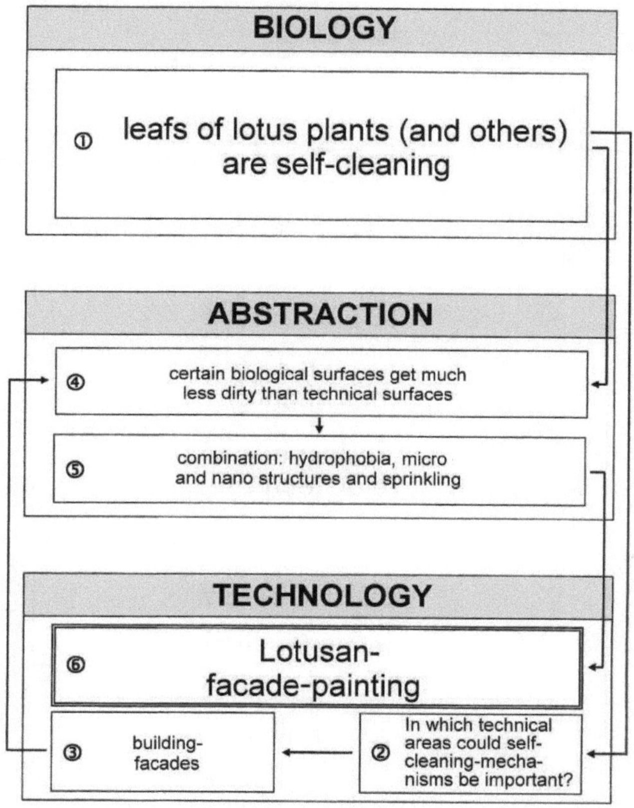

EXAMPLE FOR THE SIX-STEP-WAY:
The self cleaning-principle of the lotus leaf found a first technical application in the conception of the new facade-painting "Lotusan" (at that time the firm Ispo). Starting point was the observation that especially the surface of the leaf of the Indian lotus flower (*Nelumbo nucifera*) is self-cleaning (Bartholtt, Neinhuis 1997). So the idea was developed, one could make building-facades self cleaning.

LITERATUR: Nachtigall, W. (2010): Bionik als Wissenschaft. Erkennen → Abstrahieren → Umsetzen. Springer, Berlin. – Barthlott, W.; Neinhuis, C. (1997): Purity of the sacred lotus or escape from contamination in biological surfaces. Planta 202, 1–8.

Lu-method: Technology at the beginning – general

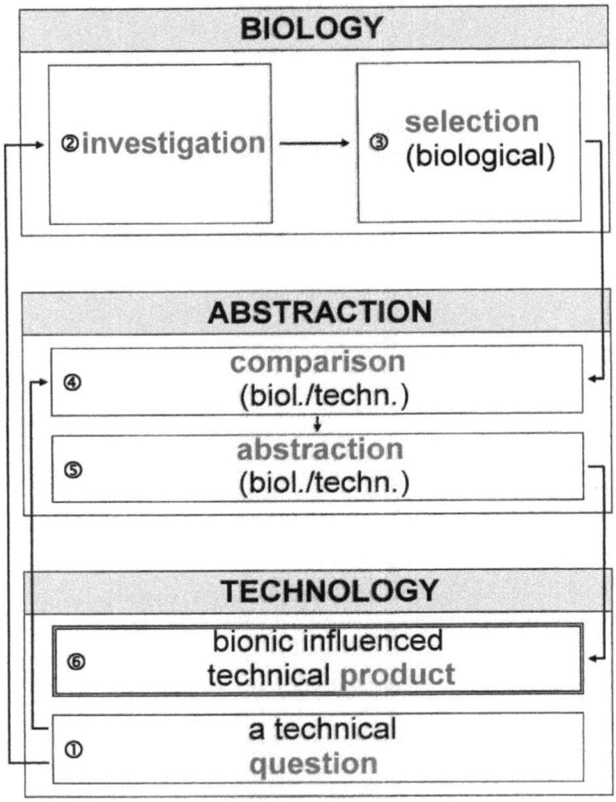

SIX-STEP-WAY:
If a problem of the technology stands at the beginning, the question is: "Which results from the biology could help to solve a technical problem?" Therefore, one researches in the biology and looks for an analogous model there, which represents the key for an existing technical lock. The accessory six-step-way is sketched in the above flow chart. An example of use is on the next page.

LITERATUR: Nachtigall, W. (2010): Bionik als Wissenschaft. Erkennen → Abstrahieren → Umsetzen. Springer, Berlin.

Lu-method: Technology at the beginning – example

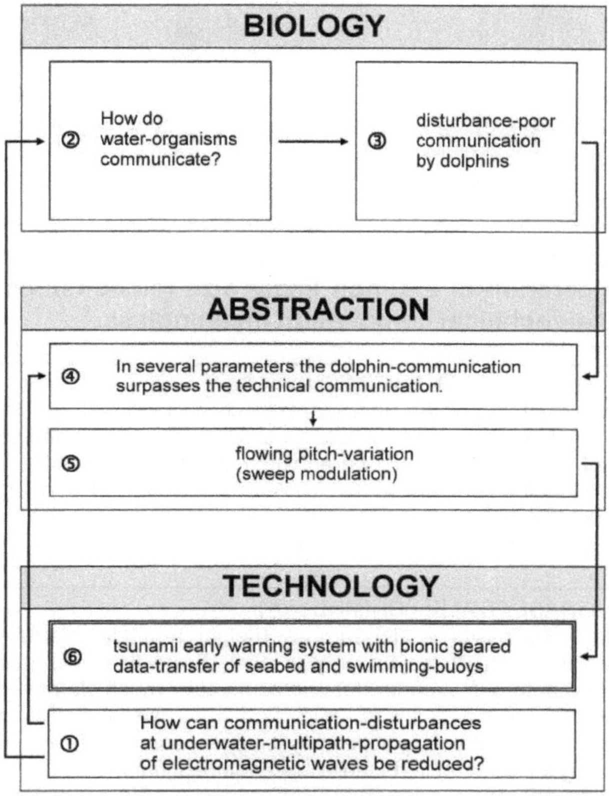

EXAMPLE FOR THE SIX-STEP-WAY:
The example refers to the removal of disturbance at wireless underwater communication by use of the "dolphin-principle". That is the way, dolphins compensate disturbances during their underwater communication, which originate due to different propagations and with it runtimes of the signals. According to this principle, a not so sensitive early warning system for tsunami was conceived by Yakovlew and Bannasch in 2006.

LITERATUR: Nachtigall, W. (2010): Bionik als Wissenschaft. Erkennen → Abstrahieren → Umsetzen. Springer, Berlin. – Yakovlew, S.; Bannasch, R. (2006): Maritime Technik. Von der Delphin-Kommunikation ... Bionik-Industriekongress. BIOKON, Berlin.

Beware of trivial bionics

Not-consideration of e.g. non linear size relationships makes a biological-technical comparison meaningless.

ANALOGY RESEARCH AT THE BEGINNING:
On p. 13, the meaning of the analogy-research was pointed out. It was said that a "courage-full comparison" should be positioned at the beginning and that every type of comparison should be admitted. Therein it is admittedly implied that – as soon as the comparison has fulfilled its heuristic function – for the further procedure one has to consider basic physical-technical relationships. If this is not taken into account, very quickly physical nonsense can result.

EXAMPLE: GRASS STALK – CHIMNEY (A)
The rye-stalk possesses a slenderness (height h/middle diameter $d_{\varnothing}$) of 400:1, a factory-chimney maybe only of 15:1. From this, it is occasionally concluded that nature is superior; for she could build much slender. However, according to the Barba Kick law of the proportional resistances for increase in size, not the linear relationship $d_{\varnothing} \propto h^1$ but the exponential $d_{\varnothing} \propto h\sqrt{h} \propto h^{1.5}$ is to be used: Higher constructions therefore must be clumsier.

EXAMPLE: JUMP CAPACIBILITIES OF THE FLEA (B)
A flea of $l = 2$ mm length can jump up approximately 30 cm. In comparison, it is called so, a 2 m big human being should have to jump over a 150 m high-rise. Active muscle power p_{mu}, however, is proportional to the cross-section ($p_{mu} \propto l^2$), passive weight force f_w, however, is proportional to the cube ($f_w \propto l^3$). Therefore it is $p_{mu}/f_w \propto l^{-1}$. So a relatively smaller force is available for larger creatures to lift off. Therefore, their jump height must be smaller.

LITERATUR: Nachtigall, W. (2001): Biomechanik. Grundlagen – Beispiele – Übungen. 2. Aufl. Vieweg, Braunschweig/Wiesbaden.

Bionics: criticism

The biggest enemy of bionics is the pseudo-bionics

Bionics – as well as other disciplines – has fallen into the "acceleration trap" and so subjected to inflationary fashions.

ALSO BIONICS IS IN THE "ACCELERATION TRAP":
Consequences of the speed increase in our world-wide network are "linear strategic and economic action full of short-sightedness, instability and intolerance to mistakes. This is also effective to the forward-looking discipline and compound science of bionics ... Whoever destroys the bases and mechanisms of the evolution step by step, takes off the fundamental bases of the bionics and their forward-looking solutions from him."

PRESENTELY, BIONIK IS FIXIED SUBOPTIMALY:
"Bionic research and development is still focused too much to the individual effect, organism or on the single, technically exemplary detail. With this often monocausal work strategy, the natural model is put artificially into an isolated "research-object", thus being cut off from its cross-linked reality. The consequence is: "Biological effect recognised, bionic model built, but improper for the market!"

PSEUDO BIONICS IST DANGEROUS:
"The current inflationary expansion of the bionics (many believe to be a bionic scientist if in a television show a bionic experiment was explained just half-heartedly to them) or the jump at the train of the partly journalistically exploited pseudo-bionics is exactly the known quantitative effect, which harms the quality of the bionics more than it is of use." A way out could be the systemic management of the science bionics (comp. p. 233 ff.).

LITERATUR: Zitate nach Küppers, U. (2010): Systemisches Bionik-Management. Eine ganzheitliche Sicht auf nachh., umweltökonomische Produktentwicklung. Wissenschaftsmanagement 1, 37–42. – Küppers, U. (2004): Bionik u. Wirtschaftlichkeit. BIUZ 5, 316–323.

Combination of bionic effects

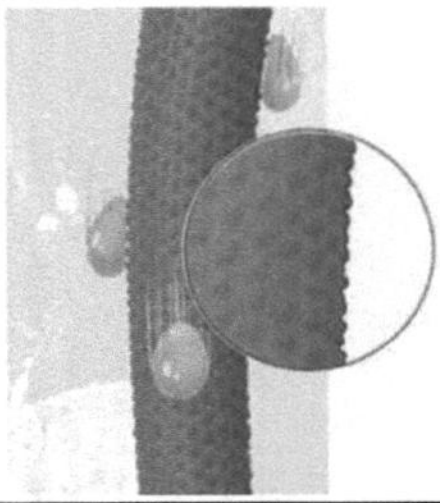

A

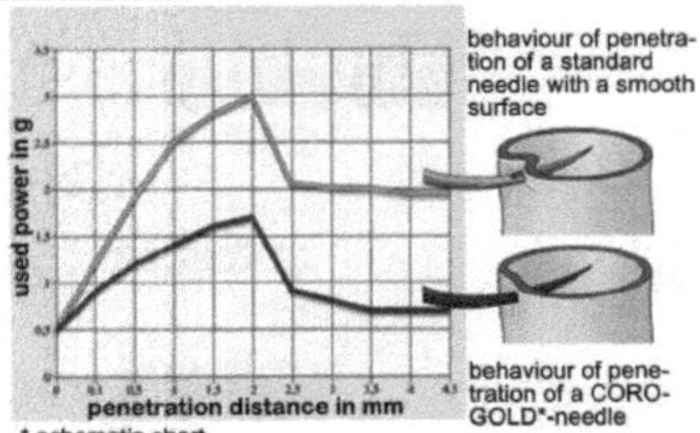

* schematic chart

B

Stimulations from the living world for the technology can be used, changed and combined in any manner.

LOTUS EFFECT:
The Indian lotus flower (*Nelumbo nucifera*) carries always clean leaves. Their surfaces are structured by papillae that are composed of protuberances from hydrophobic wax crystalloid tubules. On such surfaces, water drops roll up inevitably, by uncoiling themselves. Particles of dirt, whose adhesion to these structures is reduced, develop higher adhesion forces to the water-drops and are taken away with the uncoiling drops (comp. p. 65).

SHARK SKIN EFFECT:
The skin scales of fast swimming sharks are ordered so that the rills of one scale pass over into the rills of the next one. So, interrelated longitudinal lines are built over the streamlined shark body quasi. Therefore types of „streak lines" are built along their bodies. Among other things these prevent that transverse vortexes are formed and reduce the frictional drag by reduction of turbulent shear stress (comp. p. 210).

EXAMPLE: COMBINATION OF LOTUS AND SHARK SKIN EFFECT:
While the first effect reduces the adhesion, the latter provides less surface resistance. The Fumedica medicine-technology succeeded, in cooperation with the needle manufacturer FSSB, to combine both effects for the optimisation of operation needles (A), in fact by bringing together a special mechanical treatment (knops) and a coating. Thereby, the needle penetrates much smoother (B).

LITERATUR: Barthlott, W.; Neinhuis, C. (1997): Purity of the sacred lotus ... Planta 202, 1–8. – Bechert, D. et al. (2000): Fluid mechanics ... Naturwiss. 87, 157–171. – Fumedica Medizintechnik (2004): Datenblatt zur Entwicklung der Operationsnadel Coro-Gold®.

Value chains in biology and economy

"All countries are developing increasingly complex, intercompany value chains. The quantitative growth of these global production and logistic networks cause an increase in the global transported goods as well as the total distance transport routes. Due to the increase in transport expenses value chains and their logistic processes have an immediate relevance concerning climate and resources."

In nature biological systems exist, that can be model for a formation of value chains in economy.

BIOLOGICAL VALUE CHAIN (WKS):
Depiction of an example of the leaf-cutting ant (*Atta spec.*): 1. Fungi grow on leaf-mash. 2. Food is generated. 3. Fungus-parts serve as food. 4. Fungus-breeding takes place continuously. 5. Fungus-breeding consists of sub-processes. 6. Actors interact chemically and tactile. 7. Input factors are leaves, saliva and excrements. 8. Workers are the *Atta spec.* actors. 9. *Atta spec.* is product user *and* provider. 10. Leaf parts are carried, handled and stocked.

VALUE CHAIN IN THE ECONOMY, CHARACTERISATION:
"1. Products result from transformation of input factors. 2. Customer benefits are generated. 3. Customers use the products. 4. Products are generated in series. 5. Transformation takes place in sub-processes. 6. Information is generated and required. 7. Input factors are materials and resources. 8 / 9. Working actors / producers and customers are similar. 10. Logistical activities, as transport and handling etc., exist." (Seipold 2012)

BIONIC PROCESS MODEL FOR VALUE CHAINS:
"With the help of the specialised process models, options for action can be derived from the living nature for the formation of value chains in the economy. In three consecutive phases, biological value added chains are identified with the help of searching grids. There out, suitable organisation principles are derived and these are then transferred to concrete courses of action into the economy." (Seipold 2012)

LITERATUR: Seipold, P. (2012): Entwicklung eines bionischen Vorgehensmodells zur Gestaltung von Wertschöpfungsketten, Dissertation (unpubl.). TU Hamburg-Harburg.

Bionics and philosophy, theory of cognition

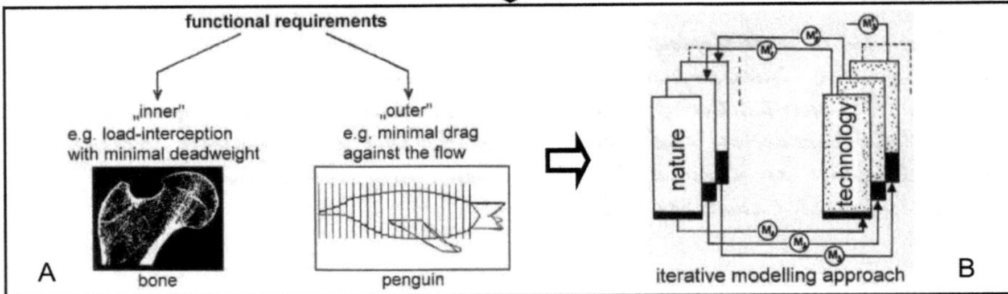

First in 2005, philosophers expressed themselves to epistemological problems in bionics – Summary 2010.

SINGLE CONTRIBUTIONS 2005:
T. Rossmann and C. Tropea published a book, in which philosophers comment in detail and critically for the first time: P. Gehring on bio-politics, A. Nordmann on techno science (change of the science culture), J.C. Schmidt on circulation theory and interdisciplinarity, G. Specht on bionic products, P. Euler on educational principles. After them bionic approach is still "epistemological mined area."

SUMMARY 2010:
In the second volume of the bionic-trilogy, W. Nachtigall has treated the "Bionics as a science" in a summarizing point of view. In fact, the sense of the bionic procedure is: A) Cognizing → B) Abstraction → C) Converting. Ad A: Biological basis; research, describing, judging. Ad B: Abstraction of biological results; working out general principles. Ad C: Technology implementation; conceptual, comparison of principles, approach.

CRITICISM OF BIONICS:
Disagreement dominates the concepts of function and design and their access to the organismic complexity. Krohs (2005) sees "biological design" as a concept coupled with the biological term of functionality (W.N. agrees with him) (A), however, a problem of intentionality rises. Also, the current model concept in biology and technology, in particular the problem of the model transfer (B) is estimated critically.

LITERATUR: Rossmann, T.; Tropea, C. (2005): Bionik. Akt. Forschungserg. ... Springer, Berlin. – Nachtigall, W. (2010): Bionik als Wissenschaft. Springer, Berlin. – Krohs, U. (2005): Biologisches Design. In: Krohs, U.; Toepfer, G. (Hrsg.): Philosophie der Biologie. Suhrkamp, Berlin.

Is the "bionic promise" kept?

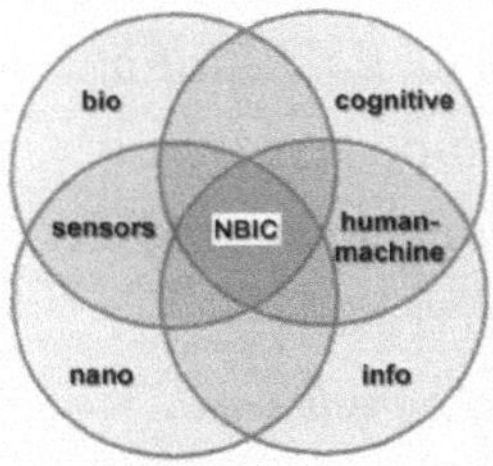

A

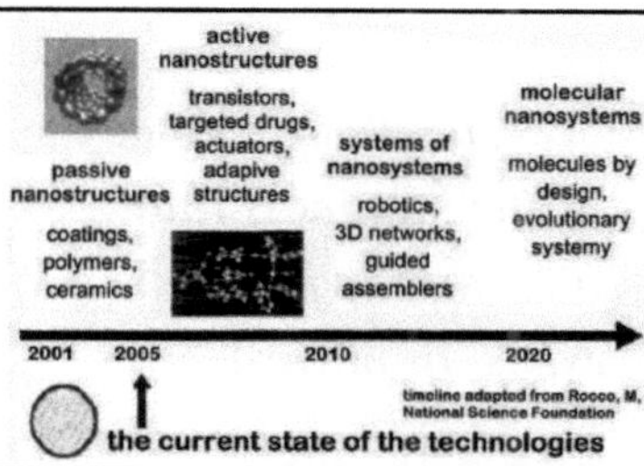

B

With the confusion of disciplines to complex techno-sciences, the responsibility of the bionics increases.

BIOLOGY:
As "bionic promise" A. von Gleich (Technological Design and Development, Univ. Bremen) formulates a type of a guideline. This marks, "how much of the ecological appropriateness and evolutionary tested awareness of the model can still be found in the respective technical bionic solutions". The problem: Bionics is in the end "applied technical biology" (Nachtigall 2002), and at the same time a typical techno science (Nachtigall 2010).

IMPACT AS TECHNO-SCIENCE:
As such, bionics has share "on the development of a quite distinctive line of new technical possibilities". So, nano-, bio-, information- and cognition-technologies converge to the interdisciplinary NBIC (A). Hereby, the danger exists that not much "of the evolutionary tried awareness and multi-dimensional optimisation can be transferred from the biological model into the technological solution". How bionically is the solution still then?

CONSEQUENCE:
With the formation of new techno sciences and their confusions also the "depth of engagement and the power" of the bionics changes. For example, nano bionics shows *novel* risks (B). "The more *synthetically* a solution is (comp. "synthetic biology"), the more one must observe the compliance of the "bionic promise". Otherwise one would speak of bionics with its "proved solutions"; however, one would not be any more bionically in the end.

LITERATUR: Gleich, A. von: (2006): Berechtigung und Reichweite des „Bionischen Versprechens". In: Kesel, A.; Zehren, D. (Hrsg.) (2006): Bionik: Patente aus der Natur. 3. Bionik-Kongress. HS Bremen 2006. . B-I-C, Bremen. S. 184–193.

Bionics: Potentials and perspectives

"A crucial advantage of the bionics is that already the result exists 'in natura'." ⇨ *„The potential of bionic research is internationally increasingly acknowledged."*

The 92-page pre-study of the German Parliament of 2006 was the basis for expertises and promotions.

CONCEPT:
With this study, the bionics was for the first time outlined more extensively by the political side. It was divided in the following 5 aspects: 1. characterization, 2. overviews, 3. bionics for new materials, 4. new bionics, 5. continuation and guidance. Several appendixes refer among others on actors and themes of bionic research, technology centres, networks, universities and research facilities, as well as also business companies in the home- and foreign country.

EVALUATION:
Beside the essential presentation of the historical development up to the differentiation into sub-zones, the study contains very essential comparative combinations. To them belong German-language definitions, BioKoN-professional groups, publications from universities, funding programmes for new materials (a main focus), analogies of technical and molecular machines, industry actors and commissioned expertises.

APPENDIX:
A good collocation of literature up to approximately 2005 includes about 150 papers. The above named appendixes are especially useful as information sources. They are structured into research and development facilities or firms – working groups and their head – topics of the working groups. Predominately characterized are the former university and also extra university research facilities in the fields of materials and substances.

LITERATUR: Bericht des Ausschusses für Bildung, Forschung und Industriefolgenabschätzung (18. Ausschuss) gem. § 56a der Geschäftsordnung. Technikfolgenabschätzung. Potenziale und Anwendungsperspektiven der Bionik. Drucksache 16/3774 vom 08.12.2006.

Bionics: Current trends and future potentials

"What causes fascination and enthusiasm in the wide public evokes scepticism in research and development and at companies from time to time?"

⇨

"Present companies are only in a restricted way capable to pursue innovation projects based on bionics."

In the year 2007, the study, quoted below, promoted by the BMBF appeared. Still today, it is valid in large parts.

CONTENT AND TRENDS:
The interesting study, published by A. v. Gleich et al., includes 4 aspects: 1. bionic trends, 2. technological perspectives of the "learning from nature", 3. topics, actors and networks; results of the literature- and patent-analyses, 4. bionics and innovation. Ad 1: Functional morphology – signal- and information-processing etc. – nano-bionics etc. – convergence of the development strands ("*converging technologies*") are summarised.

TECHNOLOGICAL POTENTIALS AND TOPICS:
Ad 2: The high status in technology prognoses is pointed out, the special affinity of the material sciences to bionics and the difficulty of its differentiation against the "non-bionics." Ad 3: The relative strong presentation of Germany is mentioned, also the fact that the rate of literature- and patent-publications world-wide arises and that bionics has an exaggerated importance for the natural scientific education and the schooling.

DEVELOPMENT-STRANDS:
Ad 4: This point bears bionics and innovations in mind. There are bionic specific innovation-boosters and -barriers. Bionic optimisation procedures, which fit to the increasing complexity of technical problems, are especially significant as "boosters". Bionics widens the field of research and is positively occupied as a strategy. On the other side the biological complexity proves to be a "barrier"; so it is difficult to take up sub-functions for examination.

LITERATUR: Gleich, A. von et al. (2007): Bionik – Aktuelle Trends und zukünftige Potentiale. S. auch http/dnb.ddb.de. Die Publikation ist kostenlos erhältlich über folgende E-Mailadresse: gleich@uni.bremen.de.

Therefore: "All bionics" – or what?

BIONICS

HAS TWO FACES

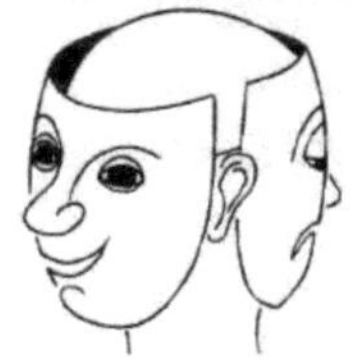

BIONICS is...

... no cure-all - however a noteworthy methodology.

... no unique research-procedure - however an essential one.

... no nature-imitation - however a nature-near system-base.

Bionics ultimately is a point of view and a tool that *can* contribute to approaches for the finding of innovations.

EXAMPLES FOR APPROACHES TO STRUCTURED INVENTIONS:

1. *Brainstorming* (*BR*): Practical ideas for solution are developed from notes with thoughts of the members of the committee at the pin board.
2. *Theory of the inventive problem solving* (TRIZ, Russ. abbrev.): G.S. Altshuller's 40-parameters-method for the solution of a "conflict of technology aim."
3. *Structured Inventive Thinking* (SIT): It is based on G. Filkowsky's simplified TRIZ; designed by C. Stephan and R. Schmierer for industrial application.

FURTHER EXAMPLES:

4. *Method of Nakayama Masakazu* (NM): Solution principles of analogues from quite different areas lead to solutions of technical problems.
5. *Method of Yoshiki Nakamura* (YN/ARIZ 02): Combines among others NM and pre-stage ARIZ 01, constricts the search area by several intermediate steps.
6. *Nature oriented invention strategy* (NAIS): B. Hill's method of catalogue sheets originates from principles of nature oriented targeting/solution finding.

7. *Contradiction oriented invention-strategy* (WOIS). H.-J. Linden; comp. p. 233.
8. "*Luscinius method*" (Lu): Bionics based. W. Nachtigall; comp. p. 255–259.

As to elaborate presentations of the methods comp. Nachtigall (2010), p. 179–199.

INCLUSION OF BIONICS:

Bionics can be included into the approaches 1-5 and 7, but it mustn't, however. The approaches 6 and 8 are actually nature based. *Bionics is a tool!*

LITERATUR: Nachtigall, W. (2010): Bionik als Wissenschaft. Erkennen → Abstrahieren → Umsetzen. Springer, Berlin. Dort detaillierte Literatur. Der Klassiker ist: Altshuller, G. S. (1984/1986/1998): Erfinden – Wege zur Lösung technischer Probleme. BTU Cottbus.

Degree of popularity and assessment of education

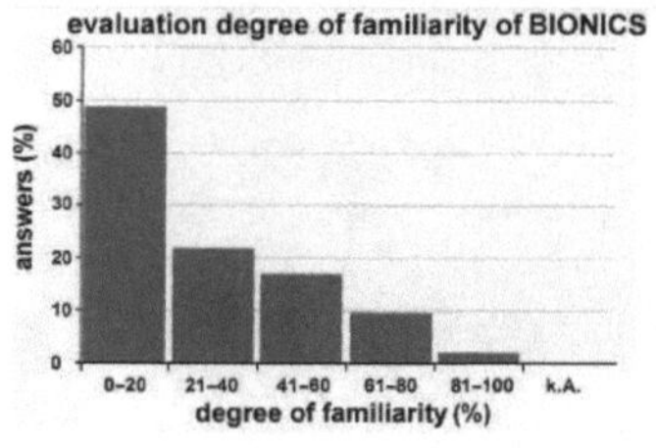

A

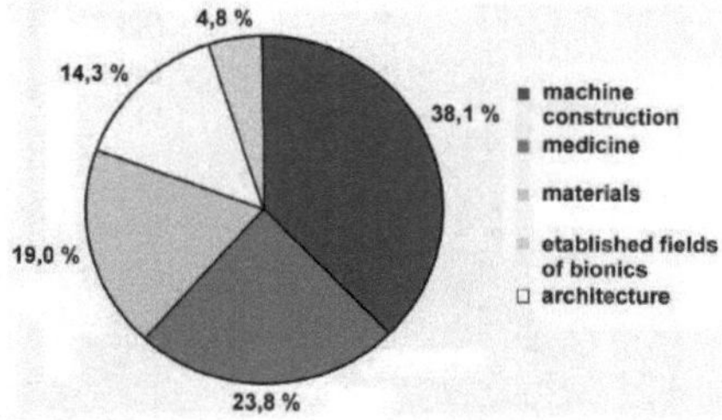

B

In the industry "bionics" is well introduced, otherwise not. Dual degree programmes usually are positively evaluated.

SURVEY:
"In the years 2009/2010, a 2-stepped examination of experts was carried out at the Bionic-Innovations-Centrum (B-I-C) of the college Bremen in order to determine general evaluations for bionics, of special interest were its publicity in the population, the carrier prospects for graduates of bionic studies as well as fields of application for bionics and bionicists." (Zehren, Kesel 2010). Scientific/technical experts were interviewed.

PUBLICITY:
Approximately half of the interviewees (n = 41) fell into the publicity class of 0–20 %, only about 2 percent into the class of 81-100 percent (A). Even within the mentioned experts, "the degree of publicity of bionics is therefore still classified as low." In a following approach (n = 96) it was asked, why this it is so. Here approximately 47 % declared that the carrier contact is missing, 33 % that the public perception is missing and 20 % that the terminology is blandly.

TRADE SPECIFIC EVALUATION AND EMPLOYMENT CHANCES:
With a dual education, one part takes place at the college, another part in a firm. This was estimated positive by 75 % of the experts (n = 21), however, this evaluation was strongly trade-specific (B). From the two possibilities, "education engineer-specifically, additionally with bionics" and "dual education leading to a '*bionicists as such*'" the first is offered in many places, the latter only at the Westphalia College of Bocholt. The carrier prospects are different.

LITERATUR: Zehren, D.; Kesel, A. (2010): Entwicklungseinschätzungen für das Ausbildungs- und Berufsfeld „Bionik" – Delphi-Studie. In: Kesel, A.; Zehren, D. (Hrsg.) (2010): Bionik: Patente der Natur. 5. Bionik-Kongress. HS Bremen 2010. B-I-C, Bremen. S. 378–383.

Books about bionics I

There are over 100 book-titles. *Non-fiction books* inform, *catalogues* accompany exhibitions, *illustrated books* entertain.

NON-FICTION BOOKS (SELECTION):

- Nachtigall, W. (2002): Bionik. Grundlagen und Beispiele für Ingenieure und Naturwissenschaftler. 2. Aufl. Springer, Berlin.
- Gleich, A. von (1998): Bionik – Ökologische Technik nach dem Vorbild der Natur. Teubner, Stuttgart.
- Rossmann, T.; Tropea, C. (2004): Bionik. Aktuelle Forschungsergebnisse in Natur-, Ingenieurs- und Geisteswissenschaften. Springer, Berlin.

CATALOGUES FOR EXHIBITION (SELECTION):

- Siemensforum München, Landesmuseum für Technik und Arbeit, Mannheim (Hrsg.) (1998): Bionik – Zukunfts-Technik lernt von der Natur.
- Bürgin, T. et al. (2000): HiTech Natur. Gemeinschaftsausstellung der Naturkundemuseen St. Gallen .... Verlag Natur-Museum Luzern.
- Ehn, F.; Seyfferth, A. (Hrsg.) (2011): Bionik – Patente der Natur. LWL Landesmuseum für Naturkunde, Landesverband Westfalen-Lippe.

ILLUSTRADED BOOKS (SELECTION):

- Nachtigall, W.; Blüchel, K. (2000): Das große Buch der Bionik. Neue Technologien nach dem Vorbid der Natur. 2. Aufl. DVA, München.
- Blüchel, K.; Malik, F. (Hrsg.) (2006.): Faszination Bionik. Die Intelligenz der Schöpfung. MCB, München.
- Umweltstiftung WWF Deutschland, Pro Futura (Hrsg.) (2005): Vision des Machbaren. Die Natur zeigt uns den Weg. Pro Futura, Waldbröl.

LITERATUR: Bücher über Bionik; z. B.: http://www.amazon.de/s?ie=UTF8&tag=firefox-de-21&index=blended&link_code=qs&field-keywords=Bionik&sourceid=Mozilla-search. – Der Natur auf der Spur Bionik - Herausforderung und Chance. Pro Futura, Waldbröl.

Books about bionics II

For some sections, there are *specific non fiction books*, also *children* and *youth books* as well as *instructions for schools*.

SPECIFIC NON FICTION BOOKS (SELECTION):

- Nachtigall, W.; Pohl, G. (2012): Bau-Bionik. 2. Aufl., Springer, Berlin.
- Nachtigall, W. (2005): Biologisches Design. Systematischer Katalog für bionisches Gestalten. Unter Mitarbeit von A. Wisser. Springer, Berlin.
- Gremmler, T. (2008): cyberBionik: Design ... Springer Spektrum, Heidelberg.
- Rheinauer, P. (2008): bionicprocess. Bionik als Vorbild für die Gestaltung von Organisationsprozessen. VDM Verlag, Saarbrücken.

CHILDREN AND YOUTH BOOKS (SELECTION):

- Zeuch, M. (2006): Was ist was? Bd. 122: Bionik. Tessloff, Nürnberg.
- Nachtigall, W. (2007): Natur macht erfinderisch. 2. Aufl. Ravensburger, Ravensburg.
- Belzer, S. (2008): Die genialsten Erfindungen der Natur. Naumann & Göbel.
- Viering, K.; Knauer, R. (2009): Bionik – Abgekupfert ... Berlin, Bloomsbury.

SCHOOL (SELECTION):

- Hill, B. (2000): Von der Natur Lernen, Unterricht Arbeit + Technik 10/, Heft mit Themenschwerpunkt Bionik im Unterricht, Friedrich Verlag, Seelze.
- Stripf, R. (2003): Biologie in der Schule – Bionik. Bd. 5/52 Aulis Verlag Deubner, Köln.
- Jelinek, C.; Wütherich, D. (2007): Prisma Projekt – Brücken und Bionik. Materialheft für Schüler. Klett, Stuttgart.

LITERATUR: Unter den einzelnen Titeln – sowie unter Stichwörtern wie Bionik; Biomimetik bzw. deren engl. Ausdrücken – sind im Internet vielfältige Informationen zu erhalten.

Series of congress-reports: IL-report

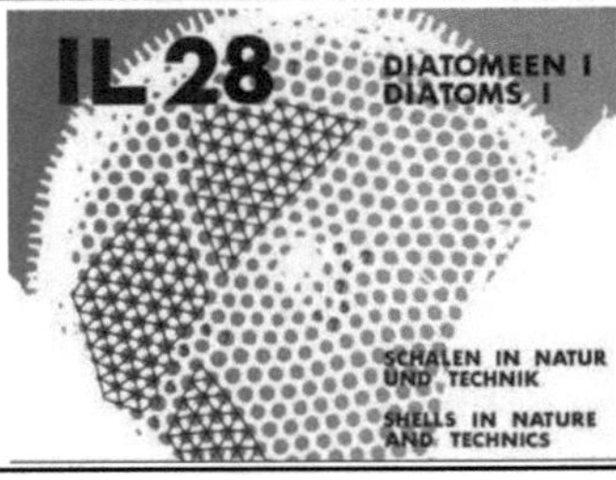

A

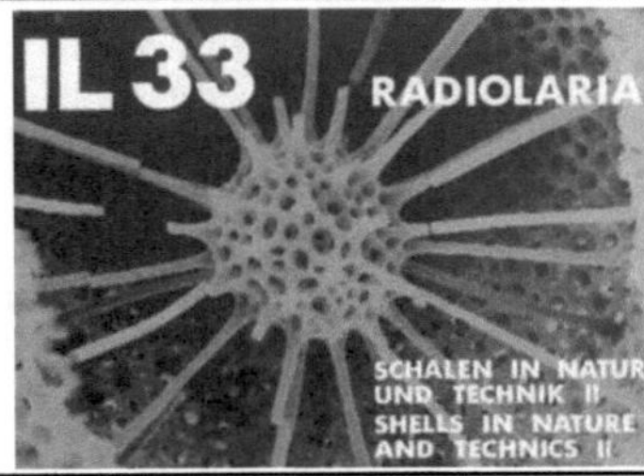

B

41 reports of the Institute for Lightweight Structures (IL) offer many info about "constructions in nature and technology."

SFB 230 OF THE DFG:
This twice extended special research field existed over 12 years and was placed in Stuttgart with the cooperation partners Tubingen and Saarbrucken. In innumerable discussions, workshops and symposia at the Institute for Lightweight Structures of the University Stuttgart (management: Frei Otto) "natural constructions" were examined by construction engineers, architects, biologists and representatives of other disciplines and were documented in reports.

TITLE SELECTION OF BIOLOGICAL-TECHNICAL TOPICS:
- Bericht Nr. 3, 4, 6 (1971, 1973): Burkhardt, B. (Hrsg.): Biology and building.
- Bericht Nr. 8 (1975): Bach, K. (Hrsg.): Nets in nature and technology.
- Bericht Nr. 9 (1977): Bach, K. (Hrsg.): Pneus in nature and technology.
- Bericht Nr. 27 (1981): Bahrdt, H.-P. (Hrsg.): Natural building.
- Bericht Nr. 28, 33 (1985, A; 1990, B): Bach, K. (Hrsg.): Shells in nature and technology.

TITLE SELECTION OF RATHER TECHNICAL TOPICS:
- Bericht Nr. 5 (1972): Burkhardt, B. (Hrsg.): Variable roofs.
- Bericht Nr. 10 (1975): Hennicke, J. (Hrsg.): Grid shells.
- Bericht Nr. 12 (1975): Bubner, E. (Hrsg.): Variable pneus.
- Bericht Nr. 16 (1976): Burkhardt, B. (Hrsg.): Tents.
- Bericht Nr. 21–25 (1979/1988/1992/1998/1990): Bach, K./Otto, F./Otto, F./Otto, F./Gaß, S. (Hrsg.): Form, force, mass.

LITERATUR: Berichte des Instituts für Leichte Flächentragwerke der Universität Stuttgart. Nr. 1– 41 (1969–1995) Unterschiedliche Herausgeber. Universität Stuttgart. Institut: Pfaffenwaldring 14, Universität, 70569 Stuttgart. In Kopien teils noch erhältlich.

Series of congress-reports: BIONA-reports

A

B

C

From the 16 BIONA-reports, 6 refer to congresses of the Saarbrucken "Society for Technical Biology and Bionics (GTBB)."

THE GTBB UND ITS CONGRESSES:
In the year 1990, the society (comp. p. 281) was founded and chaired by W. Nachtigall. Until 2002, 6 symposia and 2 workshops about technical-biological and bionic topics were held under the management of W.N. and the organisation of K. Braun and A. Wisser/Saarbrucken at different places. Results were published in reports (composition: A. Wisser). Later symposia were held in Bremen, chaired by A. Kesel/Bremen, and are published since 2007 (C).

CONGRESS 1–3:
- BIONA-report 8 (1992): 1. Bionic-Congress, Wiesbaden 1992. 168 pages, 11 publications by 11 authors (A).
- BIONA-report 9 (1994): 2. Bionic-Congress, Saarbrucken 1994. 182 pages, 13 publications by 14 authors.
- BIONA-report 10 (1996): 3. Bionic-Congress, Mannheim 1996. 215 pages, 12 publications by 18 authors, 17 Poster-short-publications by 25 authors.

CONGRESS 4–6:
- BIONA-report 12 (1998): 4. Bionic-Congress, München 1998. 337 pages, 27 publications by 25 authors, 9 Poster-short-publications by 17 authors.
- BIONA-report 15 (2001): 5. Bionic-Congress, Dessau 2000. 351 pages, 22 publications by 25 authors, 11 Poster-short-publications by 25 authors.
- BIONA-report 16 (2003): 6. Bionic-C., Saarbrucken und Bionic-Session Bistra 2002. 289 pages, publications by 18 authors, 14 Posters by 20 authors (B).

LITERATUR: BIONA-reports – Technische Biologie und Bionik 1–16 (1982–2003). Hrsg.: W. Nachtigall – W. Nachtigall und A. Wisser – A. Wisser und W. Nachtigall u. a. Akad. Wiss., Mainz, teils Fischer, Stuttgart. Teils noch erhältlich.

Bionics publications in magazines

Few specific magazines carry the term "Bionics" in the title; meanwhile, however, several accept bionic publications.

SPECIFICALLY FOR BIONIK (INCL. BIOMECHANICS) (SELECTION):

- Journal of Bionic Engineering, Elsevier
- Journal of Applied Biomechanics (*, see below)
- Journal of Biomechanics, Elsevier
- Applied Bionics and Biomechanics, IOS Press (**,see below)

ALSO TAKE BIONIC PUBLICATIONS, RATHER BROADLY (SELECTION):

- Nature
- Science
- Journal of Ecological Design Association
- Scientific American (German edition: Spektrum der Wissenschaft)
- Bild der Wissenschaft
- Biologie in unserer Zeit

ALSO TAKE BIONIC PUBLICATIONS, RATHER SPECIFICALLY (SELECTION):

- Journal of Experimental Biology
- Journal of Comparative Physiology A
- Advances in Comparative and Environmental Physiology
- Applied Physics
- Journal Fluid Mechanics
- Technik heute

LITERATUR: www.elsevier.com/wps/find/journaldescription.cws_home/707667/description#description. – http://journals.humankinetics.com/jab (*). – www.iospress.nl/journal/applied-bionics-and-biomechanics/ (**).

Television series about bionics

Good television series hold the balance between presentation with public appeal and scientific correctness.

DEVELOPMENT:
At the beginning of the 60' years, the aspect "fascinating nature → learning for technology" appeared for the first time, when W. Nachtigall with the Stölting-Film/Munich and other producers had made a series of 35-min-films. With the blooming of the bionics, also a certain "television asset stripping" started, that tried on the whole to obtain scientific solidity, however. The "big show of the natural wonders" still reaches high TV ratings.

MEANING:
Not at all, one can overestimate the broad effect of such series. One single broadcast reaches a larger audience than the lifelong lecture- and author-activity of a scientist, even if he moves into popular terrain. Is this so important? We think, yes. If one succeeds to transmit fascinatingly topics of nature and technology by maintenance of the scientific aspect, then the television fulfils a social reason task (A).

EXAMPLE:

In March 2011, a successful quadripartite documentation with bionic topics ran under the title "Bio-mimicry - naturally brilliant!" It was broadcasted by Arte and in over one dozen European countries. Titles:

1. The art of locomotion.
2. Building with efficiency.
3. Orientation in the chaos.
4. Materials of the future.

DOKU: Nachtigall, W. (ca. 1960): Fliegen müßte man können. Stölting-Film München für das ZDF – Biomimikry – Natürlich genial. 4-teilige Dokumentation von André Rehse; Gebrüder Beetz. Filmproduktion/EMS FILMS BR/ZDF/Arte Deutschland 2011, 4 x 52 min.

Exhibitions on bionics

A

B

Bionics exhibitions, designed for a wider public, took place many times in museums, also as travelling exhibitions.

EXAMPLE 1: MUSEUM FÜR TECHNIK UND ARBEIT, MANNHEIM 1996
The exhibition "Bionic – future technology learns from nature" is already classical; it achieved a large afflux. The 88-page catalogue contains contributions to flying and swimming, running and grasping, recognizing, miniaturizing, building, pleating and packing, using and preserving, as well as optimising. On this basis, the important travelling exhibition "BIOlogy and techNICS – future technology learns from nature" was worked out by the Siemens-forum (A).

EXAMPLE 2: WESTF. LANDESMUSEUM FÜR NATURKUNDE, 2011
Accompanying the excellent and generous exhibition "Bionics - patents of nature" with its important exhibits, a 156-page, readable companion book appeared, graphically and didactically beautifully done. It contains 9 chapters with detailed titles like "Bionics in medicine – tips of dr. nature" and "specialists, as far as the eye looks – the evolution makes it possibly". The book (B) with its many examples is almost a small textbook of bionics.

EXAMPLES 3 and 4: TRAVELING EXHIBITIONS 2011, 2006
A group display of the Association of Botanical Gardens is shown in several gardens from 2011 on. The 87-page accompanying book "*What technology can learn by nature*" brings essentially botanical examples.
A group display of DBU and BioKoN is on the way since 2006 as travelling exhibition with experimental possibilities, supported by a pretty 39-page companion book "*Inspiration nature – patent-shop bionics*".

LITERATUR (Titel s.o.): (1) LTA Mannheim, SiemensForum München (Hrsg.), Bappert, R. et al. (Autoren). – (2) LWL Münster, Hendricks, A. (Hrsg), Ehn, F.; Seyfferth, A.; Kriegs, O. (Autoren) – (3) Speck, T. et al. (Hrsg.), Freiburg. – (4) Brickwedde, F.; Bannasch, R. (Hrsg), Osnabrück.

Bionics at the Hannover-Messe

At the Hannover-Messe, the world's largest industry exposition, specific bionic trends were and are well represented.

FORMER YEARS (EXAMPLE):
Approximately since the 80' years, on the exhibition also bionic trends were displayed, admittedly not systematically as today, however, with selected single pieces. For example, the Berwian of I. Rechenberg was among them (comp. p. 100), a model of our Saarbrucken fin flapping pedal boat (A) or, for cars, the bionic tires of the firm Continental. Despite the large competition, such exhibits always were much in demand and very much beleaguered.

AT THE LAST TIME (EXAMPLE):
According to the researches of B. Möhl, the BioRob arm (B) is a new type of robot arms with an *antagonistic, series elastic drive concept* inspired by the elastic muscle-tendon-apparatus. This arm marks a change in the robotics from stiff to elastic systems. It is characterised by an outstanding relation of load to own weight. Its operating distance is oriented to the human arm.

PRESENCE AND LOOK-OUT (EXAMPLE):
In 2011, Festo introduced a new project from the field to bionics. The engineers of the Bionic Learning Networks were successful to develop a flight-model (C), that is autonomous and that starts without the help of further drives, flies and lands. On that occasion, the wings doe not beat only up and down but twist themselves purposefully. This happens by an active joint torsion drive (comp. p. 148). Recently a 4-winged dragonfly-model was presented.

LITERATUR: Rundschreiben 15 der GTBB (1995): Bericht zum BIONIK-Stand der UdS. – Möhl, B. (1997): Antriebskonzept patentiert, Hannover-Messe 2003. – Hannover Messe 2011: Festo lässt bionischen SmartBird abheben.

EMPHASES

AND EDUCATION

W. Nachtigall and A. Wisser, *Bionics by Examples*,
DOI 10.1007/978-3-319-05858-0_15

EMPHASES AND EDUCATION

Organizations – Competence-networks – Education.

The potentials of bionics are versatile, but they are not unlimited. If main focuses and nets emerge, this is an outer sign that the bionics has arrived at the science-business level. The VDI already developed guidelines to attain this goal. To polish up bionics, there were and are special competence-networks as well as key areas of training in schools, universities, colleges and other facilities.

Companies, networks, associations

GTBB ⇨ BIOKON BIONICS COMPETENCE NETWORK

GTBB ***BIOKON***

Bionic oriented associations like companies and networks serve to spread the ideas of bionics.

GERMANY – COMPANIES AND NETWORKS (SELECTION):
Society for Technical Biology and Bionics (GTBB): 1991 established by W. Nachtigall in Saarbrucken; 1. chairlady presently A. Kesel/Bremen.
Bionics-competence-network (Biokon): Foundation impetus 1999 through the book "Bionik" of W. Nachtigall; 1. chairmen presently T. Speck, Berlin.
International bionics-centre: Headquarter in Munich. Foundation's board presently K.-H.-K. Oeller/Munich, K. Braun/Saarbrucken.

GERMANY – LOCAL ASSOCIATIONS (SELECTION; 3 OF 12):
Bionics-Innovations-Centrum (BIC): Coordination and care of bionic projects at the college Bremen; management A. Kesel/Bremen.
Bionic engineering network (BEN): Industry-contacts. University of Applied Sciences Saarbrucken; chairman H.J. Weber/Saarbrucken.
Nieklitz Ecology- and Ecotechnology-Foundation (NICOL): Ecologically oriented installations with bionic examples; board B. Heydemann/Nieklitz.

FOREIGN COUNTRIES – REGIONAL ASSOCIATIONS (SELECTION; 3 OF 5):
England: BIONIS: The biomimetics network for industrial sustainability; University of Reading; promotes application of biomimetics results in the industry.
Austria: Bionics Austria: Austrian platform for the support of bionics research with its headquarters in Graz; 1. chairman C. Schinagl/Graz.
Sweden: Swedish Biomimetics 3000: Support of researches with the aim of taking over bionic concepts into the commercial application.

LITERATUR:
www.kompetenznetz-biomimetik.de/index.php?option=com_content&task=view&id=138&Itemid=27.

The "Bionic Learning Network" (BLN)

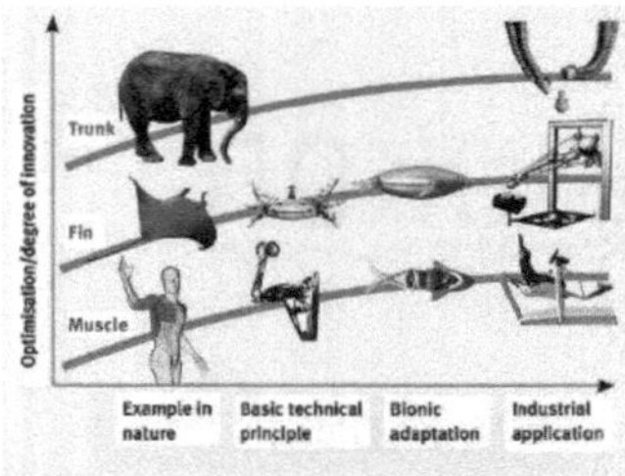

A

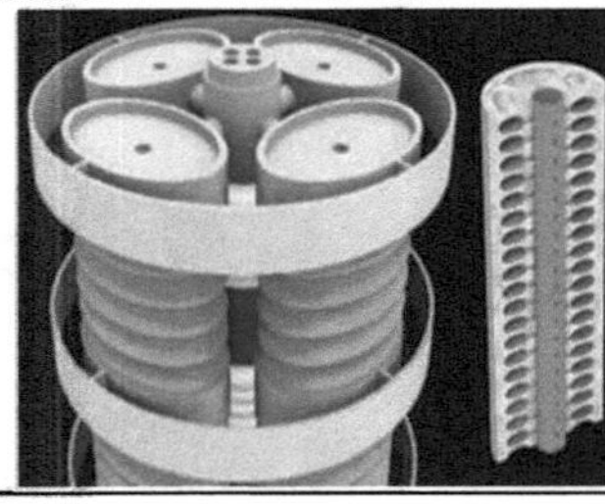

B

The BLN of the Festo INC. & Co. KG in cooperation with universities aim at solution competence by bionics.

THE IDEA:
Since beginning of the nineties, Festo is engaged in bionics. In 2006, the Bionic Learning Network was established. In 2010, Festo got the German future price for the bionic handling assistant developed according to the model of an elephant trunk. The projects of the BLN "... are expression of the solution competence and as total new technological bases ... inspire the dialogue concerning the future practicable and the wanted."

AIMS AND PREVIOUS PARTICIPANTS:
"Check-up of the market-relevance in the dialogue with customers – carrying forward the natural efficiency strategy to the automation technology – tracing of new products as well as product ideas – testing of new technologies and production procedures – developing energy-efficient and biomechatronic products." Beside internal departments, participants are universities like MIT and TU Berlin, as well as external companies and research institutes.

QUESTIONS (SELECTION):
"In the *biomechatronics* Festo tests new approaches for the control and regulation of bionic systems. – Under *biomechatronic footprint* Festo understands the documentation of (r)evolutionary developments from the BLN like for example bionic actuating elements, fins and muscles (A).– *Moving with innovative lightweight construction* (B)." The later takes place in cooperation with the Inst. for technology-oriented design innovation/HS of Art, Offenbach.

LITERATUR: Bionic Learning Network. www.festo.com/bionik. – Stoll, C.-M. (Hrsg.) (2011): Querdenken macht effizient. Wie die neue Mechatronik die Welt verändert. Sonderedition 80 Jahre Dr. Kurt Stoll. Eigenverlag.

BIONA - a supportive activity of the BMBF

BIONA

a supportive activity

Supportive activities like this allow to apply for unconventional projects, even those with a higher "innovation risk."

CONCEPT:
The supportive activity "BIONA (bionic innovations for lasting products and technologies) of the German Federal Ministry for Education and Research (BMBF) concretises the high-tech-strategy of the Federal Government in the innovation-field for environmental technologies. With altogether 30 million Euro practicable, sustainable developments and bionic approaches are promoted in the coming years". Framework-program: FONA-research for sustainability.

CLASSIFICATION OF THE BIONICS:
"Bionics pulls essential parts of the today's innovative capacity out of the connection between biological basis-research and the rapidly rising technical potentials ... The special chance of bionics lies in producing revolutionary innovations, also there, where technological improvements took place in small steps until now. Until ... it comes to a marketable product, often a long way is to be covered."

PURPOSE OF THE SUPPORTIVE ACTIVITY:
"The announcement of BIONA aims to open up the potential of natural models for generating sustainable and competitive products and technologies... The protagonists are organized in research associations from industry and science or in junior staff researcher-groups. Complementary ... cross connecting accompanying research take place" (BMBF). Also the realization of educational offers belongs to this.

LITERATUR: BIONA – Bionische Innovationen für nachhaltige Produkte und Technologien. www.bionische-innovationen.de/projekte.html. – Ansprechpartner: Projektträger im Deutschen Zentrum für Luft- und Raumfahrt, Heinrich-Könen-Str. 1, 53227 Bonn.

Awards for bionics activities

A

B

C

Beside smaller also prestigious awards have already been awarded for works in the field of bionic realizations.

GERMAN ENVIRONMENTAL AWARD:
This award, given all years by the Federal German Environment Foundation/ Osnabruck for ecologically oriented activities, awarded until now 23 people, has the highest prize money of Euro 250 000. For bionics were previously awarded:

- 1999: W. Barthlott for: "Discovery of the Lotus-effect (A)",
- 2003: C. Mattheck for: "Growth-behaviour of the trees/bionics (B)",
- 2005: B. Heydemann for: "Ecology, eco-technique, communication (C)."

FUTURE AWARD OF THE FEDERAL PRESIDENT:
The award is given to one of three applicants, which are crystallized as best representatives of forward-looking concepts out of a variety of incomings. For bionics were previously awarded:

- 2008: J. Marek, M. Offenberg, F. Metzner for: "Smart sensors ...",
- 2010: P. Post, M. Fischer, A. Grzesiak for: "Prototype elephants trunk ...",
- 2011: K. Leo, J. Blochwitz, M. Pfeiffer for: "Flexible organic photo-cells."

OTHER AWARDS (SMALL SELECTION):
International RHINELAND-AWARD OF THE TÜV:

- 2002: W. Nachtigall for issue: "Bionics and Ecology."

Treviranus-medal of the Association of German biologists:

- 2001: W. Barthlott for issue: "Bionics, Lotus-Effect",
- 2004: W. Nachtigall for: "Technical Biology and Bionics".

BIONIC AWARD VDI, DBU, *Schauenburg*, comp. p. 285.

LITERATUR: S. Homepages: DBU (z. B.: www.dbu.de/533bild16222__25994.html; Deutsche Stiftung Umwelt), Zukunftspreis Bundespräsident, VBio (Verband Biologie, Biowissenschaften und Biomedizin), VDI (Verein Deutscher Ingenieure),TÜV (Technischer Überwachungsverein).

International Bionic-Award

A

B

The international Bionic Award is addressed to junior researchers for creative results in the field of bionics.

ANNOUNCEMENT:
The International Bionic Award is remunerated with 10.000 EUR from the Schauenburg-Foundation of the Association of Sponsors of German Science and is awarded in cooperation with the Association of German Engineers (VDI) and the German Environmental Foundation (DBU) every 2 years. The submitted work can refer to a bionic product (individuals and teams can apply for it) or to a master thesis, a dissertation or a habilitation about bionics.

WINNERS OF THE AWARD 2008 (FIRST AWARDING):
M. Hermann (A), for the development of the FracTherm®-algorithm to increase the efficiency of solar-collectors and heat exchangers, being carried out with the help of a doctoral grant by the DBU. Instead of the usual serial or parallel canal structures with the disadvantage of high pressure losses and unequal shaped perfusion, fractal like multiple branched natural systems (bloodstreams, leaf-veins) were models for flat hydraulic structures.

WINNERS OF THE AWARD 2010:
M. Hollermann and F. Förster ("Die Bioniker") (B), for the development of a fastening system (dowels) according to the model of the tick. Therewith, the wall fastening of dowels should become stronger and tighter, especially in the ranges of lightweight construction and heat insulation. "With their innovative idea, the junior researchers created the basis for a marketable product, originated from a bionic development process."

LITERATUR: Presseinformation d. Fraunhofer-Instituts für Solare Energiesysteme ISE (2008): Von der Natur inspiriert. – Dr. M. Hermann erhält Internationalen Bionic-Award (I.B.A:). – Pressemitteil. d. VDI (2010): Internationaler Bionic-Award 2010: Die Technik einer Zecke als Vorbild.

Life Sciences and bionics in the VDI

VDI ⇨ analysis, abstraction, realization, application | biology, bionics, technology ⇨ BIONIC AWARD

The Association of German Engineers (VDI) links competence in the field of bionics and cooperates with other institutions.

THE VDI:
With 150 000 personal members, the VDI is Germany's biggest technical-scientific association (stand 06/2012). In 2007, beginning with the constitution of a stand-alone faculty bionics in the competence field biotechnology and in 2009 within the VDI-organisation Technologies of Life Sciences the VDI has based on first soundings (Nachtigall, 1986; Neumann, 1993) completed its competence to the topic bionics and consequently promoted Germany's leading role in bionics.

THE BIONICS INITIATIVE:
"Current studies classify Germany as an international important research location in this field. It is the aim to preserve the scientifically high level of the German research also in future and to develop it further. The international reputation must be strengthened further, and bionic approaches should be integrated sustainably and soon into the innovation processes of the industry" (citation from Finck, 2007).

ACTIVITIES:
For its realization the VDI cooperates e.g. with the German Environmental Foundation (DBU), its Centre for Environmental Communication (ZUK), the German Institute for Standardization (DIN) as well as with the competence nets BIOKON and BIOKON-international. Among others VDI guidelines and ISO-norms were prepared (p. 287–289), events carried out and co-created. Since 2008 the International Bionic-Award of the Schauenburg-Foundation was assigned (p. 285).

LITERATUR: Nachtigall, W. (1986): Konstruktionen. Biologie und Technik. VDI-Verlag, Düsseldorf. – Neumann, D. (Hrsg.) (1993): Technologie-Analyse Bionik. VDI-Technologiezentrum Physikal. Technologien, Düsseldorf. – Finck, M. (2007): Life Sciences im VDI. BIOforum 6, 19.

VDI series of guidelines for bionics

GERMAN ENGINEERS ASSOCIATION VEREIN DEUTSCHER INGENIEURE	Biomemetics Conception and strategy Differences between bionic and conventional methods/products Bionik Konzeption und Strategie Abgrenzung zwischen bionischen und konventionellen Verfahren/Produkten	VDI 6220 *Draft / Entwurf* Issue German/English Ausg. deutsch/englisch

VDI guidelines should enable and simplify the realisation of bionic developments into technical applications.

Since 2010, seven VDI guidelines for bionics were developed in the faculty bionics of the VDI, an eighth guideline is near completion. The bionics guidelines relate to technical tasks, not to systemics and management. A "framework guideline" offers a first orientation and classifies the technical guidelines:

VDI 6220 "Bionics – conception and strategy – differentiation between bionic and conventional procedures and products"

The technical VDI guidelines 6221-6226 describe the sections of the bionics:

- VDI 6221 "Bionics – functional bionic surfaces"
- VDI 6222 "Bionics – bionic robots"
- VDI 6223 "Bionics – bionic materials, structures and components"
- VDI 6224 sheet 1 "bionic optimisations – evolutionary algorithms in the application"
- VDI 6224 sheet 2 "bionic optimisations – application of biological growth-laws of the structure-mechanical optimisation of technical components"
- VDI 6225 "Bionics – bionic information processing"
- VDI 6226 "Bionics – (in preparation) architecture, structural engineering, industry design"

The VDI-guidelines can serve as basic documentations for the international standardisation. Through this leading role of the VDI, a standardised application of bionics can now be anchored as possibility for structured invention also in the international sphere.

LITERATUR: VDI – Verein Deutscher Ingenieure e. V., Fachbereich Bionik in der VDI-Gesellschaft Technologies of Life Sciences (VDI-TLS), www.vdi.de/bionik. Die Richtlinien können bestellt werden beim Beuth Verlag GmbH, Burggrafenstr. 6, 10772 Berlin, www.beuth.de.

VDI guidelines for bionics and international standardisation

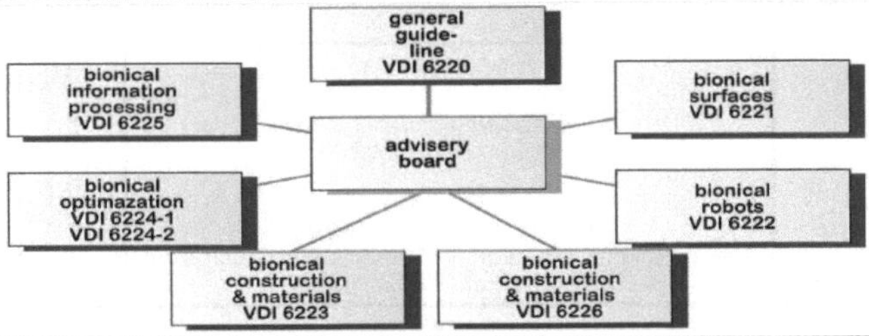

Together, the VDI and the DIN promote the international standardisation for bionics. Normalisation is the necessary

VDI FACULTY BIONICS:
The faculty covers with its guideline committees the essential technical sections of bionics: Surfaces, robots, optimisation, information processing, construction and materials, architecture and industry design. It follows approximately that of W.N. already in 1998 proposed subdivisions, however, it brings them together in a reasonable manner for industry practice. An important teamwork was the working out of the framework guideline VDI 6220.

PROJECT ISOBIONICS:
The VDI guidelines developed in Germany, which are published bilingual in German and English, can serve directly as basic documentations for the international standardisation at ISO. The establishment of a technical committee ISO/TC "Biomimetics" at ISO is passed. By the international standardization, the transfer of bionic realizations should be promoted into technical applications.

VALUATIONS:
Norms and standards (e.g. VDI guidelines) serve the quality protection of procedures and products. This on the other hand promotes the competitiveness of firms that apply norms and standards. Up to now, Germany's leading role in the field of standardisation for bionic approaches enables to apply the bionics uniformly also on international level. This will promote the acceptance of bionic approach.

LITERATUR: VDI – Verein Deutscher Ingenieure e. V., Fachbereich Bionik in der VDI-Gesellschaft Technologies of Life Sciences (VDI-TLS), www.vdi.de/bionik, www.vdi.de/isobionik).

Example: Guideline VDI 6225

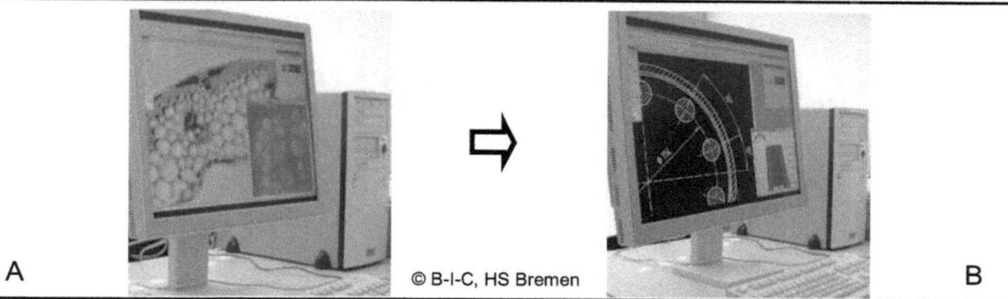

Bionics – bionic information processing; alternative approaches ► Information processing in technical Systems.

BIOLOGY:
Examples for information processing, on which technical systems can appeal, are found in the classic neuro- and sensory-physiology. At first, the information transfer over nerve-impulses interests, then the interconnection of large numbers of different nerve-cells, for example sensory and motor type, and finally the information processing from the sense organ to neural centres up to the behaviour.

PRINCIPLES:
A series of biological principles (A) finds already technical application (B), others still are in a theoretical stage. Examples: Pulse-code-modulation analogous to the transfer of nerve-impulses, parallel information processing, correlation and coincidence, prioritisation of information, pattern recognition of sensory systems, elements of memory storage, the learning and forgetting. The information bionics develops stormily.

AIM OF THE GUIDELINE:
Application related aspects are important. "It is necessary to analyse the interaction and the functioning of biological sensors, nerves and skeleton-muscle-systems (actuators), to understand and to abstract them and to make them accessible for the technology. On this occasion, algorithms can originate that are usable for software, as well as structural realizations can be won that are accessible for hardware implementation." (VDI)

LITERATUR: VDI – Verein Deutscher Ingenieure e. V., Fachbereich Bionik in der VDI-Gesellschaft Technologies of Life Sciences (VDI-TLS), www.vdi.de/bionik. Die Richtlinien können bestellt werden beim Beuth Verlag GmbH, Burggrafenstr. 6, 10772 Berlin, www.beuth.de.

Scientific didactic processing of bionics

BIONA

Federal Ministry of Education and Research

One must spread the manner of view of bionics also with didactic approaches, not only with purely scientific ones.

ADULT EDUCATION AND SCHOOLS:
The universities Bremen (A. Kesel) and Freiburg (O. and T. Speck) tried in a concerted action to implement a scientific-didactic concept with the Federal-Ministry for Education and Research as funding organisation for the further education of the named group of addressees. This should broadly forward the practice of interdisciplinary thinking, especially however the further spread of the bionics idea.

BASIC IDEAS:

- "Development of different bionic teaching and learning modules for the conveying of interdisciplinary thinking, action and problem solving."
- Workout of "offers for further education as well as teaching materials for bionics also for the adult education and for schools".
- Using of more adequate communication- and distribution-structures (e.g. also E-learning, internet portals) for publishing the modules.

PROCEDURE:

- Didactic analysis of the bionic working/development of modular teaching offers for bionics" and of "teaching and learing modules for interdisciplinary basics, necessary for bionics."
- "Development of cost-effective bionic experiments, demonstration models & teaching materials/development of standardised workshop-modules ... Aim of the project is the strengthening and the promotion of the bionic offspring."

LITERATUR: www.bionische-innovationen.de/#bildungsprojekte/bionik didaktik initiative.html. – Projektkoordination – E-Mail: akesel@bionik.hs-bremen.de.

Bionic construction kits

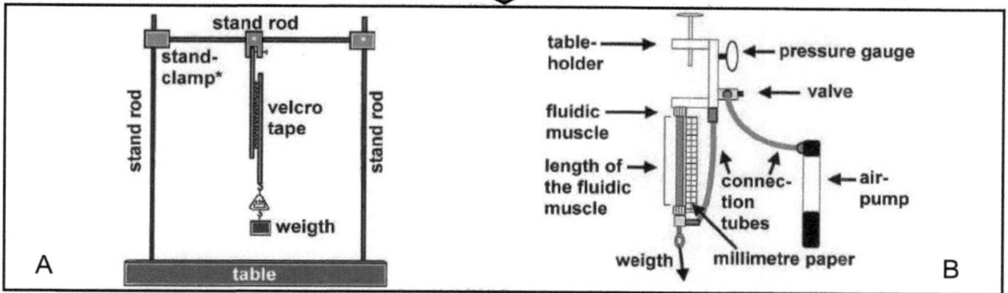

Didactically outclassing construction kit systems allow independent working, even with simple basic elements.

THE CONSTRUCTION KIT PRINCIPLE:
The didactic aspect of construction kits can hardly be overestimated. The user finds each element integrated at an identifiable place. For build-up, procedure and evaluation, he gets printed instructions for independent working. Finally he must demount a build-up into its single elements again and must sort in these again right. This system was proved for basic courses of zoophysiology by W. N. at universities (Munich, Saarbrucken).

EXAMPLE B. HILL, UNIVERSITY OF MÜNSTER:
For bionic courses, the author developed a kit system with in principle "simple" tests. One and the same test is suitable for children (qualitative statements, fun factor), pupils (simple quantitative measurements) and students (versatile quantitative and statistical evaluation). For example the adhesion of burs (A) as well as of hook and loop (adhesive) fasteners is measured and analysed under vertical and tangential tension forces.

EXAMPLE O. SPECK, FREIBURG:
To the workout for insights in biological technical contexts, the author has developed together with T. Speck and supports by the Festo INC. & Co. KG as well as the Rittal Ltd & Co. KG a "bionic-box" for 6 experiments (B). It is especially suitable for the scientific technical lessons in schools. Since the end of 2011, it is distributed by FestoDidactic under the name Bionics Lab. Didactic boxes of further authors are in development.

LITERATUR: Hill, B. (2008): Strategie naturorientiertes Lernen – Biol. Phänomene entdecken – Biol. Lösungen erfinden. Shaker, Aachen. – Speck, O.; Speck, T. (2009): Bionik-Koffer. In: Kesel, A., Zehren, D. (Hrsg.): Bionik. Patente der Natur. 4. Bionik-Kongress. Bremen. S. 97–105.

Bionics in the preschool age

A © Jörg Gräfingholt

© Heike Pfister B

In preschool, one cannot start anything with "bionics", but one can bring together nature and technology quite playfully.

NATURE AND EXPERIMENT:
Children are nosy and skilful. They learn very quickly to handle with a magnifying glass or also with a small binocular of weak enlargement (A). However, this cannot last too long. They need instruction with the focus. The right objects are not too big (one whole leaf is not of use to anything) and well structured. A small burr shows barbs. One can then experiment playfully with it: The burr sticks to the sweater.

TECHNOLOGY AND COMPARISON:
One piece adhesive fastener under the binocular shows that this is not anything else as a "technical burr." The people, who manufactured the adhesive fastener, stole this idea "from nature". Why do we not say this openly? One must certainly avoid the term "bionics" here. Also the Lotus effect® can be well converted: Produce a thin mush of flour with coloured water. The mush runs off from a kohlrabi leaf, but not from a sheet metal (B).

EXAMPLE:
The chief editor M. Janzer has put together a booklet for the preschool-magazine "*Entdeckungskiste*" (*Discovery-box*) with many kinds of practically tested examples of bionics. Chapter-titles are, for example: "Technology-theft from nature", "The engineers of tomorrow", "The inquiring mind is already there", "Skilful like the black-tailed godwit" and "Kita goes technology". So, practiced nurses can bring together technology and nature-enthusiasm.

LITERATUR: Entdeckungskiste 3/11. Zeitschrift für die Praxis in Kitas. Mit Schwerpunkten „Bionik – Von der Natur abkupfern“ – „Wissen – Das Ideenlabor Natur erforschen – „Praxisidee – Schwimmhäute mit Tüten imitieren“. Herder, Freiburg.

Bionics as school subject?

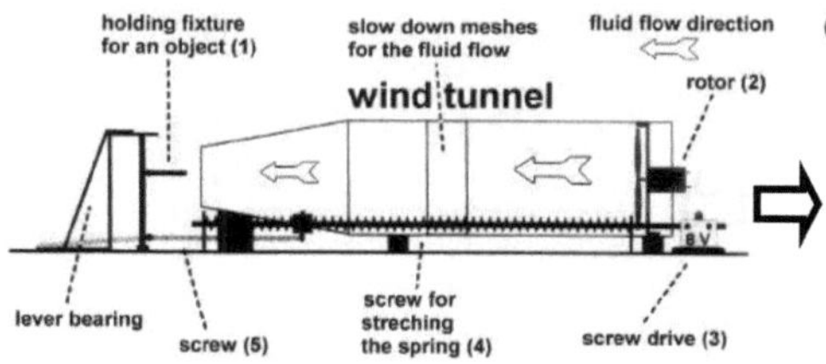

A

table of drag of different bodies
(tested in the wind tunnel with 9 V = 45 km/h wind speed)

body with A = 19,6 cm²	datas of 2 test series	difference to a disc
drob	51 / 51 mN	54 %
ball	58 / 56 mN	48 %
hemishere	61 / 63 mN	45 %
double cone	71 / 68 mN	36 %
cone	78 / 76 mN	28 %
disc	111 / 106 mN	---

B

The insight, that nature and technology are interconnected, can – by bionic approaches – not sufficiently early be awakened.

PLAYFUL PRESCHOOL LEARNING:
Here, a ("again one ...") new school subject will not be recruited. However the so important interdisciplinary bionic thinking can be introduced into the educational system already very early. This is possible already in the preschool. For example, a static feeling can be generated by using small building bricks or the flight of simple paper-gliders (cut paper sheets and staples) can be compared to that of a *Zanonia* seed (bot. gardens).

SCHOOL:
A pioneer-project (H. Birkner, A. Thanbichler a. o. from the 11. classes of the Bavarian grammar school Unterhaching with the collaboration of W.N.) took place already 1995 and proved as ideal basis for the integration of subjects. Today, in the physical lessons it is already worked frequently with bionic approaches (A, B). Or it is offered to bionic topics in advanced courses; W. N. gets every week at least 1–2 requests.

EXPERT BIONIC OFFERS:
Offers of bionicists, biologists or technicians from the universities and colleges to schools are rather rare until now. One of the few, that seeks from himself the direct way into the schools, is B. Hill, until recently professor in the faculty "Technique and their Didactics" at the University of Applied Sciences (UAS), Münster. His approaches were described on p. 291. However, the reservation changes, for pupils of today are the bionicists of tomorrow.

LITERATUR: Gymnasium Unterhaching (Hrsg.) (1995): Bionik – Lernen von der Natur. Studientag der 11. Klassen. – Hill, B. (1998): Orientierungsmodelle und ihre heuristische Nutzung ... In: Nachtigall, W., Wisser, A. (Hrsg.): BIONA-report 12. 4. Bionik-Kongress. München.

Education and study in Germany

GTBB ⇨ ***BIOKON***

At a German UAS, one can study "bionics" as a discipline; otherwise, there are supplemental bionic offerings.

HISTORY:
Berlin: Since 1972, the probably first supplemental education has been offered by I. Rechenberg at the TU Berlin, in fact with bionic lectures for listeners of all faculties. The students stayed in their traditional main subject.
Saarbrucken: Since 1990, W Nachtigall has offered a main focus "Technical Biology and Bionics for graduands in the field of biology." The students got the biologist diploma with this main focus.

FULL EDUCATION:
Bremen: Since 2003, A. Kesel offers an education to a "bionic scientist" at the UAS, outgoing from the single professorship of bionics world-wide until now. The students are taught in the principles of biology and of engineer sciences and reach the Bachelor after the 7nth semester. A 1-sem. internship abroad is implied. The education language is English. Since the SS 2008, the B. Sc. is supplemented by a 3-sem. master programme.

SUPPLEMENTAL EDUCATION:
Bionic oriented lectures and practical courses are offered as supplemental education by a row of universities, colleges and design-schools for students of life science, engineering and the design, hereby, the students stayed in their traditional main subjects.
Examples: *Darmstadt* TU (fluid engineering), *Freiburg* university (botany) *Saarbrucken* FHS (architecture) *Berlin-Weißensee*, academy of art (design).

LITERATUR: Bionik-Innovations-Centrum: http://bionik.fbsm.hs-bremen.de/pages/BIC_start.html . – Marine Bionik: Patente aus der Natur: http://bionik.fbsm.hs-bremen.de/pages/MB_start.html

Education and study at home and abroad

A

Massachusetts Institute of Technology

B

In the meantime, there are bionic offers at many European and non-European universities and other constitution.

GERMANY:
Rhineland-Westphalia Technical University Aachen (A) – University Bayreuth – Technical University Berlin – University Bonn – UAS Bremen – Technical University Darmstadt – University Freiburg – Technical University Dresden – UAS Gelsenkirchen, Bocholt – University Kiel, Technical University Ilmenau – Technical University Munich – University Munster – UAS Rhine-Waal, Kleve – Saarland University, Saarbrucken - University Stuttgart.

EUROPE BESIDES GERMANY:
Belgium: Université Libre de Bruxelles – *France*: Université de Tours – CNRS. *England*: University of Bath – University of Oxford – University of Plymouth – University of Reading. *Austria*: UAS Carinthia – Technical University Vienna – *Sweden*: Royal Institute of Technology, Stockholm – *Switzerland*: Swiss Federal Institute of Technology, Lausanne – University and Confederate Technical College, Zurich.

NON EUROPEAN:
Canada: University of British Columbia, Vancouver – University of Toronto, *Japan:* National Institute for Advanced Interdisciplinary Research, Tsukuba. *USA:* University of California, Berkeley – Duke University, Durham – Harvard University, Cambridge – University of Maryland, College Park – Massachusetts Institute of Technology, Cambridge (B) – Montana State University, Bozeman – Northeastern Univ., Boston – Pennsylvania State Univ. – Stanford Univ., Stanford etc..

LITERATUR: Datenbank des Kompetenznetzes Biomimetik (2012): www.kompetenznetz-biomimetik.de/index.php?option=com_content&task=view&id=119&Itemid=81#Datenbanken (im Aufbau).

[illegible]

In the meantime, there are some offers of [illegible] in non-European universities and other institutions

GERMANY

[illegible]

[illegible]

[illegible]

Index of names and INSTITUTIONS

W. Nachtigall and A. Wisser, *Bionics by Examples*,
DOI 10.1007/978-3-319-05858-0

S

T

Index of animals and plants

W. Nachtigall and A. Wisser, *Bionics by Examples*,
DOI 10.1007/978-3-319-05858-0

Subject index

W. Nachtigall and A. Wisser, *Bionics by Examples*,
DOI 10.1007/978-3-319-05858-0

D

R

S

MIX
Papier aus verantwortungsvollen Quellen
Paper from responsible sources
FSC® C105338

If you have any concerns about our products,
you can contact us on
ProductSafety@springernature.com

In case Publisher is established outside the EU,
the EU authorized representative is:
Springer Nature Customer Service Center GmbH
Europaplatz 3, 69115 Heidelberg, Germany

Printed by Libri Plureos GmbH
in Hamburg, Germany